Water Relations in
Membrane Transport
in Plants and Animals

Academic Press Rapid Manuscript Reproduction

Proceedings of a joint symposium held at the University of Pennsylvania, sponsored by the American Society of Plant Physiologists, the American Physiological Society, the Division of Comparative Physiology and Biochemistry of the American Society of Zoologists, and Bio-Sciences Information Service of Biological Abstracts (BIOSIS), to mark the 50th Anniversary of BIOSIS.

Water Relations in Membrane Transport in Plants and Animals

Edited by

Arthur M. Jungreis

Department of Zoology
University of Tennessee
Knoxville, Tennessee

Thomas K. Hodges

Department of Botany
and Plant Pathology
Purdue University
West Lafayette, Indiana

Arnost Kleinzeller

Department of Physiology
School of Medicine
University of Pennsylvania
Philadelphia, Pennsylvania

Stanley G. Schultz

Department of Physiology
School of Medicine
University of Pittsburgh
Pittsburgh, Pennsylvania

Academic Press, Inc.

NEW YORK   SAN FRANCISCO   LONDON   1977
A Subsidiary of Harcourt Brace Jovanovich, Publishers

QH
509
.W37

ACADEMIC PRESS, INC.
111 Fifth Avenue, New York, New York 10003

United Kingdom Edition published by
ACADEMIC PRESS, INC. (LONDON) LTD.
24/28 Oval Road, London NW1
Library of Congress Cataloging in Publication Data

Main entry under title:

Water relations in membrane transport in plants and
    animals.

    Proceedings of a symposium held during the 27th
annual fall meeting of the American Physiological
Society at the University of Pennsylvania,
Aug. 17–19, 1976.
        Includes indexes.
        1.   Biological transport–Congresses.   2.   Water
in the body–Congresses.   3.   Plant-water relation-
ships–Congresses.   4.   Plant translocation–Con-
gresses.   I.   Jungreis, Arthur M.   II.   American
Physiological Society (Founded 1887)
QH509.W37   574.8'75   77-1556
ISBN 0–12–392050–7
PRINTED IN THE UNITED STATES OF AMERICA

Contents

Session III   Epithelial Transport of Solutes and Water

CONTENTS

Session IV General Session

Contributors

Numbers in parentheses indicate the pages on which authors' contribution begin.

Ahearn, G. A. (129), Department of Zoology, University of Hawaii at Manoa, P.O. Box 1346, Coconut Island, Keneohe, Hawaii 96744

Andreoli, Thomas E. (249), Division of Nephrology, Department of Medicine, and Department of Physiology and Biophysics, School of Medicine, University of Alabama at Birmingham, Birmingham, Alabama 35294

Armstrong, W. McD. (215), Department of Physiology, School of Medicine, Indiana University, Indianapolis, Indiana 46202

Bauer, Keith (30), Department of Biological Sciences, Stanford University, Stanford, California 94305

Beames, C. G., Jr. (97), Department of Physiological Sciences, Oklahoma State University, Stillwater, Oklahoma 74074

Bisson, Mary A. (3), Department of Medicine, School of Medicine, University of North Carolina, Chapel Hill, North Carolina 27514

Blankemeyer, James T. (161), Department of Biology, Temple University, Philadelphia, Pennsylvania 19122

Boulpaep, Emile L. (355), Department of Physiology, School of Medicine, Yale University, New Haven, Connecticut 06510

Civan, Mortimer M. (187), Departments of Physiology and Medicine, School of Medicine, University of Pennsylvania, Philadelphia, Pennsylvania 19174

Conte, Frank P. (143), Department of Zoology, Oregon State University, Corvallis, Oregon 97331

Cummins, W. Raymond (30), Department of Botany, Erindale College, University of Toronto, Mississauga, Ontario, Canada

Dietz, Thomas H. (111), Department of Zoology and Physiology, Louisiana State University, Baton Rouge, Louisiana 70803

Dilley, R. A. (55), Department of Biological Sciences, Purdue University, West Lafayette, Indiana 47907

Donahue, M. J. (97), Department of Physiological Sciences, Oklahoma State University, Stillwater, Oklahoma 74074

Giaquinta, R. T. (55), Department of Biological Sciences, Purdue University, West Lafayette, Indiana 47907

Green, Paul B. (30), Department of Biological Sciences, Stanford University, Stanford, California 94305

Gutknecht, John (3), Duke University Marine Laboratory, Beaufort, North Carolina 28516

Hanson, J. B. (277), Department of Botany, University of Illinois, Urbana, Illinois 61801

Harvey, William R. (161), Department of Biology, Temple University, Philadelphia, Pennsylvania 19122

Hodges, Thomas K. (1), Department of Botany and Plant Pathology, Purdue University, West Lafayette, Indiana 47907

Jaffe, Lionel F. (15), Department of Biological Sciences, Purdue University, West Lafayette, Indiana 47907

Jungreis, Arthur M. (87, 89), Department of Zoology, University of Tennessee, Knoxville, Tennessee 37916

Kleinzeller, Arnost (275), Department of Physiology, School of Medicine, University of Pennsylvania, Philadelphia, Pennsylvania 19174

Kregenow, Floyd M. (291) Laboratory of Kidney and Electrolyte Metabolism, National Heart and Lung Institute, National Institutes of Health, Bethesda, Maryland 20014

Maddrell, S. H. P. (303), Agricultural Research Council Unit of Invertebrate Chemistry and Physiology, and Department of Zoology, Cambridge University, Cambridge CB2 3EJ, England

Maginniss, L. A. (129), Department of Zoology, University of Hawaii at Manoa, Coconut Island, Keneohe, Hawaii 96744

Merz, J. M. (97), Department of Physiological Sciences, Oklahoma State University, Stillwater, Oklahoma 74074

Ort, D. R. (55), Department of Biological Sciences, Purdue University, West Lafayette, Indiana 47907

Patlak, Clifford S. (249), Theoretical Statistics and Mathematics Branch, National Institute of Mental Health, National Institutes of Health, Bethesda, Maryland 20014

Phillips, John E. (333), Department of Zoology, University of British Columbia, Vancouver, British Columbia, V6T 1W5, Canada

Prochaska, L. J. (55), Department of Biological Sciences, Purdue University, West Lafayette, Indiana 47907

Raschke, Klaus (47), ERDA Plant Research Laboratory, Michigan State University, East Lansing, Michigan 48824

Riegel, J. A. (121), Department of Zoology, Westfield College, University of London, London, NW3 7ST, England

Robinson, Kenneth R. (15), Department of Biological Sciences, Purdue University, West Lafayette, Indiana 47907

Schafer, James A. (249), Department of Physiology and Biophysics, School of Medicine, University of Alabama at Birmingham, Birmingham, Alabama 35294

Schultz, Stanley G. (183), Department of Physiology, School of Medicine, University of Pittsburgh, Pittsburgh, Pennsylvania 15261

Slayman, Clifford L. (69), Department of Physiology, School of Medicine, Yale University, New Haven, Connecticut 06510

Song, Y. K. (129), Department of Zoology, University of Hawaii at Manoa, Coconut Island, Keneohe, Hawaii 96744

Spangler, Robert A. (315), Department of Biophysical Sciences, State University of New York at Buffalo, Buffalo, New York 14203

Tormey, John McD. (233), Department of Physiology, School of Medicine, University of California at Los Angeles, Los Angeles, California 90024

Tornquist, A. (129), Department of Zoology, University of Hawaii at Manoa, Coconut Island, Keneohe, Hawaii 96744

Wright, Ernest M. (199), Department of Physiology, School of Medicine, University of California at Los Angeles, Los Angeles, California 90024

Preface

The chapters of this book were prepared for presentation in a symposium dealing with Water Relations in Membranes in Plants and Animals, during the 27th Annual Fall Meeting of the American Physiological Society held at The University of Pennsylvania, August 17-19, 1976. This symposium was cosponsored by the American Physiological Society, American Society of Plant Physiologists, the Division of Comparative Physiology and Biochemistry of the American Society of Zoologists, and BioSciences Information Service of Biological Abstracts (BIOSIS) to mark the 50th anniversary of BIOSIS.

Water is a ubiquitous molecule upon which life as we know it is dependent. To be living also required the evolution of phosphate linked polymers and biomembranes, which served to husband the molecules whose syntheses were associated with both positive and negative free energies, as well as the molecules required for the synthesis of these other molecules. Thus water, compartmentation, and life are intimately intertwined. Compartmentation via membranes also necessitated restricting the movement of water across these synthetic barriers, and therein begins our story. Eucaryotic cells have evolved a wide range of mechanisms to cope with problems of water transport, functions that can be likened to gates, barriers, and revolving doors. In addition, the incorporation of cells into tissues resulted in the formation of additional pathways for water movement, namely between and across (epithelial) cells.

The purpose of this symposium is to explore the common modes of water regulation in plants and animals. In these proceedings, mechanisms employed to restrict water flow across plant and metazoan animal cells are described. Putative differences in mechanisms of water regulation retained by plant versus animal cells become inconsequential in the light of the numerous similarities: dependence upon bioelectric potentials maintained across cell membranes, energy dependence of uphill water movement, and solute coupling during water transport.

Chapters 1-7 deal with specific mechanisms of water transport in plants. Chapters 8-15 with specific mechanisms in invertebrates, Chapters 16-21 with specific mechanisms in vertebrates, and Chapters 22-28 with generalized mechanisms common to plants and animals.

This symposium was immensely successful in effecting its desired goals. Physiologists studying plants discovered that, despite their small numbers, significant progress (both relative and absolute) has been made in elucidating the mechanisms of water flow in algae and higher plants. Physiologists studying invertebrates found their work to be of the same high caliber as that dealing with vertebrate tissues. Moreover, they discovered that the observed transport properties could readily be related to the ecology of the organisms, in turn permitting a deeper understanding of homeostatic mechanisms in poikilotherms. Physiologists working with vertebrates came away with the knowledge that all living cells must solve basic problems of water metabolism via comparable mechanisms, and that tissues other than toad bladder, frog skin, rabbit kidney and gall bladder, and rat intestine might better be employed to answer fundamental questions of water regulation in the metazoa.

I wish to thank Miss Carole A. Price for her infinite patience in handling the correspondence associated with this symposium, and Mrs. Phyllis Bice for typing final drafts of manuscripts that the authors and the Senior Editor were all too eager to change. I wish to thank Drs. Thomas K. Hodges, Arnost Kleinzeller, and Stanley G. Schultz for organizing excellent sessions and for the initial editorial work on manuscripts from authors in their respective sessions. I also wish to thank Dr. Kleinzeller for help above and beyond the call of duty in making this symposium a reality.

Arthur M. Jungreis
Senior Editor

WATER AND SOLUTE TRANSPORT IN PLANT CELLS:
CHAIRMAN'S INTRODUCTORY COMMENTS

Thomas K. Hodges

Purdue University

On behalf of the American Society of Plant Physiologists, we
are very proud to participate in this symposium commenorating the
50th anniversary of Biological Abstracts.
Water relations of plants have intigued scientists for centuries.
The reasons for this are obvious. Different plants grow under a
variety of aquatic conditions. Some spend their entire life completely submerged, others live in the near absence of water, and
still others live in environments where the available water supply
fluctuates widely. Furthermore, certain plants can live under
very saline conditions and others can live in very infertile areas.
During the last 50 years, great strides have been made in measuring
the internal water potential of plant cells, the hormonal control
of cell water potentials, the energetic aspects of ion transport,
the electrical currents through growth points of cells, and the
unique and diverse anatomical features of plants which enable them
to cope with these widely different environmental conditions. Unfortunately, only a few of these topics will be considered in to-
days papers.
In organizing this session I had to continually remind myself
that the symposium was being held during the annual meeting of the
American Physiological Society, which is primarily "animal" oriented. For this reason, I felt a special obligation to invite plant
scientists who were not only doing the most exciting and important
research in this field but who could also put their work into
proper perspective in the allotted 20 minutes. Although I have
received some static about the 20 minute limitation, I am confident
that each of the speakers will provide both an overview and some
specifics of their own work in this time period.
In selecting the speakers, I also sought diversity with respect
to the systems studied and the approaches taken. There will be

1

two papers dealing with algal cells, one with a fungus, one with
the coleptile of young seedlings, one with stomata, and one with
chloroplasts. Two of the papers are concerned primarily with
water relations, another two deal mostly with ion transport, and
the remaining two are concerned with both of these aspects of plant
physiology.

ION TRANSPORT AND OSMOTIC REGULATION
IN GIANT ALGAL CELLS

John Gutknecht and Mary A. Bisson

Duke University

INTRODUCTION

All plant and animal cells possess the ability to regulate the
volume and composition of their cytoplasms. In most plants, bacteria and fungi the problem of volume regulation is "simplified"
by the presence of a cell wall which prevents uncontrolled osmotic
swelling during exposure to changes in environmental osmotic pressure. The presence of a cell wall allows the development of a turgor pressure, which is due to the hyperosmolality of intracellular
fluids and the consequent uptake of water by osmosis across the
semipermeable plasma membrane. The hyperosmolality of intracellular fluids may be caused either by the active uptake of salts or
by the biochemical synthesis of organic solutes or, more often, by
a combination of the two. Under steady-state conditions the turgor pressure is approximately equal to the difference between the
osmotic pressures of intracellular and extracellular fluids. Turgor pressures vary widely among walled cells, ranging from about
0.5 to 25 bar.
Turgor pressure serves a variety of important functions in the
life of a plant. Turgor is the driving force for expansion growth,
and turgor is responsible for maintaining the form and mechanical
rigidity of plant cells and tissues. Controlled changes in turgor are also responsible for the diurnal movements of plants, the
rapid movements of sensitive and carnivorous plants, and the opening and closing of stomatal pores.
Our review will discuss mainly the regulation of turgor in some
marine and estuarine algae which are subjected to large fluctuations in external osmotic pressure due to tidal cycles, rain show-
ers and/or evaporation. We will show first that turgor is regulated in several species of giant-celled algae, and we will then

discuss the effectors, signals, and possible transduction mechanisms involved in turgor regulation. For more comprehensive reviews of ionic and osmotic regulation in algae see Hope and Walker (1975), Hellebust (1976), Cram (1976), Raven (1976) and Gutknecht, Hastings and Bisson (1977).

TURGOR REGULATION IN *Valonia*, *Codium* and *Halicystis*

Figure 1 shows the relationship between steady-state turgor

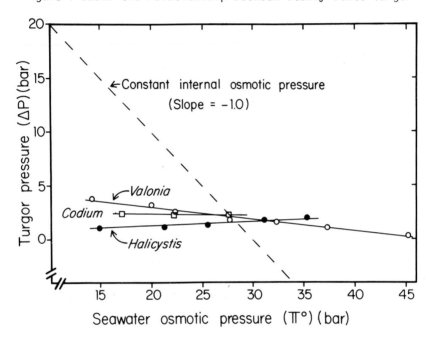

Fig. 1. *Steady-state turgor pressures of three marine algae adapted to various salinities. An osmotic pressure of 26 bar corresponds to full-strength seawater (35 ppt salinity). Turgor was estimated by measuring the difference between the osmotic pressures of intracellular and extracellular fluids, as indicated by freezing point depression. For details see Bisson and Gutknecht (1975) and Hastings and Gutknecht (1976).*

pressure (ΔP) and external osmotic pressure (π^o) in three species of giant-celled marine algae which were adapted gradually to a wide range of salinities. If intracellular osmotic pressure did not change in response to changes in salinity, then the slope of ΔP vs. π^o would be -1.0 (dashed line in Fig. 1). On the other hand, if turgor were maintained constant at all salinities, then the slope of ΔP vs. π^o would be 0, i.e., turgor regulation would be 100% effective. By this criterion, the effectiveness of turgor regulation is 90% in *Valonia,* 100% in *Codium* and 105% in *Halicystis*. Although these results demonstrate the capacity of these algae to maintain a constant turgor over a wide range of external salinities, they provide no information about the mechanisms involved in turgor regulation.

IDENTIFICATION OF THE EFFECTOR IN TURGOR REGULATION

By "effector" we mean the solute transport or metabolic synthesis-degradation system which is primarily responsible for controlling intracellular osmotic pressure. The ionic compositions of the cell saps of *Valonia, Codium* and *Halicystis* are shown in Table 1.

TABLE 1

Ionic composition of cell saps, vacuole potentials and turgor pressures in three species of giant-celled marine algae.

Species or external solution	Ion concentration (mM)						Turgor pressure (bar)	Vacuole potential (mV)	Reference
	Na^+	K^+	Ca^{2+}	Mg^{2+}	Cl^-	SO_4^{2-}			
Seawater (salinity = 36 ppt)	494	10.4	10.6	55	575	29			Barnes (1954)
Valonia macrophysa	195	431	1.0	1.1	600	0.3	1.5	+7	Hastings and Gutknecht (1976)
Codium decorticatum	302	368	11	21	568	74	2.3	-76	Bisson and Gutknecht (1975)
Halicystis parvula	415	9.2	42	65	579	0.6	0.5	-82	Graves and Gutknecht (1976)

The saps of these three algae show a remarkably wide range of ionic compositions. The cation concentrations range from very low K^+ in *Halicystis parvula* to very high K^+ in *Valonia macrophysa,* with Na^+ making up most of the cation balance. In all three algae Cl^- is the predominant vacuolar anion, and in all three the sap osmolality is at least 95% accounted for by the inorganic ions shown in Table 1. Thus, organic solutes and trace elements contribute at most only 5% to the total sap osmolality. Although the ionic compositions of the cytoplasms may differ greatly from those of the vacuolar sap, the osmotic relations of the whole cell are dominated by the large vacuole. Thus, in this review we will deal only with transport processes which control the content and composition of the cell sap.

A comparison of the calculated equilibrium potentials with the measured vacuole potentials suggests that K^+ is actively absorbed in *Valonia* and that Cl^- is actively absorbed in *Codium* and *Halicystis* (for details, see references listed in Table 1). Although there is evidence for active extrusion of Na^+ in all three species, we are primarily interested in the transport processes which are responsible for generating and maintaining the hyperosmolality of the sap, since these processes are most likely to be the key elements in the turgor regulation systems. By studying net salt transport during osmotic stress we have shown that the K^+ transport system in *Valonia* and the Cl^- transport system in *Codium* and *Halicystis* are mainly responsible for the controlled changes in sap osmolality which occur in response to salinity changes (Hastings and Gutknecht, 1976; Bisson and Gutknecht, 1975, 1976, and unpublished data).

IDENTIFICATION OF THE PRIMARY SIGNAL FOR TURGOR REGULATION

The fact that a cell controls its turgor pressure (ΔP) does not prove that the primary signal for regulation is a change in ΔP, because a variety of physical and chemical parameters are closely related to turgor and any one of these might serve as the primary signal for regulation. First, since turgor pressure is caused by water flowing down its osmotic gradient, $\Delta \Pi$ rather than ΔP might be the primary signal for regulation. Second, if the environmental osmolality is constant, as in fresh water, then in order to control either $\Delta \Pi$ or ΔP a cell needs only to regulate intracellular osmotic pressure (Π^i) or the intracellular concentrations of specific solutes. Third, since changes in environmental hydrostatic pressure are usually small compared to the turgor pressure, a cell could conceivably control ΔP by regulating only P^i, the internal hydrostatic pressure. Fourth, since changes in turgor pressure always cause changes in cell volume, cell volume rather than turgor pressure may be the regulated parameter. Fifth, since changes in turgor pressure can cause changes in the partial pressures of dissolved gases such as O_2 and CO_2, these parameters must also be considered as possible signals for turgor regulation.

In order to clearly identify the primary signal in a turgor control system, we must be able to control all of the physical and chemical parameters related to turgor pressure. Although this is a difficult task, the method of continuous intracellular perfusion provides the best opportunity to achieve this goal. Figure 2 shows the experimental apparatus used to study the effects of turgor pressure on ion transport and electrical properties of *Valonia* and *Halicystis* (Hastings and Gutknecht, 1974; Graves and Gutknecht, 1976). Both sides of the intracellular perfusion system can be pressurized by means of compressed gas, while continuous perfusion is maintained by the small difference between the levels of the inflow and outflow reservoirs. In the perfused cell, membrane current and voltage can be controlled and recorded, and solute and water fluxes between vacuole and external solution can be easily measured. The major advantage of the perfused cell for studying osmotic regulation is that ΔP, $\Delta \Pi$, Π^i, Π^o and specific ion concentrations can all be varied independently, which allows us to determine which, if any, of these parameters is the primary signal for turgor regulation. The major limitation of the method is that only giant cells (> 4 mm diameter) can be easily studied. For controlling and measuring turgor in smaller (nonperfused) cells (> 500 μm diameter) a micropressure probe has been developed by Zimmermann, Raede and Steudle (1969) and Steudle, Luttge and Zimmermann (1975).

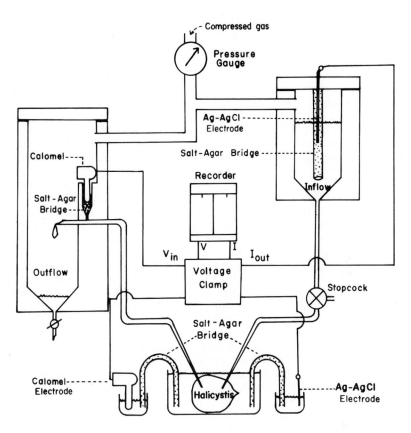

Fig. 2. A pressurized perfusion system for studying the effects of turgor pressure on ion transport and electrical properties of giant algal cells. The system includes two closed plexiglas reservoirs which are pressurized by a common gas supply. The heights of the inflow and outflow reservoirs are adjusted so as to maintain a gravity driven flow of 0.05 to 0.25 ml/min, depending on the cell size and species. Samples of perfusate are taken every 5-15 min from the outflow port. (Modified after Graves and Gutknecht, 1976).

TURGOR-SENSITIVE K^+ TRANSPORT IN *Valonia*

For the remainder of this paper we will discuss primarily *Valonia*, the alga which has been most intensively studied with regard to turgor regulation. We will discuss together several species - *V. macrophysa*, *V. ventricosa* and *V. utricularis* - all of which are characterized by a large cell size (up to several cm diameter), high K^+ sap, and a slightly positive vacuole potential. As shown in Fig. 1, the steady-turgor in *V. macrophysa*, which is normally about 1.5 bar, changes only 3 bar in response to a 30-bar change in seawater osmotic pressure. Thus, turgor regulation in *V. macrophysa* is 90% effective. Similar results have been obtained with *V.utricularis* (Zimmermann and Steudle, 1974).

In *V. macrophysa* the salts important in turgor regulation are KCl and NaCl. When salinity changes, changes in vacuolar KCl concentration account for 85% of the change in sap osmolality, and changes in NaCl concentration account for the remaining 15% (Hastings and Gutknecht, 1976). The two ions which are most important in turgor regulation in *Valonia* are K^+ and Cl^-, but only K^+ is actively transported into the vacuole. The evidence for active K^+ uptake includes (1) a large net influx of K^+ under short-circuit conditions (2) a flux ratio (J_K^{in}/J_K^{out}) which is 100 times larger than that predicted for simple diffusion (3) inhibition of K^+ influx by cyanide and azide and (4) maintenance of a steady-state vacuolar K^+ concentration which is about 100 mV out of equilibrium at all salinities (Gutknecht, 1966, 1967; Hastings and Gutknecht, 1974, 1976 and unpublished data). Chloride appears to be passively transported as the counter ion to K^+. This statement is based on (1) the absence of a net Cl^- flux under short-circuit conditions in internally perfused cells and (2) agreement between the observed and predicted flux ratios for simple diffusion in normally growing cells (Gutknecht, 1966; Hastings and Gutknecht, 1976, and unpublished data).

K^+ is the ion most directly involved in turgor regulation in *Valonia*. The active trasnport of K^+ into the vacuole is stimulated by a decrease in the turgor pressure (Fig. 3), both in internally perfused cells in which the turgor pressure is controlled directly and in intact cells in which the turgor pressure is reduced by raising the external osmolality with an impermeant nonelectrolyte (Gutknecht, 1968; Cram, 1973; Hastings and Gutknecht, 1974). In contrast, the K^+ permeability is not significantly affected by turgor pressure over the range of 0-1.5 bar (Gutknecht, 1968; Hastings, 1975). However, K^+ efflux is markedly increased by high turgor pressures, a phenomenon which Zimmermann and Steudle (1974) have attributed to a reversal in the direction of the K^+ pump.

Before we can conclude that the primary signal for turgor regulation is a change in turgor pressure *per se* we must consider the alternative possibilities. First, there is the possibility that the error signal is a change in the internal pressure (P^i), rather than a change in turgor (ΔP). This question can only be answered by varying independently the external and internal hydrostatic pressures during K^+ flux measurements. By working in a large hyperbaric chamber, Hastings and Gutknecht (1974) showed that changes in ambient pressure over the range of 1 to 2 bar has no effect on K^+ transport, whereas decreasing the turgor from 1 to 0 bar greatly stimulates active K^+ uptake (Fig. 3). In these hyperbaric experiments pO_2 was held constant by using appropriate gas mixtures in order to rule out possible effects of changing oxygen tension on ion transport.

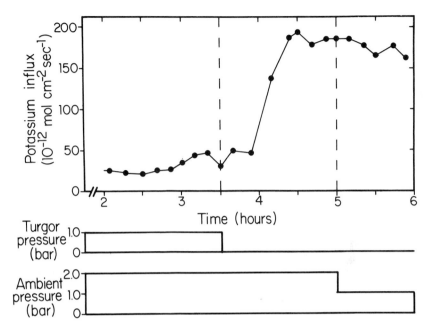

Fig. 3. Effects of turgor pressure and ambient hydrostatis pressure on the one-way K^+ influx (measured with $^{42}K^+$) in an internally perfused cell of Valonia macrophysa. The cell was impaled at t = 0 hr, and the turgor was held at 1 bar until t = 3.5 hr. Ambient hydrostatic pressure was held at 2 bar from about 0.5 hr to 5 hr, at which time the ambient pressure was reduced to 1 bar. The perfusate was an artificial cell sap (K^+ = 454 mM, Na^+ =145 mM, Cl^- = 602 mM, pH - 5.5), and the external solution was sea water with a composition similar to that shown in Table 1. The vacuolar potential ranged from +5 to +11 mV. Under these conditions about 98% of the K^+ influx occurs by active transport. (From Hastings and Gutknecht, 1974.)

Another important variable in osmotic regulation experiments is the external ion concentration. For example, when salinity changes, proportional changes in C_K^0 occur, and the active transport system may respond primarily to this change in C_K^0 or C_K^i/C_K^0, rather than to a change in ΔP. This possibility has been ruled out by several experiments which show collectively that (1) the K^+ pump is stimulated by decreased turgor at constant C_K^0 and (2) large changes in C_K^0 do not affect steady-state turgor pressure when external osmolality is held constant (Gutknecht, 1968; Hastings and Gutknecht, 1974; Cram, 1974; Zimmermann and Steudle, 1974).

Possible direct effects of osmolality changes on turgor regulation and K^+ transport have also been tested. Only in an internally perfused cell can both variables, ΔP and $\Delta \Pi$, be controlled independently during ion flux measurements. Again, several studies show that the K^+ pump in *Valonia* responds to a change in $\Delta\pi$ (Gutknecht, 1968; Zimmermann and Steudle, 1974). Furthermore, no coupling between water fluxes and ionic fluxes has been observed in *Valonia*, either during tests for "solvent drag" or electroosmosis (Gutknecht, 1968; Zimmermann and Steudle, 1974).

The last parameter which must be considered as a possible signal for turgor regulation is the change in cell volume which occurs when ΔP changes. In *Valonia* the large volumetric elastic modules of $1-2 \times 10^2$ bar insures that when ΔP changes, the fractional volume change will be very small, about 0.5%/bar (Villegas, 1967; Zimmermann and Steudle, 1974). Thus, a macroscopic volume change can probably be excluded as a primary signal for turgor regulation. However, as we discuss in the next section, microscopic deformations in the plasmalemma caused by turgor may indeed be involved in the turgor response.

THE TURGOR-PRESSURE TRANSDUCER

Small changes in turgor pressure (< 1 bar) cannot directly cause large changes in either the electrochemical gradients for ions or the rate of active ion transport. Even the most pressure-sensitive biological processes known change by only about 0.3% per bar change in hydrostatic pressure (Johnson and Eyring, 1970). Thus, the change in turgor is probably transduced and amplified in order to modify the rate of ion transport. The fact that a change in the pressure gradient alone is sufficient to modify the rate of ion transport implies that a turgor-pressure transducer must reside along the pressure gradient. Since the central vacuole, the tonoplast and the cytoplasm are all fluid or semifluid structures which are unable to support a significant pressure gradient, we infer that virtually all of the pressure gradient, and thus also the pressure transducer, must be located in the plasmalemma-cell wall complex.

The interface between the cell wall and the plasmalemma is the obvious starting point in the search for insights into the nature of the pressure transducer. Although cell walls of different algae vary widely in composition and structure (see Preston, 1974), one of the best characterized cell walls is that of *Valonia*. The wall of *Valonia* is composed largely of cellulose which exists in several hundred sheets of cylindrical microfibrils. Successive sheets of parallel microfibrils are laid down at roughly $90°$ angles to one another. Our preliminary freeze-fracture electron micrographs on *Valonia macrophysa* show that the microfibrils are 28-33 nm in diameter, and the spaces between the microfibrils range from 0-50 nm (D.F. Hastings, unpublished). Although the permeability properties of the *Valonia* cell wall have not yet been characterized, the electrical conductance (0.125 mhos cm^{-2}) is only 1.5% of the conductance of an equivalent thickness of seawater (Hastings, 1975). From a knowledge of the microfibril diameter and the electrical conductance of the wall, relative to seawater, and assuming a tortuous diffusion path twice as long as the wall thickness, we estimate the gap between the microfibrils to be roughly 0.9 nm.

In the absence of a detailed knowledge about the turgor transduction process, a large number of plausible mechanisms have been proposed, most of which involve interactions between the cell wall and plasmalemma. Some of the possibilities include (1) blocking of transport sites as the membrane is appressed against the cellulose microfibrils (Hastings, 1975: Zimmermann, Steudle and Lelkes, 1976) (2) changes in membrane folding or stretching (Steudle, 1974) (3) electrostatic interactions between fixed charged groups in the cell wall and plasmalemma (Hastings and Gutknecht, 1974) (4) membrane compression (Coster, Steudle and Zimmermann, 1976) and (5) anisotropic membrane stretching and/or appression, as shown schematically in Fig. 4.

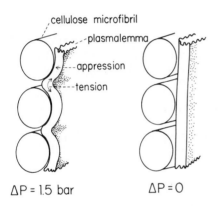

cellulose microfibril

plasmalemma

appression

tension

$\Delta P = 1.5$ bar $\Delta P = 0$

Fig. 4. Diagram of hypothetical turgor-induced membrane stretching in Valonia. The plasmalema lies adjacent to the innermost layer of cellulose microfibrils. Changes in turgor pressure causes changes in the degree of intercalation of the plasmalemma between the microfibrils, as well as changes in the degree of appression of the membrane against the microfibrils. These changes in strain may cause redistribution of membrane components, release of chemical messengers, etc.

When turgor pressure forces the plasmalemma against the cell wall, anisotropic membrane tension or stretch might cause local changes in membrane structure, e.g., a redistribution of membrane components. This hypothesis is based partly on the observed asymmetrical distribution of lipids in small single walled vesicles about 20 nm in diameter (see Roseman, Litman and Thompson, 1974; Huang, Sipe, Chow and Martin, 1974). When the radius of curvature of a lipid bilayer is very small, asymmetrical physical forces on the two monolayers tend to cause a redistribution of membrane components in order to minimize the strain produced by packing lipids of various sizes and shapes into a curved bilayer (see also Evans, 1974; Israelachvili and Mitchel, 1975). The redistribution might occur either within the plane of the membrane (lateral diffusion) or across the membrane (flip-flop). These hypothetical changes in membrane structure could affect K^+ transport directly, e.g., by changing membrane permeability, or indirectly by causing the release of a chemical messenger which then diffuses to the K^+ pump site, which may be the tonoplast.

On the basis of rather limited data we propose the following working model for turgor-sensitive K^+ transport in *Valonia*. A pressure transducer resides in the plasmalemma which is normally intercalated between the cellulose microfibrils, as shown in Fig. 4. When turgor decreases, the membrane "relaxes" into a more nearly planar configuration, which causes the release of a chemical messenger. The chemical messenger diffuses to the tonoplast where it stimulates the active transport of K^+ into the vacuole. The passive K^+ permeability of the tonoplast, which is assumed to be the rate limiting barrier for K^+ diffusion, is not affected by the chemical message. Because the permeability of the plasmalemma to K^+ is relatively high, the increase in inward K^+ pumping at the tonoplast does not cause a drastic reduction in the K^+ concentration of the cytoplasm. The inward K^+ pump at the tonoplast is electrogenic, and the stimulation of this K^+ pump causes a positive shift in the vacuole potential.

This model of hypertonic regulation is consistent with the known ion transport and electrical properties of *Valonia* over the turgor range of 0 to 1.5 bar (see Gutknecht and Dainty, 1968; Hastings and Gutknecht, 1974, 1976). However, the model is incomplete and is not the only one which is consistent with the available data. One major defect in the model is that a simple inhibition of the inward K^+ pump by high turgor is not sufficient to explain the rapid excretion of KCl which occurs during hypoosmotic (i.e., hyperturgor) stress (see Zimmermann and Steudle, 1974; Hastings, 1975; Gutknecht, Hastings and Bisson, 1977). Thus, we suspect that *Valonia* may have separate mechanisms for hypertonic and hypotonic regulation. Ion flux measurements during hypoosmotic stress are needed to resolve this question.

SUMMARY AND CONCLUSIONS

Several species of giant-celled marine algae maintain a nearly constant turgor in spite of large fluctuations in environmental osmotic pressure. In *Valonia* the primary signal in the negative feedback system is a change in turgor *per se*, and the primary effector is a turgor-sensitive K^+ transport system. In other giant-celled algae our understanding of the mechanisms of turgor regulation is less complete. However, the available data suggest that *Chaetomorpha* may have a turgor-sensitive K^+ transport system similar to that in *Valonia* (Steudle and Zimmermann, 1971), and *Codium* and *Halicystis* may possess turgor-sensitive Cl^- transport systems (Bisson and Gutknecht, 1976; Graves, 1974). *Nitella*, a fresh-water alga which never experiences large changes in external osmolality, apparently regulates its turgor indirectly by regulating π^i, the internal osmotic pressure (Nakagawa, Kataoka and Taxawa, 1974; Cram, 1976).

The prospects for future work in this relatively unexplored area of cell physiology are very exciting. Of the specific problems which need further study, perhaps the most intriguing, as well as the most challenging, is the elucidation of the turgor transducing mechanism and the control reactions which link the pressure transducer to the ion transport system. In no cell has the turgor-pressure sensor or the mechanism of the information transfer been clearly identified. In this regard we think it is especially important to learn more about physical and chemical interactions which occur between the plasmalemma and cell wall.

The existence of turgor regulation has been demonstrated in only a few species, and comparative studies are needed to provide phylogenetic perspective on turgor regulation. Although our work on giant algal cells is motivated largely by convenience, we believe that insights obtained from giant-celled algae can be extended to more complex organisms. However, because of our small sample size any generalizations implied at this point must be considered as speculative. Since osmotic regulation has been carefully studied in only a few species of plants, bacteria and fungi, it is likely that new and different types of osmotic regulation await the attention of future investigators.

ACKNOWLEDGEMENTS

This work was supported in part by U.S.P.H.S. Grant # HL 12157. We thank Drs. W.J. Cram, D.F. Hastings, E. Steudle and U. Zimmermann for sending us unpublished manuscripts and/or data, Dr. D.F. Hastings for helpful discussions throughout the course of this work.

REFERENCES

Barnes, H. (1954) *J. Exp. Biol. 31:* 582.
Bisson, M.A. & Gutknecht, J. (1975) *J. Memb. Biol. 24:* 183.
Bisson, M.A. & Gutknecht, J. (1976) *In* Transmembrane Ionic Exchanges in Plants. (Tellier, M. *et al.*, eds.) Centre National de la Recherche Scientifique, Paris In Press.
Cram, W.J. (1973) *J. Exp. Bot. 24:* 328.
Cram, W.J. (1976) *In* "Encyclopedia of Plant Physiology" Vol. 2A Luttge, U. & Pitman, M.G. eds. New York: Springer-Verlag.
Coster, H.G.L., Steudle, E. & Zimmermann, U. (1976) *Plant Physiol.* In press.
Evans, E.A. (1974) *Biophys. J. 14:* 923.
Graves, J.S. (1974) Ph.D. Thesis, Duke University.
Graves, J.S. & Gutknecht, J. (1976) *J. Gen. Physiol. 67:* 579.
Gutknecht, J. (1966) *Biol. Bull. 230:* 331.
Gutknecht, J. (1967) *J. Gen. Physiol. 50:* 1821.
Gutknecht, J. (1968) *Science 160:* 68.
Gutknecht, J. & Dainty, J. (1968) *Oceanogr. Mar. Biol. Ann. Rev. 6:* 163.
Gutknecht, J.W., Hastings, D.F. & Bisson, M.A. (1977) *In* "Transport Across Biological Membranes" Giebisch, G. *et al.*,eds. New York: Springer-Verlag. In press.

Hastings, D.F. (1975) Ph.D. Thesis, Duke University. University Microfilms, Ann Arbor, Mich.

Hastings, D.F. & Gutknecht, J. (1974) *In* "Membrane Transport in Plants" (U. Zimmermann & J. Dainty, eds.) New York:Springer-Verlag.

Hastings, D.F. & Gutknecht, J. (1976) *J. Memb. Biol.* 28: 363.

Hellebust, J.A. (1976). *Ann. Rev. Plant Physiol.* 27: 485.

Hope, A.B. & Walker, N.A. (1975) "Physiology of Giant Algal Cells" New York: Cambridge University Press.

Huang, C.H., Spie, J.P., Chow, S.T. & Martin, R.B. (1974) *Proc. Nat. Acad. Sci., USA, 71:* 359.

Israelachvili, J.N. & Mitchel, D.J. (1975) *Biochim. Biophys. Acta 389:* 13.

Johnson, F.H. & Eyring, H. (1970) *In* "High Pressure Effects on Cellular Processes" (Zimmermann, A.M. ed.) New York: Academic Press.

Nakagawa, S., Kataoka, H. & Tazawa, M. (1974) *Plant Cell Physiol.* 15: 457.

Preston, R.D. (1974) "The Physical Biology of Plant Cell Walls". London: Chapman and Hall.

Raven, J.A. (1976) *In* "Encyclopedia of Plant Physiology," Vol. 2A. (U. Luttge & Pitman, M.G. eds.) New York: Springer-Verlag.

Roseman, M., Litman, B.J. & Thompson, T.E. (1975) *Biochemistry* 24: 4826.

Steudle, E. & Zimmermann, U. (1971) *Z. Naturforsch.* 26: 1276.

Steudle, E., Luttge,U. & Zimmermann, U. (1975) *Planta 126:* 229.

Steudle, E. & Zimmermann, U. (1974) *In* "Membrane Transport in Plants" (Zimmermann, U. & Dainty, J., eds.) New York:Springer-Verlag.

Villegas, L. (1967) *Biochim. Biophys. Acta 136:* 590.

Zimmermann, U., Raede,H. & Steudle, E. (1969) *Naturwissenschaften* 56: 634.

Zimmermann, U. & Steudle, E. (1974) *J. Memb. Biol. 16:* 331.

Zimmermann, U., Steudle, E. & Lelkes, P.I. (1976) *Plant Physiol.* In press.

TRANSCELLULAR ION MOVEMENTS AND GROWTH LOCALIZATION
IN FUCOID EGGS AND OTHER CELLS

Kenneth R. Robinson and Lionel F. Jaffe

Purdue University

INTRODUCTION

Our interest in electrolyte transport in plant cells is a result of our efforts to understand the highly epigenetic process of
axis formation in the zygotes of certain maine algae, *Fucus* and
Pelvetia. The unfertilized eggs of these plants are radially symmetric; some hours after fertilization (but before first cell division) this symmetry is visibly broken by a rhizoidal bulge which
can appear at any region of the surface. The location of the rhizoid can be determined by unilateral light, nearby eggs, imposed
electric fields, and gradients of a number of substances including
K^+, auxin, and pH. The discovery some years ago by one of us that
developing fucoid eggs drive an electrical current through themselves (Jaffe, 1966) focused our attention on the cell surface
both for its likely role as a receiver of external stimuli and as
an amplifier of those stimuli into the gross asymmetry of rhizoid
growth.

Before the localization problem could be attacked in these
terms, it was necessary to study the physiology of the whole egg.
It was found that the membrane potential (E_m) of the unfertilized
egg is about -30 mV (Weisenseel & Jaffe, 1972, 1974); after fertilization, Em increases rapidly to about -70 mV. These findings
were confirmed by tracer flux measurement of the major ions of sea
water which indicated that the unfertilized egg is permeable K^+,
Na^+, and Cl^- but that after fertilization, a high degree of
selectivity for K^+ develops (Robinson & Jaffe, 1973). These post-
fertilization changes are accompanied by the development of about
5 atm of turgor pressure, largely due to the accumulation of KCl
(Allen, Jacobsen, Joaquin and Jaffe, 1972). One mechanism for
this accumulation is the activation of a pump to move Cl^- inward

15

and the virtual elimination of the passive Cl⁻ leak (Robinson & Jaffe, 1973). (It should be mentioned that these eggs do not have a large vacuole or any large inclusions, so microelectrode and tracer flux studies are relatively easy compared to those of most plant cells.)

ULTRASTRUCTURE OF CELL WALL FORMATION

The development of turgor pressure after fertilization coincides with and is dependent on the secretion of a cell wall, since the unfertilized <u>Fucus</u> egg is remarkable among plant cells in that it lacks a wall (Levring, 1952). H. Benjamin Peng and Klaus Schroter in this laboratory have recently conducted an electron microscopic study of this process using freeze-fracture and conventional thin-section techniques.

Prior to fertilization, numerous vesicles about 0.2 µm in diameter can be seen subjacent to the plasma membrane (Fig. 1) in

Fig. 1a. Electron micrograph of a replica of a fracture through an unfertilized <u>Pelvetia</u> egg. Note the numerous cortical vesicles (CV) adjacent to the plasmalemma (PL).

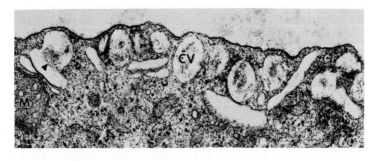

Fig. 1b. Electron micrograph of a thin section through an unfertilized <u>Pelvetia</u> egg. Micrographs by Dr. Klaus Schroter.

both freeze-fracture replicas and in thin sections. Within a few minutes after fertilization, nearly all of these cortical vesicles have fused with the plasma membrane and a thin (0.1 µm) cell wall has formed. This process is analogous to the well-known formation of a fertilization membrane in animal eggs, which also depends on the release of preformed secretion vesicles.

Peng and Jaffe (1976a, 1976b) have developed a method for obtaining large numbers of freeze-fracture replicas of one pole or the other of polarizing <u>Pelvetia</u> eggs. This method involves filling the small holes of nickel screens (as in Robinson and Jaffe, 1975) with eggs and orienting the population with unilateral light. A metal "hat" is then applied to the monolayer eggs on the screen and the resulting sandwich quickly frozen. When the hat is pulled off of the eggs the membrane leaflet is exposed as well as fractures through the cell wall and cytoplasm.

The fractures through the cell wall show that there is a characteristic organization of the wall microfibrils (Fig. 2a). These

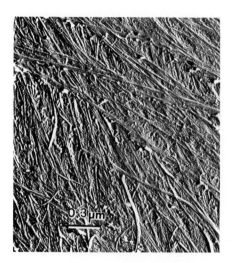

Fig. 2a. A fracture through the wall of a <u>Pelvetia</u> zygote showing cell wall microfibrils.

fibrils are layered, with the orientation of each subsequent lamella being turned about 35° from the previous one. When obliquely cross-fractured, this produces a bow-arc pattern (Fig. 2b) similar to that seen in insect cuticles (Bouligand, 1972). The innermost lamellae of the wall leave imprints on the plasma membrane that can be clearly seen as grooves in the protoplasmic face (PF) or as ridges on the endoplasmic face (EF) (see Fig. 3). These imprints are a result of the cell's turgor pressure, which forces the membrane against the wall, and of adhesion to the wall.

Also seen in Fig. 3 are 10 nm intramembranous particles. In the fertilized egg, these particles are often seen in linear arrays of 5 to 40 particles. Peng and Jaffe (1976) propose that the particle strings are responsible for orienting the initial microfibrils that are deposited immediately after fertilization.

When freeze-fracture replicas of presumptive rhizoid and thallus ends of a fertilized egg are compared, more vesicle fusion sites are seen at the shaded (presumptive rhizoid) end than at the lighted

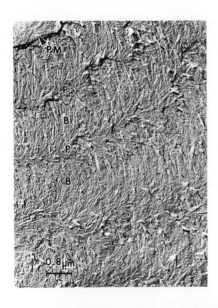

Fig. 2b. An oblique cross-fracture through the wall showing the bow-shaped arcs. These micrographs as well as the ones in Fig. 3 and 4 are by Dr. H.B. Peng.

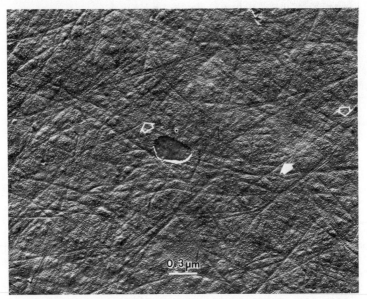

Fig. 3. The P-face of the plasma membrane of a 9-hr embryo. Imprints of the wall microfibrils are seen as grooves (solid arrow). Open arrows point to particle strings.

end (Peng, 1976). These vesicle fusion sites are identified by amorphous membrane patches which lack fibrillar imprints and appear to be pushed inward. The differences between the two poles of these cells are obvious shortly after unilateral illumination is begun and constitutes a second phase of cell wall secretion, the first being the release of the cortical vesicles upon fertilization (Fig. 4a). A third phase of secretion begins about 5 hr after

Fig. 4a. Secretion on E-face at 1½ hr after fertilization at the presumptive rhizoid. The material secreted between the plasma membrane and the wall separates the membrane and the fibrillar wall, resulting in patches (lower right portion of photograph) devoid of fibrillar imprints (curved arrow). Numerous secretion sites (open arrows) are also obvious.

fertilization. Secretion at the presumptive thallus end ceases and numerous imprint-free patches appear at the rhizoid end of the cell. These patches vary in size between 0.3 and 0.9 μm, and they persist after germination (Fig. 4b). Presumably, this third phase of secretion is involved in the export of material required for the

Fig. 4b. E-face of the plasma membrane at rhizoid end of 6-hr embryo. Numerous dome-shaped imaginations are characteristic of the third phase of secretion.

formation of new wall and membrane associated with germination.
    Finally, a substantial difference in the number of isolated intramembrane particles is seen between the two poles. By 8.5 hr after fertilization, the particle density on the P face at the presumptive thallus end is nearly twice that at the rhizoid end.

In summary, there are three distinct phases of secretion observed during the early development of the early Pelvetia embryo. The first is the release of preformed cortical vesicles just after fertilization and the rapid formation of zygote wall. The second is presumably involved with wall thickening and anchoring the cell to the substratum. It also may be involved with the initial, tentative polarization of the zygote since it is more intense at the presumptive rhizoid than at the presumptive thallus. The third phase is associated with irreversible polarization and rhizoidal growth.

ELECTRICAL CORRELATES OF DEVELOPMENT

As was mentioned above, the developing fucoid egg produces a transcellular current such that current enters the rhizoidal pole and leaves the thallus. This current was first measured by placing hundreds of eggs in a capillary tube, polarizing the population with unilateral light and measuring the potential difference between the two ends of the tube (Jaffe, 1966). In order to study the current pattern around a single egg, the "vibrating probe" has been developed (Jaffe and Nuccitelli, 1974). The critical element of this instrument is a solder-filled glass microelectrode at the tip of which is a 25-μm platinum black ball. This electrode is attached to a piezoelectric element which vibrates in response to an applied a.c. signal. The vibrating electrode can thus be used to explore the electric current pattern around eggs since it will record a voltage difference between the two extreme positions of its vibration (about 30 μm apart) if there is in fact a current flowing. It should be emphasized that these measurements are done extracellularly in the conductive medium (sea water in the present case) that surrounds the cells. Because of the extremely low resistance of the platinum ball at the a.c. frequencies at which it is vibrated, the noise of this probe is very low and voltages of 1-2 nV can be measured in sea water. This is at least 1000 times better resolution than can be achieved with static, saline-filled micropipettes of similar dimensions.

Using this probe it has been found that the current pattern around developing Pelvetia eggs and embryos has two components, a steady one and a pulsating one. The pulses first appear soon after germination and consist of current entering the growing tip and leaving the rest of the embryo (see Fig. 5). They last about 100 sec and occur at a rate of 1-5 per hour. At their peak, the current density of these pulses at the surface of the embryos is about 3-10 μA/cm^2. The pulses are monophasic in nature, which indicates that they are not propagated around the embryo (Nuccitelli and Jaffe, 1974). By changing the composition of the surrounding sea water, it has been found that Cl^- efflux at the rhizoid tip is responsible for most of the entering current. The nature of the evidence for this is twofold: first, the current pulse amplitude increases 60% when external Cl^- is reduced 7-fold, and second, when the membrane potential is varied by changing the external K^+

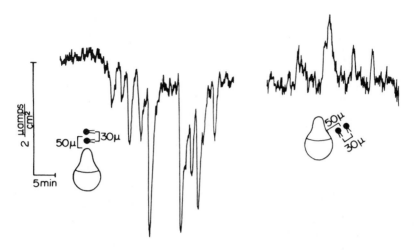

Fig. 5. The pulses on the left were recorded at the tip of a 1-day-old Pelvetia embryo. They are inward in direction. The pulses on the right were normal to the base of the rhizoid and were outward.

concentration, the direction of the current pulses reverses at a potential that could only be the Cl⁻ equilibrium potential. Incidentally, this latter measurement implies that the cytoplasmic Cl⁻ concentration is about 110 mM, which is less than half of the value obtained if the total chloride content is divided by the volume of the cell water (Allen, Jacobsen, Joaquin and Jaffe, 1972); thus large amounts of chloride must be sequestered within the cell.

There are at least two possible developmental roles for the current pulses. One is the creation of an internal electric field which can act electrophoretically on charged entities in the cytoplasm and membrane. This possibility will be discussed more fully below. A second role that the pulses might play is in the regulation of turgor pressure. That this is so has recently been shown by Nuccitelli and Jaffe (1976). They found that the current pulses may be modulated by the osmotic strength of the surrounding medium. Increasing the external osmotic pressure suppressed pulsing, while decreasing osmotic pressure stimulated pulsing (Fig. 6). A decrease of external osmolarity of only 3% produced a measurable increase in the size of the pulses; further decreases caused pulse size to increase nonlinearly. Of course, the electrically measured efflux may be only a fraction of the total loss of ions upon decreasing osmolarity since the loss of KCl from the same area of the membrane would be electrically indetectable. Nuccitelli and Jaffe (1976) therefore determined the total Cl⁻ efflux by preloading a large population of cells with ^{36}Cl and then monitoring the change in efflux in response to lowered external osmolarity. A 17% decrease in osmolarity resulted in a transient 33-fold increase in Cl⁻ efflux (Fig. 7). Further, they calculated the concomitant loss of

21

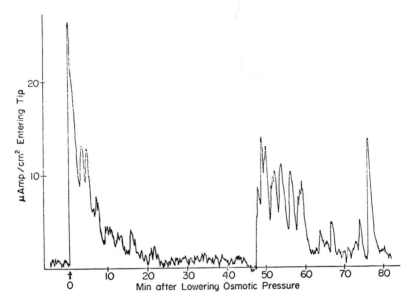

Fig. 6. *The electrical response of a 1-day-old* Pelvetia *embryo to a 20% decrease in external osmotic strength.*

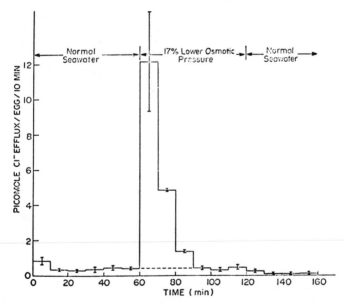

Fig. 7. *Cl⁻ efflux from a 1-day-old* Pelvetia *embryo as a result of a 17% decrease in external osmotic strength. The area above the dotted line is taken as the excess Cl⁻ efflux in response to the changed osmolarity and represents 17 pmoles during 40 min.*

KCl that this increased efflux represented was just enough to restore the cells' turgor pressure to its original value.

The picture that emerges is that the rhizoidal tip acts as a pressure transducer. When it senses an excess turgor pressure, its passive Cl^- leak is increased and Cl^- leaves. K^+ follows since electroneutrality must be maintained. This KCl release is electrically detectable because the Cl^- leak is localized at the tip while K^+ efflux is spread out more uniformly over the whole cell. Nuccitelli and Jaffe believe that the Cl^- permeability is controlled by Ca^{++}, which they suggest enters the Pelvetia embryo when the rhizoid membrane is stretched as a result of increased turgor pressure.

Cl^- action potentials have also been noted in fresh water algae, Nitella and Chara (Gaffey and Mullins, 1958; Mullins, 1962) and the marine alga, Acetabularia (Gradman, Gottfried and Blasel, 1973).

The steady currents in Pelvetia have been most extensively studied in the pregermination zygote (Nuccitelli, 1975). It was found that the current entered the cell in a localized area at various regions, starting as early as 1 hr after fertilization. The position of the current entry shifted around with respect to the direction of the unilateral light before finally stabilizing at about 5 hr after fertilization on the shaded side where germination would eventually occur. The currents were on the order of 1 $\mu A/cm^2$ at the surface.

LOCAL CATION ENTRY AND ELECTROPHORESIS

The intracellular effect of the extracellularly observed currents will depend greatly on which ions are responsible for the current. To answer this question, we have developed a method which allows us to determine the ion fluxes at either end of population of polarized fucoid eggs (Robinson and Jaffe, 1975). This involves filling the 75-μm-diameter holes of a nickel screen with 90-μm Pelvetia eggs and illuminating the screen from one side or the other. Since the eggs secrete a sticky glue, they seal themselves into the holes and effectively isolate the two sides of the screen from each other. Tracer ions can then be added to either side and the influx of ions into either the rhizoid or thallus halves of the egg can be measured. Using these methods, we have found that calcium enters the presumptive rhizoidal region of Pelvetia eggs and is pumped out the thallus end (Robinson and Jaffe, 1975). This calcium current is maximal at the earliest time that it can be measured (about 5 hr after fertilization) and it steadily declines as the time of germination approaches (see Fig. 8).

The involvement of calcium in the currents detected around developing fucoid eggs is central to our idea of the developmental significance of these currents. We believe that the currents have as their target the interior of the egg and that the intracellular field that they generate moves charged entities to the surface at one end of the cell or the other, thereby polarizing it. The size of the intracellular field that can be generated depends on the

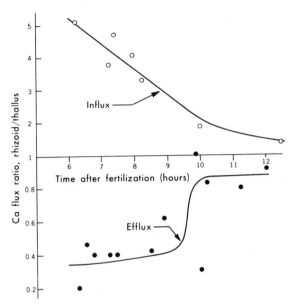

Fig. 8. Calcium influx and efflux ratios in developing Pelvetia eggs. Each point is the ratio of two simultaneous flux measurements on half-eggs made as described in the text.

mobility of the ions which are carrying the current and calcium is surely the most immobile of all small ions in cytoplasm (see Baker, 1972 for a discussion). In simplest terms, the voltage generated by a given ionic current will be proportional to the resistance of the medium to the ion. We calculate (Jaffe and Nuccitelli, 1974) that the measured calcium current in these 90-μm-diameter fucoid eggs could produce voltage differences on the order of a few millivolts across their diameter. Since the polarization process takes place over a period of hours, such intracellular fields could easily move vesicles or molecules with typical electrophoretic mobilities of 1 μ/sec per volt/cm the few micrometers that might be necessary for them to fuse with the plasma membrane.

If this idea is correct, it should be possible to polarize Pelvetia eggs by putting them in a gradient such that more calcium is driven in one part of the eggs than the other. We have tried such experiments by putting sea water with different concentrations of calcium on the two sides of egg-bearing nickel screens as described above and examining the direction of the rhizoid outgrowths to see if they appear on the high or low calcium side. We assumed that since calcium entry is a passive, downhill process that more calcium would enter the sides of the eggs that were exposed to the higher concentration of calcium. To our surprise, however, we found that the rhizoids tended to grow toward the lower calcium side. Before abandoning our calcium gradient hypothesis, we did influx measurements on whole cells, using [45]Ca in sea waters with various calcium concentrations. Again, to our surprise, we found that more calcium entered the cells at lower concentrations than at the normal concentration of 10 μm. Indeed, at 1.5 μm Ca, the influx was

24

7 times greater than at 10 μm (Fig. 9). We have begun a systematic study of polarization of _Pelvetia_ eggs by various calcium gradients and our preliminary results are that the eggs tend to germinate toward the side where more calcium enters.

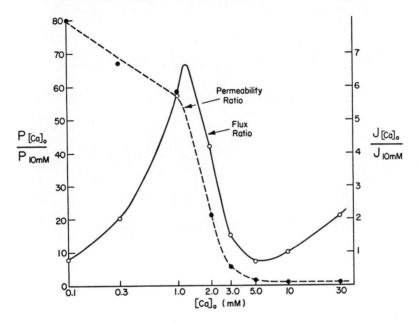

Fig. 9. _Ca flux and permeability ratios as a function of external Ca concentration. The solid curve shows the Ca flux divided by the flux at 10 mM external Ca. The permeability ratios (filled circles) were calculated by dividing the flux ratios by the corresponding concentration ratios._

So far, we have concentrated on the cytoplasm as the target of transcellular electric fields. A second possible target is the plasma membrane. We have recently engaged in a theoretical and experimental study of the effect of small applied electric field on charged, mobile particles within the plane of the plasma membrane (Jaffe, 1976; Poo and Robinson, 1976). Theoretical considerations indicate that a potential difference of a millivolt across a cell could produce a 10% gradient of a population mobile, charged particles in the membrane. Experiments on cultured embryonic muscle cells show that the concanavalin A receptor is grossly redistributed by a voltage drop of 10 mV across the 30-μm width of the cells in 4-5 hr (Fig. 10) and that a detectable asymmetry occurs with a voltage drop of less than a millivolt across the cell after 10-20 hr. These experiments were done by passing a current via agar bridges through the medium in which the cells were growing for various times. The field was then removed and FITC-Con A added; after washing off the FITC-Con A, the cells were fixed in cold acetone

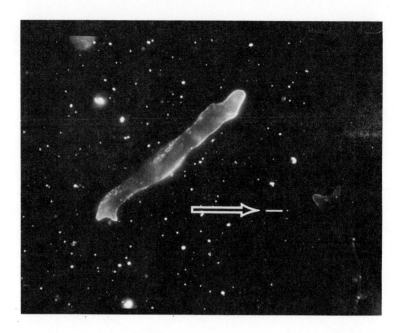

Fig. 10. An embryonic muscle cell which was stained with FITC-Con A after it had had a voltage of 12 mV across it for 4½ hr. The negative electrode was to the right. Cells which were not placed in an electric field showed uniform fluorescence at the periphery.

and examined in a fluorescent microscope. In control cells, a bright, uniform ring of fluorescence was seen at the edge of each cell, but after the field was applied, the ring faded on the positive side and became brighter on the negative side of each cell. We calculate that the electrophoretic mobility of the Con A receptor in the membrane is about 5×10^{-4} μm/sec per V/cm. In another experiment, the return to a symmetric distribution was followed after the receptors had been moved to one side with an electric field. From this information, we estimate the diffusion coefficient of the Con A receptor to be $1-2 \times 10^{-11}$ cm²/sec. It should be emphasized that both of these numbers, the mobility and the diffusion coefficient, are estimates for the unlabeled receptor, not the receptor-Con A complex.

ELECTRICAL CURRENTS AND CALCIUM ACCUMULATIONS IN POLLEN

The vibrating electrode has been used to explore the electric fields around lily pollen germinating in vitro (Weisenseel et al, 1975). Since the pollen does not stick itself to the substratum as fucoid eggs do, it was necessary to isolate the pollen from the mechanical disturbance of the vibration. This was accomplished by growing the pollen on the underside of an ion-permeable, 3-μm thick cellulose membrane. The pollen was held to the membrane by surface

26

tension while the electrode was in the medium above the membrane.
The currents associated with germinating pollen resemble the ones found around fucoid eggs. The pollen current, considered as a flow of positive charge, enters the prospective growth site of the ungerminated grain and leaves the opposite end (see Fig. 11).

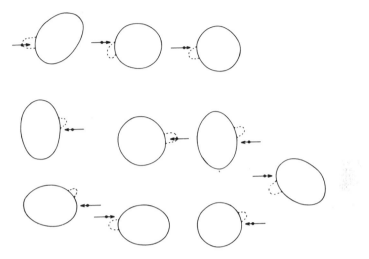

Fig. 11. *The point of entry of current in ungerminated lily pollen grains indicates the point at which germination will occur. The arrows show the location of the maximum inward current before germination and the dotted lines show the location of the subsequent outgrowth.*

After germination, the current enters the growing tip and the region behind it while leaving the part of the tube nearest the grain itself (see Fig. 12.) The current also has two components, a steady one and a pulsed one. The steady current at the tip of tube is typically 250-300 nA/cm^2. The current pulses appear only after the tube reaches a length of 1 mm. The entry for the pulses in concentrated more toward the growing tip than is the entry region for the steady current.
Since these pollen grains will grow in a simple medium whose only ions are K^+, H^+, and Ca^{++}, it is easy to determine the ionic nature of the current. Weisenseel and Jaffe (1976) have found that increasing K^+ stimulates the steady current while decreasing K^+ immediately inhibits it. Comparable changes in H^+ have the opposite effect while changes in Ca^{++} have little effect. They conclude that the steady growth current is carried in by a potassium leak and out by a proton pump and that a minor, but controlling, element of the inward current is Ca^{++}. That the current is necessary for growth is indicated by their finding that growth is rapidly and reversably blocked when the current is stopped by removing extracellular K^+. Weisenseel and Jaffe argue that this effect may

27

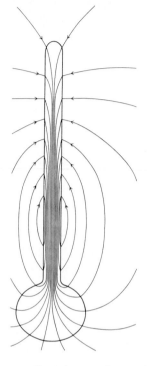

Fig. 12. The current pattern around and inside growing lily pollen. The internal current pattern is inferred from the external one, which was measured with the vibrating electrode.

be due to the change in membrane potential caused by the change in potassium, which in turn reduces calcium entry.

The most dramatic evidence of the involvement of calcium in pollen tip growth is the finding by L.A. Jaffe (Jaffe, Weisenseel and Jaffe, 1975) that calcium is localized at the growing tip. They grew lily pollen in ^{45}Ca and then made autoradiographs of the whole grains at -78°C. Photometric measurements of these autoradiographs showed that the calcium concentration at the tip was 2-4 times as great as in the rest of the tube.

SUMMARY

The unfertilized fucoid egg is a naked protoplast, lacking a cell wall or any protective coating beyond the plasma membrane. Fertilization triggers the release of preformed cortical vesicles which leads to the rapid formation of a cell wall. Two further phases of cell wall secretion occur; the first involves secretion over the entire zygote (although somewhat localized at the prospective rhizoid), and the second is highly localized at the pre-rhizoid region.

As the cell wall is formed, a turgor pressure is developed and the zygote begins to drive an electrical current through itself. A substantial part of this early current in unilaterally illuminated cells is localized calcium entry at the presumptive rhizoid region. We believe that the function of the current is the establishment of an internal calcium gradient which, in turn, results in a cytoplasmic electric field. This field, we think, is a key link between the external light signal and the ultimate response of the cell: the formation of a rhizoid at the darker end where more calcium enters. There are two potential targets of the electric field -- charged particles or molecules in the cytoplasm or charged entities in the plasma membrane.

Germinating lily pollen is found to have electrical currents associated with its growth similar to the fucoid egg. Blockage of this current by reducing extracellular potassium also blocks elongation.

REFERENCES

Allen, R.D., Jacobsen, L., Joaquin, J., & Jaffe, L.F. (1972) *Dev. Biol. 27:* 538.

Baker, P.R. (1972) *Progr. Biophys. Mol. Biol. 24:* 185.

Gaffey, C.T. & Mullins, L.J. (1958) *J. Physiol. (London) 144:* 505.

Gradmann, D., Gottfried, W. & Glasel, R.M. (1973) *Biochem. Biophys. Acta 323:* 151.

Jaffe, L.A., Weisenseel, M.H. & Jaffe, L.F. (1975) *J. Cell Biol. 67:* 488.

Jaffe, L.F. (1966) *Proc. Natl. Acad. Sci. 56:* 1102.

Jaffe, L.F. (1976) Submitted to *Nature.*

Jaffe, L.F. & Nuccitelli, R. (1974) *J. Cell Biol. 63:* 614.

Jaffe, L.F., Robinson, K.R. & Nuccitelli, R. (1975) *Ann. N.Y. Acad. Sci. 238:* 372.

Levring, T. (1952) *Physiol. Plant. 5:* 528.

Mullins, L.J. (1962) *Nature 196:* 986.

Nuccitelli, R. (1975) Ph.D. Thesis, Purdue University, West Lafayette, IN.

Nuccitelli, R. & Jaffe, L.F. (1974) *Proc. Natl. Acad. Sci. 71:* 4855.

Nuccitelli, R. & Jaffe, L.F. (1976) *Planta,* In press.

Peng, H.B. & Jaffe, L.F. (1976a) *J. Cell Biol.,* In press.

Peng, H.B. & Jaffe, L.F. (1976b) *Planta,* In press.

Poo, M.M. & Robinson, K.R. (1976)- submitted to *Nature.*

Robinson, K.R. & Jaffe, L.F. (1975) *Science 187:* 70.

Weisenseel, M.H. & Jaffe, L.F. (1972) *Dev. Biol. 27:* 555.

Weisenseel, M.H. & Jaffe, L.F. (1974) *Exp. Cell Res. 89:* 55.

Weisenseel, M.H. & Jaffe, L.F. (1976) *Planta,* In press.

Weisenseel, M.H., Nuccitelli, R. & Jaffe, L.F. (1975) *J. Cell Biol. 66:* 556.

BIOPHYSICAL MODEL FOR PLANT CELL GROWTH: AUXIN EFFECTS

Paul B. Green, Keith Bauer, and W. Raymond Cummins

Stanford University

INTRODUCTION

In the most general terms the growth of the plant cell is to be viewed as the yielding of the cell wall to the pressure of the vacuole, the gain in volume being mainly due to the uptake of water. The relation describing the yielding process is

$$\underline{r}_s = \underline{m}_s \ (\underline{P} - \underline{Y}_s) \tag{1}$$

where $\underline{r}$ is rate, $\underline{P}$ is turgor pressure, $\underline{Y}$ is a yielding threshold below which the wall is inextensible, and $\underline{m}$ is an operational "extensibility" coefficient relating rate to turgor in excess of $\underline{Y}$. The subscript refers to steady rate. At the same time the uptake of water is governed by

$$\underline{r}_s = \underline{L} \ (\Delta\pi - \underline{P}) \tag{2}$$

where $\underline{L}$ is hydraulic conductivity, and $\Delta\pi$ is the osmotic value of the cell contents above its environment. Comparison of (1) and (2) presents the paradox that high $\underline{P}$ promotes rate in (1) and inhibits it in (2). Combining these two relationships into a single expression (Lockhart, 1965; Ray, Green and Cleland, 1972), however, provides a resolution of the paradox.

$$\underline{r}_s = \underline{m}_s \ \underline{L} \ (\Delta\pi - \underline{Y}_s)/(\underline{L} + \underline{m}_s) \tag{3}$$

The form of (3) is such that if $\underline{m}$ or $\underline{L}$ is very large, rate will be relatively insensitive to it and (3) will reduce to (1) or (2). If $\underline{m}$ is large, a small excess of $\underline{P}$ over $\underline{Y}$ suffices to extend the cell

30

rapidly via (1), and this same low $\underline{P}$ favors water entry via (2). If $\underline{L}$ is large, then $\underline{P}$ need only be slightly smaller than $\Delta\pi$ to give a good growth rate via (2) while this same high pressure acts to extend the cell via (1). There is evidence that in some plants at least $\underline{L}$ is large relative to $\underline{m}_s$. In <u>Nitella</u> (Green, Erickson and Buggy, 1971) a given osmotic shift in the bathing medium is reflected almost immediately in a corresponding change in turgor as measured inside the cell. Also, grass coleoptiles that have been germinated in penta-erythritol solutions and studied in a continuous perfusion chamber, can be shown to be nearly in osmotic equilibrium, a condition where $\underline{L}$ is large relative to $\underline{m}_s$ (Green and Cummins, 1974). This has the benefit of reducing the general expression (3) back to an approximation of (1) where $\underline{P} \approx \Delta\pi$:

$$\underline{r}_s = \underline{m}_s \ (\Delta\pi - \underline{Y}_s) \tag{4}$$

Further physiological simplification comes from the findings that growth stimulation by auxin is not accompanied by an increase in $\Delta\pi$, nor is the slowing of growth in aging (-IAA) accompanied by a decrease in $\Delta\pi$ (Ordin, Applewhite and Bonner, 1956). Hence changes in the steady growth rate presumably relate mainly to changes in $\underline{m}_s$ or $\underline{Y}$. In the case of auxin stimulation, batch culture experiments (Cleland, 1959) and flow chamber experiments (Green and Cummins, 1974) indicate that the change is mainly in $\underline{m}_s$, apparent wall extensibility (Fig. 1).

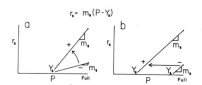

Fig. 1. Graphical representations of alternate explanations for the observation that both high turgor (P) and IAA (+) promote steady rate (r_s). Action of the hormone, arrow, could take either of two extreme forms, a vs. b. Case a is favored, the slope of the relation between steady rate and P is changed, rather than the intercept (Y). Present work shows that the change in m_s is the result of a resetting of a self-stability system. The physical extensibility of the tissue is not necessarily measured in m_s.

IS WALL-YIELDING A COMPLEX PROCESS?

The simplest possible interpretation of extensibility is that it is a simple reciprocal viscosity. One indication that it is at least not an ordinary reciprocal viscosity is that growth is very closely dependent on metabolism. Ray and Ruesink (1962) showed that rate fell essentially immediately upon permeation of the tissue by metabolic inhibitors. Thus despite the presence of sizable turgor pressure, the growth process does not "coast" as one might expect from the simple interpretation. A second indication comes from <u>Nitella</u> where a direct test for simplicity was applied. If equation (4) is valid as such, then known variations in $\Delta\pi$, readily imposed by varying the tonicity of the bathing medium, should bring on strictly corresponding changes in growth rate. In <u>Nitella</u> this could be done under conditions where the turgor pressure inside the cell was continuously monitored. It was found that a small drop in turgor brought on an immediate cessation of growth, then a gradual partial recovery of the original rate. Response to a sharp turgor increase was similar: a large immediate rise in rate, then downward adjustment to a new somewhat higher steady value (Green, Erickson and Buggy, 1971). This revealed self-stabilization in this alga cell where adjustment clearly did not involve compensatory/changes in turgor pressure. Thus either $\underline{m}_s$ or $\underline{Y}_s$ (or both) appears to be complex.

With a precedent set in an alga, and with the view that a thin yielding structure subjected to pressures of many bars could not extend consistently without some type of control, one can ask whether self-stability is general in cell walls.

IS THERE SELF-REGULATION IN HIGHER PLANTS?

A study relating steady growth rate to turgor pressure (Green and Cummins, 1974) employed abrupt turgor shifts to determine steady rate at a variety of turgor pressures. It was noted that, as in <u>Nitella</u>, a small shift in turgor brought on an "exaggerated" initial response which was later largely compensated. Only the <u>net</u> effects of this behavior are embodied in equation (4), hence the need for the subscript referring to steady rate. Since these results indicated a self-stabilization system in higher plants, it follows that any effective growth regulator must reset the equilibrium value for the system. To study how this is done one must shift his frame of reference, temporarily, from equation (4) and consider the general features of self-regulatory kinetics. A general treatment is appropriate because it is not clear whether $\underline{m}_s$ or $\underline{Y}_s$ in equation (4) is the term displaying regulatory behavior.

WHAT TYPES OF SELF-REGULATION COULD APPLY?

The most generalized approach to the <u>in vivo</u> analysis of regulation is to study the time-course of the response of a system to perturbation.

The simplest possibilities for rate stabilization involve the concept that change in rate ($\underline{dr/dt}$) is a function of the difference between a tendency to accelerate and one to decelerate. When the difference between the two tendencies is zero, the system is stabilized. When a difference exists, this imbalance tends to change rate toward a defined steady rate, $\underline{r}_s$. One possibility can be stated:

$$dr/dt = \underline{A} - \underline{D} \cdot \underline{r} \qquad\qquad (5)$$

where $\underline{A}$ is a constant tendency to accelerate and $\underline{D}$ is a coupling factor relating a tendency to decelerate to the rate, $\underline{r}$, which the system displays at any given time. If $\underline{r}$ is zero, $\underline{A}$ alone acts on the system to accelerate it (dr/dt is positive); if $\underline{r}$ is infinite, the term $\underline{D} \cdot \underline{r}$ will predominate and the system will decelerate (dr/dt is negative). When $\underline{A} = \underline{D} \cdot \underline{r}$, stability has been achieved, dr/dt being zero. This steady rate, $\underline{r}_s$, has the value $\underline{A}/\underline{D}$. A convenient feature of this kind of stability is that the return to a stabilized rate following perturbation is such that the "error" (departure from $\underline{r}_s$) diminishes exponentially with time. See Appendix.

An alternate extreme possibility, equally simple, is

$$dr/dt = \underline{A}'/\underline{r} - \underline{D}' \qquad\qquad (6)$$

where $\underline{A}'$ is a coupling factor relating acceleration to rate, and $\underline{D}'$ is a constant tendency to decelerate. This relation yields the same solution for $\underline{r}_s$, $\underline{A}/\underline{D}$. It embodies a remarkable and testable prediction for the response to extreme reduction in rate, namely extremely high acceleration. (The tendency to accelerate is measured by the quotient $\underline{A}'/\underline{r}$). Hybrid behavior, with $\underline{r}$ involved with both $\underline{A}$ and $\underline{D}$ is possible, of course.

The purposes of the present study are

a. To determine if either extreme type of self-regulation applies, at least approximately:

b. To see which of the two major terms, $\underline{A}$ or $\underline{D}$ ($\underline{A}'$ or $\underline{D}'$) is changed when steady growth rate is altered by (1) auxin addition (2) natural variation and (3) aging.

c. To see if a model suitable for the above can be made compatible with the phenomenon of stored growth. When a tissue is held at low turgor for a period, then returned to full turgor, an ensuing "burst" of growth often yields sufficient extra length so as to almost fully compensate for the length "lost" during the slow growth at low turgor (Fig. 2a-c). This phenomenon, described here in terms of turgor inhibition, is called stored growth (Ray, 1961).

PHYSIOLOGICAL PROCEDURES.

Rye coleoptile sections were the material used. Their preparation as well as the details of the automated growth-measuring apparatus have been described (Green and Cummins, 1974). The most important consideration is the rapid attainment of osmo-elastic equilibrium. Very rapid length changes occur in response to turgor shifts even when the tissue is not growing. These decay with a half-time of about one minute. Comparable growing tissue shows the same type of fast response to be superimposed upon a slower (half-time about 5 min) adjustment in rate which is taken to be the

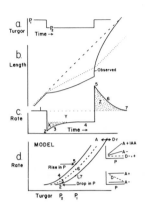

Fig. 2. *Diagram illustrating "stored growth" (a, b, c) and a general biophysical model for growth (d). a. Time course of turgor. b. Concurrent inelastic length change, solid line. The acceleration following a turgor drop leads to a new slower steady rate. The "growth burst" after the turgor step-up takes length back to the trajectory it would have remained on had no turgor drop occurred. This is "100%" stored growth. If no regulation or storage had occurred, the system would show the behavior of the dotted line. c. Same events* as in b, but shown as rate data. The length "lost" during adjustment to step-down is the area X, that gained upon step-up is Z. When Z = X + Y, there is 100% stored growth. Numbers sequence these events for application of the model in d. d. The self-stabilizing equation (5) is considered to act on a concave-upward curve relating rate to turgor, as in Fig. 8. This relation moves left or right so as to keep a balance between A and D·r, the opposing terms of the self-stabilizing relation. At steady rate, at high turgor (P_1) the curve is stationary as at 1. A drop in turgor to P_2 causes only a modest drop in rate, 2, because the slope of the curve is gentle. Subsequent upward adjustment in rate, 3, occurs during the rebalancing of the equation. At 4, balance is achieved in terms of new values for A and D. The abrupt rise in P at 5 leads to a large increase in rate because the part of the curve involved has a steep slope. Subsequent adjustment, 6, to restore the original state, 7, involves a displacement of the curve to the right. Since the part of the curve used is steeper, A and D both take on higher values. Inserts show two possibilities for values of A and D that would give the steady rate behavior shown in Fig. 7a.

growth response. In the analysis of the response to turgor shift we therefore ignore the first six minutes so as to study the growth response proper. All solution changes were automated, as was the conversion of length changes to rate changes. Perfusion fluid was 1% sucrose, 10 mM phosphate buffer pH 5.6. Turgor was altered (by nominal amount) with mannitol.

ANALYTICAL PROCEDURES

In equation (5) dr/dt is linear with rate; in (6) it is linear with $1/r$. Thus the conversion of rate as a function of time, to "rate of change of rate" (dr/dt) was necessary. This was accom-

34

plished by measuring rate at about one-minute intervals during a response to turgor shift, and fitting the data with a four-term polynomial, by computer. The derivative of the fitted curve was then plotted against r (or $1/r$). For the determination of $\underline{A}$, $\underline{D}$, ($\underline{A}'$, $\underline{D}'$) a straight line was fitted to the curve of dr/dt vs. r (or $1/r$). In equation (5), $\underline{D}$ is the negative of the slope of the relation, $\underline{A}$ is the intercept on the $\underline{Y}$-axis. See Fig. 3a, b.

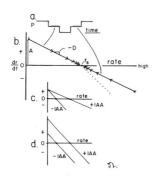

Fig. 3. *Idealized kinetic test for correlating the action of auxin with one of two terms in a self-stability function. The turgor regime in* a *is applied and in* b *one follows the course of rate change during the dotted times in* a. *After turgor step-down the system accelerates,* dr/dt *being positive, but does so in ever decreasing fashion. Time course is symbolized by arrowheads approaching steady rate,* r_s, *exponentially. After turgor step-up, the system* decelerates. *Following equation* (5), A *is determined by the intercept,* D *by the slope, of the line. Hence if mean rate, A/D, is increased by IAA then either D may be decreased, as in* c, *or A may be increased, as in* d *(or both). Dotted line shows that data from step-ups tend to have larger A and larger D than data from step-down. This feature is explained in Fig. 2.*

It is a special feature of (5) that the return from an unusually high (or low) rate to the stabilized value is an exponential function (see Appendix). Thus $\underline{D}$ is the negative slope of a plot of the absolute value or $(r_s - r)$ against time. Analysis of the responses to turgor step-up and step-down, were done separately. Because of the higher rates involved, analysis was mainly on the former response.

NATURE OF SELF-STABILITY

The Pertinent Equation. Of the two extreme possibilities present in equations (5) and (6) our data gave the better fit with (5) as shown in Fig. 4. The extreme acceleration behavior at low rate expected with (6) was not seen. A reasonably good linear decay of $\ln | r_s - r |$ favoring (5) was also confirmed. This has been reported by Green and Cummins (1974). Using equation (5), observed steady rate was reasonably close to calculated steady rate (A/D) as shown in Fig. 5 and Table I. While equation (5) is preferred, the fit is far from perfect. The plots of dr/dt/vs. r were not straight; often an oscillatory response was noted (Fig. 4b). Nonetheless it was felt that the plots were sufficiently diagnostic to allow us to see if a gross change in steady rate

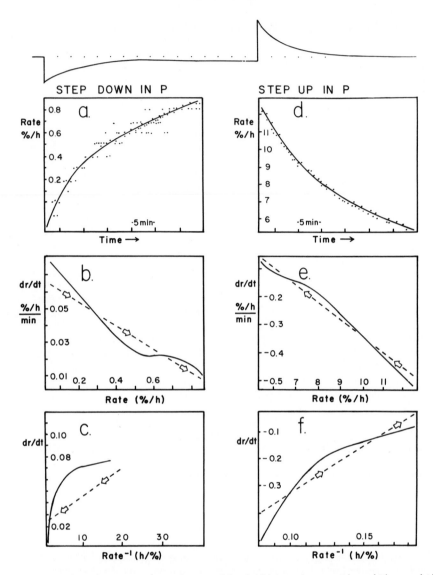

Fig. 4. Kinetic test for the applicability of equation (5) or (6) as the self-stability system. Data on response to turgor drop are in the left column, those to step-up are at right. a, d. Raw data are taken and a polynomial fit is applied (solid line). b, e. The derivative of the above curves is taken and, to test (5), is plotted against rate (solid line). Again a straight line is fitted, dashed line, to give values for A and D as in Fig. 2. Time course of stabilization is shown by the open arrows. c, f. As before but with dr/dt plotted against 1/rate to test equation (6). The fit is worse; hence equation (5) is used in analysis.

caused by IAA had been brought on by a significant change in either A or D.

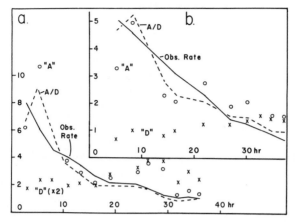

Fig. 5. Time course of the loss of rate with aging, in IAA. The quotient A/D correlates fairly well with observed rate. Values of A tend to fall with time, a slow rise in D is seen particularly after 20 hr. Both tendencies reduce A/D and hence reduce steady rate.

Table 1. Analysis of IAA Action. All sections were 14mm and were preincubated for 1 hr without IAA.

	A	D	A/D	obs. r_s*	n
a) -IAA	0.12	0.13	0.9	1.1	13
b) +IAA	0.54	0.08	6.7	5.5	9
c) +IAA for 5 hr. then -IAA	0.17	0.20	0.8	0.75	12

*Rates are expressed as %/hr after turgor step-up.

Evaluation of A and D in Response to IAA. Following the scheme to measure A and D independently as shown in Figure 3 we found that the increase in the value of r_s, manifest in A/D, was due mainly to an increase in A rather than a decrease in D. Data are summarized in Table 1. In all cases the tissue was depleted of IAA at the start. When we compare the values of A and D of such tissue (a) with comparable tissue that had responded to IAA (b), we find a roughly fivefold increase in rate. The ratio of values of D is 1.6, that of the values of A is 4.5. Hence the effect is due mainly to an increase in A, but also, in part, to a fall in D. When the -IAA values are taken from tissue that had spent several hours growing in the apparatus with IAA before being depleted, the

37

differences are less striking (c). The ratio of values of D̲ is now 2.5, that of A, about 3.3. The ratio of steady rates was 7.3. Of the two types of experiment the former is better for comparison because aging of the tissue is not involved; i.e., the experiments are done in parallel rather than in series.

Natural Variation (+ IAA). Since we studied individual coleoptiles it was possible to characterize naturally "slow" vs. "fast" ones. This was done in the presence of IAA. A given coleoptile would be given repeated shifts in turgor and several values of A̲ and D̲ were obtained. When such values for many coleoptiles are plotted against steady rate, as in Figure 6, it is seen that, as more rapidly growing tissue is examined, the value of A̲ goes up relatively proportionately, while there is a small but significant decline in D̲.

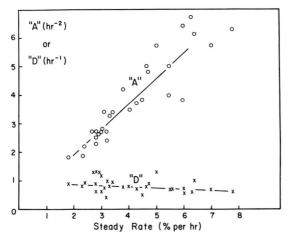

Fig. 6. *Independent measurements of A and D are plotted against increasing steady rate for a variety of rye coleoptiles growing at different rates. Rapidly growing tissue has considerably higher values of A, somewhat lower values of D.*

Aging. Since some of the variation above could be due to the tissue aging while in the chamber, an independent study of aging was carried out. Single coleoptile sections were left in the apparatus for more than a day, + IAA, until their growth rate approached zero. The decline in rate for the first 15 hr (approximately) was due almost entirely to a fall in A̲; thereafter a rise in D̲ was seen to accompany the continued decline in A̲ (Fig. 5).

It is concluded that growth rate can be analyzed kinetically as a self-stabilizing system following equation (5). The resetting of the stabilizing system by auxin (when compared with -IAA tissue studied in parallel) and by simple aging of excised tissue involves primarily the A̲ term. This involves changes in the tendency for the rate to accelerate, as against changes in D̲, the coefficient coupling deceleration to rate. Studies of the basis of rate regulation by gross level of turgor are planned.

A MORE DETAILED MODEL

The analysis so far, and the finding that parameters influencing
steady rate (aging, auxin) mainly influence the acceleration term,
is based on a single aspect of rate kinetics: the decay characteristics following turgor shift. Unexplained observations are that
a) the system shows damped oscillations indicating higher order
differential equations may be appropriate; b) the values of $\underline{A}$ and
$\underline{D}$, while giving about the same ratio for turgor step-up vs. turgor
step-down, are typically lower in absolute value following the
step-down (Fig. 3); c) the model in equation (5) predicts the
course of the responses in the two directions to be reciprocal in
character whereas in fact they are not. The area shown as $\underline{X}$ on
the rate vs. time curve in Figure 2c associated with the response
to step-down, is less than that associated with a step-up ($\underline{Z}$ in
Fig. 2c). This means that the length "lost" during adjustment to
step-down is less than that "gained" during step-up; hence one can
say that some potential for growth has been stored during the period of low turgor and released upon return to high turgor; d) the
relation of the regulatory behavior in (5) to the phenomenology in
(4), which described the steady state rates, needs clarification.
We have not studied the oscillatory phenomena (a), but further study
of the rate transients has led to a more detailed model which
accounts for the behavior in (b), (c) and which covers item (d).

Each response to a turgor shift is composed of an initial
immediate rate shift followed by the slow rate adjustment which has
already been analyzed. The initial response, the "true" physical
extensibility that the tissue must have, can also be studied. The
two responses, studied in combinatation, clarify many points.

The magnitude of the initial growth rate shift cannot be directly read off the trace because the observed abrupt rate change is
composed of both an osmoelastic component and the "irreversible"
or growth component. The apparent initial shift in the irreversible
component can be estimated by treating the latter stages of the
rate decay, made up only of the growth component, as an exponential
function and back-extrapolating rate to the time of the turgor
shift (Fig. 7). When this is done, it is found that the initial
shift in rate is less for step-down than step-up, from the same
starting value of turgor (data not shown). The same nonlinearity
is evident in the fact that larger turgor step-ups, made from the
same basal level of turgor, yield disportionately larger shifts
in the immediate rate response. This nonlinearity is easier to
measure and is shown in Figure 8. This is the instantaneous rate
vs. turgor relation; it is presumably the physical "extensibility"
of the walls in the tissue. This is distinct from the r_s vs. $\underline{P}$
relation which reflects only stabilized values of extensivility.

Examination of the reason for the greater area under the step-
up transient ("stored growth") shows that much of it is due to the
larger initial shift in rate (Fig. 2). In a nonregulating system
this would be mysterious. Reciprocal turgor shifts on a fixed
curve, however nonlinear it might be, should give reciprocal

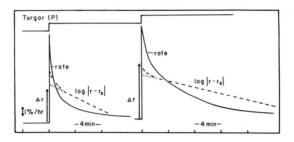

Fig. 7. Characteristization of initial growth rate responses to turgor shift despite the presence of large osmo-elastic effects. The slow phase of the rate adjustment is fitted as a linear exponential function. This (dashed) line is back extrapolated to the time of the shift and the anti-logarithm of the value is taken as the initial growth rate (vertical arrow). That consecutive step-ups in turgor show equivalent, or even increasing, immediate rate responses indicates that the suppression of rate occurs by a change in the intercept, rather than the slope, or an instantaneous rate/turgor relation.

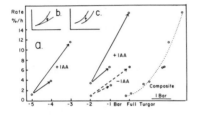

Fig. 8 Rough characterization of the apparent non-linear relation between rate and turgor. The initial rate response (Fig. 7) is shown for two sizes of turgor up-shift (1 and 2 bars), both with and without IAA, at high and low gross levels of turgor. Assuming a single rate/turgor relation applies to all such changes, independent of initial turgor and - or + IAA, the points are displaced to the right to make a composite curve (dotted). Note that in all cases, including the composite, a disproportionate rate increase accompanies a doubling of the size of the turgor shift.

immediate shifts in rate. That the immediate responses are not equivalent for the two directions of shift, reveals that the instantaneous rate vs. turgor has been somehow modified during the self-stabilization between shifts.

Information as to the nature of this modification, and hence the basis of the hysteresis manifest in stored growth, comes from an examination of the immediate rate response to consecutive step-ups in turgor. Considering the rate vs. turgor relation as being concave upward (Fig. 8a), the slow compensatory reduction in rate following the first step-up could be due to a reduction in slope (flattening) of the relation (Fig. 8b) or to a lateral shifting of the relation to the right toward higher turgor values (Fig. 8c).

If it were due to the former, one would expect the initial rate shift in the second step-up to be reduced due to the previous curve flattening. Actually the second rate shift tends to be somewhat larger, so it is concluded that the second type of modification, a lateral shifting, is more likely to occur.

We can now restate equation (5) in more explicit terms, as is done in Figure 2. The opposing accelerating and rate-dependent decelerating processes are manifest as opposing tendencies to shift laterally, a nonlinear instantaneous extensibility curve. This model will display hysteresis of the type seen in stored growth. As shown in Figure 2 an initial high turgor, P_1, is associated with a steady rate shown as $\underline{1}$ on both the rate trace in Fig. 2c and on the instantaneous extensibility relation in Fig. 2d. When turgor is abruptly dropped to P_2, rate drops abruptly, down a gentle part of the extensibility curve. The immediate new value is shown as $\underline{2}$ in Fig. 2c and 2d. This leads to an imbalance in the right-hand members of the stability equation because of the change in $\underline{r}$. At the new low turgor the values of $\underline{A}$ (or $\underline{D}$) are somewhat different so that $\underline{A/D}$ has a slightly lower value. A new balance is approached by a displacement, to the left, of the instantaneous relation, shown as $\underline{3}$ in $\underline{c}$ and $\underline{d}$. A new, lower, equilibrium rate is reached at $\underline{4}$. Because the curve in $\underline{d}$ has been shifted to the left, the return to the original turgor, P_1, involves a steeper part of this curve. Hence the initial shift upward in rate is greater than the previous shift downward. This means that the course of regulation will involve a relatively large adjustment of rate. Hence it will be associated with a great gain in length (area Z). This gain in length upon turgor step-up will exceed the "loss" in length associated with the adjustment to turgor step-down (area X). The net gain is the turgor-related "stored growth".

Other investigars, particularly Cleland (1972) have been unable to detect stored growth, whereas we find it consistently. In terms of the above model, stored growth is a dependent on the self-stability system making large adjustments after being suddenly placed far from its equilibrium rate by abrupt turgor shift. If the turgor shift is gradual, however, as is usually the case in other laboratories where extraordinary measures are not taken to hasten it, the self-stabilization system will "keep up" with it and hence significantly different parts of the instantaneous relation will not be involved. So there will be no hysteresis, and no stored growth.

The final interrelation between equation (4) and (5) cannot be made at present because we are uncertain as to effects of gross level of turgor on $\underline{A}$ and $\underline{D}$. Assume for the moment, however, that reduced turgor lowers $\underline{A}$, just as lack of auxin lowers it. Then the two equations become unified as shown in Figure 2d, upper inset, with the equilibrium behavior of rate reflecting the equilibrium (and instantaneous) values of $\underline{A}$. An alternate possibility, lower inset, is that $\underline{A}$ is insensitive to $\underline{P}$ while $\underline{D}$ falls with $\underline{P}$ (and is insensitive to $\overline{IAA}$). The final valid reconciliation of the two equations will require knowledge of $\underline{A}$ and $\underline{D}$ as functions of turgor

41

as well as of the presence/absence of IAA. Study of this is complicated by the possibility that, during long periods at low mean
turgor, compensatory changes in turgor may occur. Correction for
this in higher plant material is difficult.

At any event the different steady rates seen at different turgor,
which define M_s, appear to be equilibrium values of a self-regulatory system involving lateral displacement of a nonlinear instantaneous extensibility function. This lateral displacement involved in
self-stability (Fig. 2) is independent of the issue of whether or
not a lateral displacement is associated with the stimulation of
steady rate by auxin, the issue in Figure 1.

DISCUSSION

The major qualitative conclusion of this paper is that the yielding cell wall is not properly viewed as a simple visco-elastic structure which is either "softened" or "loosened" when growth is
accelerated. Rather the wall should be analyzed as a self-regulatory system. The system in this case has the properties of a continuous tendency to accelerate rate opposed by a rate-dependent
tendency to decelerate. Several changes in steady rate appear related mainly to changes in the former term. That the regulation
is carried out by the cell wall, rather than a compensatory change
in turgor, is based on observations that the compensation is carried
out rapidly compared to estimates of rates of change in $\Delta\pi$ (see
Green and Cummins, 1974). Also, in <u>Nitella</u>, analogous compensation
is carried out demonstrably by changes in wall properties (Green,
Erickson and Buggy, 1971).

In <u>Nitella</u> the chronic acceleration component of the system,
equivalent to $\underline{A}$ in this study, was sensitive to azide. It is thus
possible that the near immediate sensitivity of coleoptile growth
to such inhibition (Ray, 1961) is mediated by the sensitivity of
the $\underline{A}$ term. The physiological meaning of changes in the $\underline{A}$ term
is not addressed in this work. However, since the term has the
features of a continuous activity which can vary in rate, it could
represent the flux of vesicles to the cell wall. The accelerating
action could be related to more rapid delivery of enzymes, or wall
polymers into the wall, presumably by vesicles (Ray, 1969). The
synthesis of all matrix polymers is promoted by IAA.

The nature of the apparently more conservative $\underline{D}$ term can only
be speculated on. The simplest possibility is that it reflects
some physical property of the wall. The wall's chemical structure
does not change grossly with auxin stimulation (Ray and Baker, 1964).
The physical property of "strain hardening", common to many materials, could be manifest in a rate-dependent tendency for the cell
wall to resist extension. In fact this was implied by Cleland
(1968) in a model for plant growth that had implied self-regulatory
properties. His was an <u>in vitro</u> study, using killed cell walls,
where "strain-hardening" (analogous to $\underline{D}$) was the only parameter
directly measured. The results indicated growth rate was increased
by a lowering of $\underline{D}$, rather than by a raising of $\underline{A}$, in terms of our

analysis. It is possible that our samll decline in D, see in vivo, relates to his in vitro observations.

Ideally the self-regulation changes should be studied in systems where one can monitor turgor continuously (as one can in Nitella) and where there is a good hormone response (lacking in Nitella). Further work should clarify the physiological nature of "A" and possibly confirm the suspicion that "D" is a physical property of the wall.

The relation of the current pH concepts of auxin action (e.g. Cleland, 1976) to the present system is not clear. Abrupt pH changes in our system lead to exceptionally rapid changes in rate which immediately begin to decline. Thus there is no steady state for analysis. Kinetically the effect is as if the instantaneous rate vs. turgor curve were shifted abruptly to the left by low pH, thus accelerating rate dramatically, and then the apparently unaltered stabilizing system brings the curve back to the right. That the rapid effects of auxin, including the pH shift, may be physiologically distinct from the long-term change in steady state brought on by auxin is clearly indicated by the work of Vanderhoef and Stahl (1975).

The present type of in vivo analysis dissects rate into measurable instantaneous and self-stabilizing features. This intricacy allows more detailed examination of the action of regulators, chemical or physical, and presumably reflects the intricacy of the growth process in nature. Many of the described complexities have a potential simple explanation in an IAA mediated flux of polymer to the cell wall, the polymer having an accelerating effect, like grease, when it arrives, but stiffening, like taffy, upon being extended. The observed self regulatory behavior would be manifested, These opposing properties in a single polymer mixture could serve, in principle, just as well as the more populat duality of lysis and (bond) synthesis.

SUMMARY

The rate-limiting component in a general equation for the biophysics of plant growth appears to be the wall extensibility term. If extensibility were a simple physical property of the wall, growth rate (r) should directly reflect variations in turgor pressure, the driving force for growth. Abrupt modulation of turgor pressure in coleoptiles shows, however, that rate changes do not accurately parallel changes in turgor. The departure from expectation reveals the operation of a self-stabilizing system for wall extensibility. Kinetic analysis shows this to consist of a chronic tendency for rate to accelerate (A) normally balanced by a rate-dependent tendency to decelerate ($D \cdot r$). Steady rate A/D. These two factors can be measured independently at different steady rates. Increasing the rate by addition of auxin is associated with a major increase in A, a minor decrease in D. This suggests that the hormone acts mainly to increase the rate of some chronic process acting on the wall ($\underline{A}$), rather than to change a presumably structural coupling of rate to deceleration ($\underline{D}$). Additional kinetic analysis can account for hysteresis in the rate response to turgor shift, the phenomenon of stored growth. The previously simple concept of "wall loosening" is shown to reflect the operation of a non-linear self-stabilizing system.

APPENDIX

The time course of return of rate to a stabilized value, following equation (5), is exponential.

$$d r/d t = \underline{A} - \underline{D} \cdot \underline{r} \tag{5}$$

Steady rate, $\underline{r}_s$, equals $\underline{A}/\underline{D}$ hence $\underline{A}$ is $\underline{r}_s \cdot \underline{D}$ and we may write

$$d r/d t = \underline{D} \cdot (\underline{r}_s - \underline{r}) \tag{7}$$

Rearranging, $1/(\underline{r}_s - \underline{r}) \cdot d r/d t = \underline{D} \tag{8}$

Since $\underline{d}(\underline{r}_s - \underline{r})/d t = -d r/d t \tag{9}$

we may substitute for $d r/d t$ to get

$$[1/(\underline{r}_s - \underline{r})] \cdot \underline{d}(\underline{r}_s - \underline{r})/d r = -\underline{D} \tag{10}$$

which integrates to

$$(\underline{r}_s - \underline{r}) = (\underline{r}_s - \underline{r})_0 \; e^{-\underline{D} \cdot \underline{t}} \tag{11}$$

REFERENCES

Cleland, R.D. (1959) *Physiol. Plant. 12:* 809.
Cleland, R.E. (1968) In *Biochemistry and Physiology of Plant Growth Substances* (F. Wightman & G. Setterfield, eds.) p. 613. Runge Press, Ottawa.
Cleland, R.E. (1972) *Planta 106:* 61.
Cleland, R.E. (1976) *Planta 128:* 201.
Green, P.B. & Cummins, W.R. (1974) *Plant Physiol. 54:* 863.
Green, P.B., Erickson, R.O. & Buggy, J. (1971) *Plant Physiol. 47:* 423.
Lockhart, J.A. (1965) *J. Theoret. Biol. 8:* 264.
Ordin, L., Applewhite, T.H. & Bonner, J. (1956) *Plant Physiol. 31:* 44.
Ray, P.M. (1961) In *Control Mechanisms in Cellular Processes* (D.M. Bonner, ed.). P. 185. New York: The Ronald Press.
Ray, P.M. (1969) *Devel. Biol. Suppl. 3:* 172.
Ray, P.M. & Baker, D.B. (1964) *Plant Physiol. 40:* 353.
Ray, P.M., Green, P.B., & Cleland, R.E. (1972) *Nature 239:*163.
Ray, P.M. & Ruesink, A.W. (1962) *Dev. Biol. 4:* 377.

Vanderhoef, L.N. & Stahl, C.A. (1975) *Proc. Nat. Acad. Sci. U.S.A.* *72:* 1822.

THE OSMOTIC MOTOR OF STOMATAL MOVEMENT

Klaus Raschke

MSU/ERDA Plant Research Laboratory

INTRODUCTION: Function of Stomata

Stomata are small pores in the epidermis of plants, surrounded by pairs of specialized cells called guard cells. These pores open when the guard cells swell; swelling is caused by an osmotic intake of water. The osmotic pressure in the guard cells increases when a low partial pressure of CO_2 in the leaf signals a demand for CO_2; the osmotic pressure decreases when the CO_2 concentration rises or when water stress develops in the plant. In the latter case, a plant hormone, (+)-abscisic acid (ABA), appears to act as a messenger. This hormone is rapidly formed in stressed leaf tissue, presumably in response to a loss in turgor, and then travels to the guard cells. Within minutes, at lower concentrations of ABA after about half an hour, the guard cells begin to lose solutes and the pores narrow. Stomata are parts of a feedback system whose function it is to admit CO_2 to photosynthesizing tissue while preventing loss of turgor. I wish to describe that part of the system which acts as the effector in the regulation of gas exchange; I shall try to summarize how transport and metabolism of osmotica are used by the plant to adjust stomatal aperture. (Stomatal action has been treated more extensively than is here possible. See recent review - Raschke, 1975a).

CHANGES IN STOMATAL VOLUME, PRESSURE AND SOLUTE CONTENT

Knowledge of the volume and pressure changes occurring in guard cells is a requisite for an assessment of the solute requirement for stomatal opening. By taking microphotographs of stomata at various levels of focus and at various stages of stomatal opening we found that the lumina of a pair of guard cells of <u>Vicia</u> <u>faba</u>

had, on an average, a volume of 6.3 pl when stomata were closed, and a volume twice as large when the maximal width of the stomatal aperture measured 15 μm. The relationship between pore width, A (μm), and guard cell volume, V (pl) was linear: $V = 6.3 + 0.43 A$ (Raschke and Dickerson, 1973). The hydrostatic pressure in guard cells was derived from simultaneous measurements of guard-cell volumes and osmotic pressures (by short-term incipient plasmolysis) in epidermal strips bathed in solutions of known water potential. The turgor in guard cells of widely opened stomata of V. faba was often as high as 50 atm (possibly not all species are able to build up such high pressures in guard cells). To our surprise, the increment in turgor needed to produce a unit increment in cell volume decreased with stomatal aperture. The relationship between guard cell turgor, p, volume modulus of elasticity, ε; and the relative increase in guard cell volume, $v = (V-V_0)/V_0$, (with V_0 = guard-cell volume at zero turgor) could be described by $p = \varepsilon v^n$; n was found to have the value 0.67 (Raschke, Dickerson & Pierce, 1973). An n of 0.67 happens to result in a nearly linear relationship between solute content, s (in osmoles), and volume increase, v, of guard cells (Raschke 1976):

$$\frac{ds}{dv} = \frac{\leftarrow V_0}{RT} \{ nv^{n-1} + (n+1)v^n \} \qquad (1)$$

A linear relationship between solute content and stomatal aperture was not only found by us but also by Fischer (1972) on V. faba and by Sawhney and Zelitch (1969) in stomata of Nicotiana tabacum. This linearity appears to be an evolution facilitating the functioning of the stomata as regulators: the amount of osmoticum required to produce a unit change in stomatal conductance for gases remains the same over the whole range of stomatal apertures.

Grasses and sedges have stomata which appear to use their osmotica more efficiently than stomata with kidney-shaped guard cells. The middle portions of these cells have been reduced to narrow tubes, and most of the volume changes occur only in the bulbous ends of the cells. Grass stomata are able to open and close fast (particularly those of maize). The rapidity of stomatal movement in grasses is further enhanced by the presence of two subsidiary cells which function as storages for osmotica and possibly also water.

TRANSPORT OF IONS

The solutes involved in stomatal movement are salts of potassium. The amounts of K^+ taken up by guard cells during stomatal opening are often massive; we found a $[K^+]$ of 0.9 eq l^{-1} in guard cells of open stomata of Vicia faba (Humble and Raschke, 1971). In some species, the whole epidermis serves as a diffuse source of K^+. In others, individual epidermal cells adjoining stomata are found filled with K^+. In still other species, these neighboring cells are also morphologically distinct from ordinary epidermal

48

cells and can be recognized as subsidiary cells.

Measurements with the electron-probe microanalyzer showed that the amounts of K^+ taken up by guard cells of opening stomata are sufficient to account for the measured increases in guard-cell pressure and volume (Table 1). As a matter of fact, there was too

Table 1. K content, volume, osmotic pressure and solute content of guard cells from open and closed stomata of Vicia faba. (From Raschke et al. 1973, based on data by Humble & Raschke 1971).

Stomata	Aperture µm	K content p eq stoma	Guard cell volume, pl*	Osm. pressure at incipient plasmolysis, atm	Solute content, p osmol stoma
open	12	4.24	4.8	80	7.2
closed	2	0.20	2.6	20	1.8
difference between open and closed	10	4.04	2.2	60	5.4

*guard cell volume at incipient plasmolysis: 2.2 pl.

much K^+ in guard cells of open stomata of V. faba if K^+ was associated with a monovalent anion. But a balance between the components determining the chemical potential of water in guard cells was obtained when a divalent anion was assumed. The electron microprobe did not show any changes in the content of guard cells in N, P, or S, and only small changes in Cl. The divalent anions balancing the positive charges of K^+ had, therefore, to be organic. Malate was suspected, and Allaway (1973) showed that malate indeed appeared in the epidermis of V. faba when stomata opened. We now know that aspartate, citrate, fumarate, succinate, and glycerate may also occur, although in smaller amounts than malate (Dittrich and Raschke, unpublished).

There exist conditions which enable guard cells of V. faba to use also Cl^- as counter ion for K^+. Availability of Cl^- appears to be one of them; a low level of Ca^{++} is another (Pallaghy 1970). But there are other conditions affecting the use of Cl^- by guard cells; we do not know them yet. An important role is played by chloride in stomata of Zea mays. On the average, Cl^- accounts for 40% of the negative charges in this species. When stomata are closed, Cl^- is stored in the subsidiary cells, together with K^+. After initiation of stomatal opening, both ions move rapidly into the guard cells. This transfer becomes detectable 1 or 2 min after the opening stimulus has been given. When stomata close again, K^+ as well as Cl^- return to the subsidiary cells. The

total content of K^+ and Cl^- of a stomatal complex of Z. mays does
not change during this shuttle, only the distribution of the two
ions between guard and subsidiary cells does (Raschke and Fellows,
1971).

Maintenance of electroneutrality during stomatal opening is no
problem as long as K^+ and Cl^- are migrating at a ratio of 1:1.
This is, however, rarely the case; transport of K^+ usually exceeds
that of Cl^-. We saw, that in stomata of V. faba, the participation
of Cl^- in stomatal movement can be negligible. In this case, guard
cells secrete H^+ when they take up K^+. This can be shown by placing
epidermal strips with broken epidermal cells but intact guard cells
in solutions of K^+ salts. When stomata are opening, a decline of
the pH of the solution can be monitored. The H^+ excreted by the
guard cells can be titrated. The amount of H^+ released approximately equals the amount of K^+ accumulated during opening (Raschke
and Humble, 1973), as long as a concomitant transport of Cl^- does
not occur.

METABOLISM OF ORGANIC ACIDS

It is an old observation that the starch grains in the chloroplasts of guard cells become smaller or even disappear during
stomatal opening. Possibly, glycolysis, followed by carboxylation
of phosphoenolpyruvate ultimately yields malic and other organic
acids. Their anions are transported into the vacuoles of the
guard cells; the hydrogen ions are exchanged for K^+ with the tissue
surrounding the guard cells. Epidermal tissue possesses phosphoenolpyruvate carboxylase activity in proportion to the stomatal
frequency (Willmer, Pallas and Black, 1973). Epidermal tissue in-
corporates labelled bicarbonate into malate and aspartate, in the
light as well as in darkness (Willmer and Dittrich, 1974), and the
power of epidermal tissue to fix CO_2 resides in the guard cells,
as the early autoradiograms of Shaw and Maclachlan (1954) prove.
These findings demonstrate that carbon-4 of malate is derived from
CO_2. In other experiments, radioactive sugars and glucose-1-phos-
phate were supplied to epidermal strips (Dittrich and Raschke,
unpublished). Again, malate was the metabolite most heavily labelled. We conclude, that sugars can be the source for carbons 1
through 3 of malate; the hypothesis of acid formation from carbohydrates during stomatal opening can be maintained. Further experiments showed that during stomatal closure, malate is disposed of
by three processes occurring simultaneously, (i) leakage from guard
cells, (ii) oxidation in the tricarboxylic acid cycle, and (iii)
gluconeogenesis after decarbonxylation.

EFFECTS OF CO_2 AND ABSCISIC ACID

Some information on stomatal responses to CO_2 and ABA may be
necessary to develop a hypothesis of the mechanism which functions
as the osmotic motor of stomatal movement. Stomatal pores become

narrower in response to an increase in the $[CO_2]$ in the intercellular spaces of leaves. In Zea mays, this response was found to
follow saturation kinetics, with half-saturation of the velocity of
closing occurring near 200 μl l^{-1} in the air (Raschke 1972). ABA
also causes closure. The relationship between $[ABA]$ and stomatal
aperture does, however, not follow a hyperbola; it rather resembles
a Freundlich absorption isotherm (Raschke 1975b). In some species,
there is a strong interaction between responses to CO_2 and those
to ABA. In Xanthium strumarium for instance, and also in cotton
plants (Gossypium hirsutum) before they flower, open stomata will
not close if only one of the two substances is present. ABA and
CO_2 are required simultaneously for the modulation of stomatal
aperture (Raschke 1975b). These observations give clues to a possible mechanism of regulation of stomatal movement by CO_2 and ABA.

THE OSMOTIC MOTOR OF STOMATAL MOVEMENT:
SUMMARY, HYPOTHESIS AND CONCLUSION

Stomata open when the guard cells inflate. The increase in
pressure and volume of guard cells is the osmotic effect of an im-
port of K^+ and Cl^- and formation from carbohydrates and CO_2 of
organic acids, mainly malic. The relationship between stomatal
aperture, guard-cell volume and solute content is linear. The H^+
of the organic acids formed during opening is excreted by the guard
cells in exchange for K^+; the anions accompany K^+ into the vacuole.
During stomatal closure, K^+, Cl^-, and part of the malate are released from the guard cells; the other part of the malate enters
glyconeogenesis or is oxidized in the citric acid cycle. The osmotic pressure in the guard cells decreases in the presence of absci-
sic acid, a plant hormone formed under stress; it also decreases
if the $[CO_2]$ in the intercellular spaces is high. The last statement seems to be in conflict with the earlier one referring to the
requirement of CO_2 for the production of malate during stomatal
opening. This conflict can be resolved, as will be shown below.
Guard cells have a negative transmembrane potential (Pallaghy
1968). It would be the result of an active expulsion of H^+ from
the guard cells followed by (i) an active transport of anions into
the vacuole, (ii) an uptake of K^+ by the cell along the electrochemical potential gradient, (iii) formation of organic acids in
response to alkalinization of the cytoplasm, and (iv) Cl^- may be
exchanged for OH^-, for the same reason. Production of organic
acids should be self-limiting through feedback control of phosphoenolpyruvate carboxylase (Osmond 1976). The cytoplasm will acidify
and its malate content will increase if the rates of removal of
malate and explusion of H^+ cannot keep pace with the rate of malate
formation. I envisage this to happen when the $[CO_2]$ in intercellular spaces increases; this could explain control of stomatal aperture by $[CO_2]$ in the air. Initiation of stomatal closure by high
$[CO_2]$ requires the additional assumption that a low cytoplasmic
pH and a high malate concentration lead to the leakage of ions from
the vacuoles. Jackson & Taylor (1970) reported that nondissociated

aliphatic acids, including malic, caused leakage of ions from roots.
The interaction between CO_2 and ABA points to the possibility that ABA acts on guard cells by inhibiting the expulsion of H^+ (Raschke 1975a,b). Enhancement of malate formation by ABA could be assumed as an alternate explanation, although it would be more difficult to interpret all relevant observations (e.g. inhibition of the opening movement by ABA).

Comparing the view gained on the stomatal mechanism with the knowledge accumulated on other membrane transport (and metabolism) in algae serves homeostasis with respect to hydrostatic pressure or cell volume, guard cells use transport (and metabolism) of ions to vary pressure and volume for the purpose of regulating gas exchange. In engineering terms: the regulator has become a servo mechanism, accomplished by the addition of sensing mechanisms responding to signals from the plant body. The plant appears to be designed in a modular fashion. Insights gained at one front of plant research may, therefore, help to make transparent opaque areas on other fronts.

Acknowledgements. Supported by U.S. Energy Research and Development Administration Contract E(11-1)-1338 and the Deutsche Forschungsgemeinschaft. Part of the work was performed while the author held a John Simon Guggenheim Fellowship.

REFERENCES

Allaway, W.G. (1973) *Planta 110:* 63.
Fischer, R.A. (1972) *Aust. J. Biol. Sci. 25:* 1107.
Humble, G.D. & Raschke, K. (1971) *Plant Physiol. 48:* 442.
Jackson, P.C. & Taylor, J.M. (1970) *Plant Physiol. 46:* 538.
Osmond, C.B. (1976) In: *Transport in Plants.* (Encyclopedia of plant physiology; Vol. 2, pt. A, V. Luttge & M.G. Pitman, eds.) Berlin-Heidelberg: Springer-Verlag.
Pallaghy, C.K. (1968) *Planta 80:* 147.
Pallaghy, C.K. (1970) *Z. Pflanzenphysiol. 62:* 58.
Raschke, K. (1972) *Plant Physiol. 49:* 229.
Raschke, K. (1975a) *Ann. Rev. Plant Physiol. 26:* 309.
Raschke, K. (1975b) *Planta 125:* 243.
Raschke, K. (1976) In: *Transport and Transfer Processes in Plants* (I.F. Wardlaw & J.B. Passioura, eds.) New York: Academic Press.
Raschke, K. & Fellows, M.P. (1971) *Planta 101;* 296.
Raschke, K. & Dickerson, M. (1973) *Plant Research '72, MSU/AEC Plant Res. Lab.,* Mich. State Univ. 153.
Raschke, K. & Humble, G.D. (1973) *Planta 115:* 47.
Raschke, K., Dickerson, M. & Pierce, M. (1973) *Plant Research '72, MSU/AEC Plant Res. Lab.,* Mich. State Univ. 149.
Sawhney, B.L. & Zelitch, I. (1969) *Plant Physiol. 44:* 1350.
Shaw, M. & Maclachlan, G.A. (1954) *Gen. J. Bot. 32:* 784.
Willmer, C.M., Pallas, J.E. Jr. & Black, C.C. Jr. (1973) *Plant Physiol. 52:* 448.

Willmer, C.M. & Dittrich, P. (1974) *Planta 117:* 123.

CONTROL OF PROTON TRANSLOCATION IN THE CHLOROPLAST WATER OXIDATION SYSTEM

R.A. Dilley, R.T. Giaquinta, L.J. Prochaska and D.R. Ort

Purdue University

INTRODUCTION

Energy transduction (ATP synthesis) in cell organelle membranes such as chloroplasts and mitochondria seems to involve the generation and utilization of ion gradients, particularly proton gradients. On the surface, one might think that this topic is somewhat peripheral to the theme of this symposium. However, we shall see that water is quite centrally involved in the membrane transport phenomena associated with the bioenergetics of ATP formation in chloroplasts. In particular, the oxidation of water by the photosynthetic electron transport chain with the concomitant release of H^+, is intimately involved with chloroplast energy transduction.

Mitchell (1966) has provided a conceptual scheme, the chemiosmotic hypothesis, postulating that electron transport generates a transmembrane protonmotive force that is a obligatory stage in the transduction of oxidation-reduction potential energy into chemical work such as ATP synthesis. A further postulate is that the membrane bound coupling factor, or ATPase, interacts in some way with the protons so that, as the proton gradient is dissipated, the energy available is coupled to the formation of ATP. Many, if not most, workers in bioenergetics accept the basic tenets of this conceptual scheme, because it has been so fruitful in providing a rationale framework for formulating and testing hypotheses. The richness of this hypothesis notwithstanding, it is still a mystery as to how the flux of protons through the membrane can be mechanistically linked to ATP formation. What molecular events occur that permit the transduction of the electrochemical potential energy of a proton gradient into the chemical energy stored in ATP?

This report deals with our recent experiments that attempt to probe into one possible class of molecular events, i.e. membrane conformational changes which could be involved in the transduction of energy, Green and Ji (1972) and Boyer and coworkers (1973) have proposed completely different models in which membrane conformational changes are suggested to be essential intermediate energy transduction steps linking electron transport to phosphorylation. Our viewpoint concerning the possible role of conformational changes is presented in papers by Dilley (1967) and Dilley and Giaquinta (1975).

We will present evidence for the hypothesis that electron transport in a certain segment of the chloroplast electron transport chair (between photosystem II and plastoquinone) controls the directionality of protons released during the photochemical oxidation of water. Our model proposes that in the normal course of events the water protons are deposited within the membrane, where they interact with amino acid functional groups, leading to conformational changes in the membrane components. There seems to be a close correlation between this proton-dependent conformational change and the ATP formation coupled to electron transport from water to plastoquinone, although the molecular mechanisms are not yet understood.

CHEMICAL MODIFIERS AS PROBES FOR CONFORMATIONAL CHANGES

The experimental approach we are using to monitor membrane conformational changes involves covalently binding chemical modification reagents. It is expected that major membrane conformational changes will alter the exposure and/or the chemical reactivity of certain amino acid functional groups of membrane proteins, just as has been found in some cases with purified enzyme systems (Timasheff and Gorbunoff, 1967). Implicit in this approach is the requirement that the modification reagent not penetrate to the interior of the proteins or the membrane. Thus, changes in protein or membrane conformation should alter the exposure of certain functional groups to the external media, permitting an enhancement of functional group derivatization. The use of a radioactive reagent provides a convenient measure of incorporation, although flourescent probes have also been used.

A CHLOROPLAST MEMBRANE CONFORMATIONAL CHANGE

We first became aware of an electron transport dependent conformational change in the chloroplast membrane from observations of the differential binding of (^{35}S) diazonium benzene sulfonate (DABS) to chloroplast membranes that are kept in darkness or exposed to light (Giaquinta and Dilley, 1973). The binding of DABS is typically increased about four-fold by conditions permitting electron transport (Table 1). The extra binding is inhibited by the electron transport inhibitor DCMU. An unexpected finding was that cyclic electron transport mediated by phenazine methosulfate (PMS) in the presence of DCMU did not potentiate the extra DABS binding, even though the cyclic electron transport supported the generation of H^+ acculmation (formation of the energized state).

56

Conditions	(^{35}S) DABS $\frac{\text{nmoles DABS}}{\text{mg chl}}$
Light	31.0
Dark	7.0
Light + 20 μM DCMU	7.7
Light + 30 μM PMS	30.6
Light + 20 μM DCMU + 30 μM PMS	8.2
Light + 10 μM Nigericin	36.0

Table 1. Diazonium binding to chloroplast membranes under various conditions. Reaction conditions were as described in Table 3 of Giaquinta and Dilley (1973). Chloroplasts (1 mg chl/ml) were exposed to 2 mM (^{35}S) DABS in either darkness or light for 30 seconds with the additions made as indicated.

In the absence of DCMU, PMS cyclic electron flow is expected to be accompanied by a component of electron flow from water oxidation by photosystem II. The extra DABS binding was not inhibited by uncouplers such as nigericin (Table 1) or chlorocarbonyl cyanide phenylhydrazone (CCCP) as shown in further detail by Giaquinta, Dilley, Selman and Anderson (1974a).

Evidence that the electron transport-induced extra binding of DABS is due to a membrane conformational change is indicated by the fact that salt addition or pH changes (in the dark), induce additional DABS binding (Giaquinta and Dilley, 1974). The salt and pH changes are known to induce membrane conformational changes (Dilley and Rothstein, 1967). Fig. 1 shows that addition of 25 mM $MgCl_2$ or 300 mM KCl to low salt-suspended chloroplasts induces a 1.5-fold increase in DABS binding (Giaquinta and Dilley, 1974).

Other chemical modification reagents show the same pattern of extra binding under photosystem II electron transport conditions, the same response to DCMU inhibition, the last of enhanced binding when cyclic electron flow occurs in the presence of DCMU, and the same insensitivity to nigericin uncoupling. Table II shows typical data obtained with (^{3}H)-iodoacetate, a chemical modifier that reacts with sulfhydryl groups and amines (Means and Feeney, 1971). The level of incorporation is much less than is found with the diazonium reagent, as expected because of the broader reactivity of DABS, but a similar pattern is found.

Likewise, the modification of carboxyl groups by carbodiimides plus a nucleophile showed about a two-fold increase in binding in light compared to dark or light + DCMU conditions (Table III). In this experiment carboxyl groups were activated by 1-ethyl-3(3-dimethylaminopropyl) carbodiimide in the presence of the nucleophile (^{14}C)-glycine ethylester. The nucleophile forms a stable, covalent derivative with the activated carboxyl group (Means and Feeney, 1971).

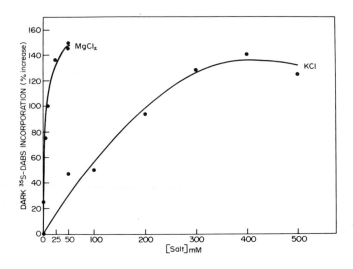

Fig. 1. *Diazonium benzene sulfonate binding to chloroplast membranes affected by salt addition. Chloroplasts were prepared as described by Giaquinta and Dilley (1973) and resuspended in 200 mM sucrose and 10 mM NaH₂PO₄ pH 7.2. Chloroplasts at 0.1 mg chl per ml were suspended in 200 mM sucrose, 2 mM NaH₂PO₄ pH 7.2, and 5 mM (³⁵S)-DABS. Various MgCl₂ or KCl additions were made on separate reaction mixtures and the amount² of DABS incorporated into the membrane was measured.*

It is possible that some part of the extra DABS binding induced by electron transport could be due to a highly reactive species of photochemically reduced DABS. Lockau and Selman (1976) have shown that DABS can act as an electron acceptor and when it is reduced there is apparently a transient state induced having greater chemical reactivity than the parent compound. In our experiments methylviologen or PMS are present as electron acceptors, and these compounds appear to compete quite effectively with DABS for the electron coming from photosystem I. This is evident in the data of Table IV in the case of I⁻ electron donation and methylviologen reduction. In that case, the electron transport does not elicite extra DABS binding (for reasons explained below) nor does the DABS become activated by reduction so as to show the greater chemical reactivity.

The failure of uncouplers to inhibit the extra DABS or iodoacetate binding as well as the inability of cyclic electron transport in the presence of DCMU to elicit the extra binding suggest that the phenomenon which we term a conformational change is not a straight forward aspect of the energized state. In the former case (presence of uncouplers) the energized state (as defined by the capacity to form ATP) is inhibited and in the latter case, (PMS mediated cyclic electron flow in the presence of DCMU) the energized state is attained.

Conditions	$\dfrac{\text{nmoles Iodoacetate}}{\text{mg chl}}$
Exp. 1	
Light	1.65 ± 0.15 S.E.
Dark	1.36 ± 0.12
Light + 20 µM DCMU	1.29 ± 0.19
Exp. 2	
Light + 30 µM PMS	1.38 ± 0.10
Light + 30 µM PMS + 12 µM DCMU	1.17 ± 0.06
Dark	1.22 ± 0.04
Exp. 3	
Light	1.44 ± 0.12
Dark	1.18 ± 0.11
Light + 10 µM Nigericin	1.40 ± 0.04
Dark + 10 µM Nigericin	1.14 ± 0.17

Table II. Iodoacetate incorporation into chloroplast membranes under various conditions. Exp. 1. chloroplasts (0.30 mg chl/ml) were suspended in 25 mM NaCl, 20 mM HPO$_4$, pH 7.6, 0.5 mM methylviologen, 2.5 mM Na ascorbate and 5 mM (^{3}H)-iodoacetate at 20 mCi/mmole. The exposure to iodoacetate was for 30 sec in the various treatments followed by quenching with 0.1 M cysteine. The chloroplasts were washed three times to remove non-membrane bound iodoacetate and the radioactivity determined by scintillation counting. Exp. 2. Reaction conditions were similar to Exp. 1 except phenazine methosulfate (PMS) was the electron transport cofactor. Exp. 3. Reaction conditions were similar to Exp. 1. Valinomycin (5 µM) was also present with nigericin to assure complete dissipation of any K$^+$ gradients that might have been induced by nigericin.

PARTIAL ELECTRON TRANSPORT REACTIONS AND DABS BINDING

Additional experiments showed that the extra DABS binding reflects an aspect of photosystem II function and has no relationship to photosystem I function. These studies are explained in detail in Giaquinta et al. (1974b, 1974c, 1975) and Giaquinta and Dilley (1975). The important results of these experiments can best be represented by considering the partial reactions of photosystem II. Normal electron transport proceeds in the sequence shown below:

$$H_2O \rightarrow PS\ II \rightarrow Q \rightarrow PQ \rightarrow cyt\ f \rightarrow PC \rightarrow PS\ I \rightarrow electron\ acceptor \quad (1)$$

where PS II indicates the photochemical reaction, Q the primary electron acceptor, PQ is plastoquinone, cyt f is cytochrome f and PC is the copper protein plastocyanin. By suitable treatments and the use of particular reagents, it is possible to isolate certain portions of the electron flow as shown:

Treatment	$\dfrac{\text{nmol}(^{14}\text{C})\text{-glycine ethylester bound}}{\text{mg chl}}$	$\dfrac{\mu\text{Eq e}^-}{\text{mg chl-hr}}$
Light	12.5 ± 0.8	350
Dark	6.0 ± 1.2	-
Light + 12.5 µM DCMU	5.2 ± 0.4	15
Dark + 12.5 µM DCMU	6.9 ± 0.6	-

Table III. The effects of light and DCMU upon the 1 ethyl-3(3 dimethyl aminopropyl) carbodiimide mediated incorporation of (^{14}C)-glycine ethyl ester. Chloroplasts were incubated at a concentration of 0.3 mg/ml at 20°C for one minute in 20 mM Imidazole, NaOH pH 7.2, 50 mM NaCl, 1 mM MgCl$_2$, 0.5 mM methylviologen, and 0.5 mM NaN$_3$. Dual red light sources (light intensity = 0.4 kiloergs/cm-sec^2) were projected onto.the chloroplasts for 30 sec prior to chemical modification to activate electron transport. Then 20 mM EDC and 50 mM (^{14}C)-GEE (0.4 mCi/mmol) were added simultaneously to the reaction medium and allowed to react for 1 min. The reaction mixture (2 ml) was then removed with a syringe and immediately injected into 30 ml of 0.2 M Tris-succinate, pH 6.0, to quench the reaction. After centrifugation at 20,000 x g for 10 minutes, the chloroplasts were resuspended in 0.2 M Tris-succinate, pH 7.2. After a similar centrifugation step, the chloroplasts were washed twice in 5 mM HEPES, NaOH, pH 7.6, 200 mM sucrose, 5 mM MgCl$_2$, 0.01% BSA. After a final centrifugation step, the chloroplasts were resuspened in distilled water and 60-100 µgrams of chlorophyll material were placed into liquid scintillation vials. 1 ml of 30% H$_2$O$_2$ was added and the vials were heated at 60°C for 5-7 hours. 10 ml of tritosol liquid scintillator was added to each vial and radioactivity incorporated into the chloroplasts was determined on a Beckman LS-100 liquid scintillation counter.

$$\begin{matrix} \text{exogenous} \\ \text{electron donors} \end{matrix}$$

$$H_2O \qquad\qquad \rightarrow \ PS\ II \rightarrow Q \rightarrow PQ \ \rightarrow \ \rightarrow \ \ldots \qquad\qquad (2)$$

$$\begin{matrix} \text{inhibition by Tris} \\ \text{or NH}_2\text{OH treatment} \end{matrix}$$

$$H_2O \rightarrow PS\ II \rightarrow Q \overset{DCMU}{\underset{\downarrow}{\rightarrow}} \ \ldots \qquad\qquad (3)$$
$$SiMo$$

In diagram (2) the normal H$_2$O oxidation is inhibited by washing the membranes in 0.8 M Tris-HCl buffer or 10 mM NH$_2$OH. Alternate electron donors such as I$^-$ can substitute for H$_2$O as a source of electrons or donors such as catechol (Izawa and Ort, 1974; Ort and Izawa, 1973) can donate both electrons and H$^+$ upon oxidation (see the review by Trebst (1974) for further details and references). Diagram 3 indicates that silicomolybdate intercepts electrons prior to the DCMU block (Giaquinta and Dilley, 1975).

Using these various partial electron transport reactions it was possible to show that the conformational change requires (a) proton release upon oxidation of the photosystem II donor and (b) electron flow beyond Q. The requirement for proton release is shown in Table IV. In the control chloroplasts normal electron transfer from H_2O to methylviologen gave nearly a three-fold increase in DABS binding. With hydroxylamine-treated chloroplasts, H_2O oxidation was inhibited almost completely and the level of DABS binding was reduced to nearly the dark level. When I^- addition restored electron transfer to more than a 10-fold higher rate, there was no additional DABS binding. When Mn^{++} was the exogenous donor to photosystem II, there was about a three-fold increase in DABS binding (see Table 1 of Giaquinta et al. 1974a). The oxidation of Mn^{++} to Mn^{+++} results in the liberation of a H^+ because of the high affinity of Mn^{+++} for OH^- ions. Thus Mn^{++} mimics water as an electron and proton source upon oxidation.

System	Reaction Conditions	Electron transfer rate prior to DABS treatment μeq $(Hr\ mg\ chl)^{-1}$	(^{35}S)DABS Binding $\dfrac{nmoles}{mg\ chl}$
Control chloroplasts			
1. $H_2O \rightarrow$ methylviologen	Light	84	38
2. $H_2O \rightarrow$ methylviologen	Dark	-	14
Hydroxylamine-treated chloroplasts			
3. $H_2O \rightarrow$ methylviologen	Light	10	18
4. $I^- \rightarrow$ methylviologen	Light	105	18

Table IV. Incorporation of $(^{35}S)DABS$ during electron transport to methylviologen in hydroxylamine-treated or tris-treated chloroplasts. For details for lines 1-4 see Table III of Giaquinta et al. (1975).

A qualitatively similar pattern was obtained with iodoacetate binding under conditions of I^- electron donation and methylviologen reduction. In this case the light and dark control treatments ($H_2O \rightarrow MV$) gave 1.47±0.2 and 0.98±0.04 nmoles incorporated per mg chl respectively; and the $I^- \rightarrow MV$ light and dark treatments gave 0.93± 0.05 and 0.94±0.4 nmoles incorporated/mg chl.

When silicomolybdate is used as the electron acceptor in the presence of DCMU, electrons from water do not reach segments of the electron transport chain beyond Q. In the absence of DCMU, silicomolybdate accepts electrons in two places, one site before or at Q just as in the presence of DCMU, and the second site is at or near plastoquinone (Giaquinta and Dilley, 1975). The reduction of SiMo in the presence of DCMU does not result in any light-dependent DABS binding, as shown in Table V. The lower incorporation of DABS in the $H_2O \rightarrow SiMo$ case [90 compared to 174 nmoles $(mg\ chl)^{-1}$] and the lack of any additional binding in the $H_2O \rightarrow SiMo$ (+DCMU)

System	Reaction Conditions	Electron transport Prior to DABS Treatment μeq (mg chl hr)$^{-1}$	(^{35}S)DABS Binding nmoles mg chl
$H_2O \rightarrow$ Ferricyanide	Light	151	174
$H_2O \rightarrow$ Ferricyanide	Dark	-	36
$H_2O \rightarrow$ SiMo	Light	166	90
$H_2O \rightarrow$ SiMo + DCMU	Light	61	39
$H_2O \rightarrow$ SiMo + DCMU	Dark	-	37

Table V. Incorporation of DABS into chloroplast membranes with silicomolybdate as the electron acceptor. Reaction conditions were as given for Table II of Giaquinta et al. (1975).

reaction is consistent with the premise that part of the electron transport in the $H_2O \rightarrow$ SiMo is probably going to SiMo before the DCMU block. The remainder of the electron flow goes through to the plastoquinone region, and that portion apparently is linked to the conformational change.

H^+ UPTAKE, PHOSPHORYLATION AND CONFORMATIONAL CHANGES

The partial electron transport sequence from H_2O to SiMo (+DCMU) does not support phosphorylation or H^+ ion accumulation or the extra DABS binding. Fig. 2 shows that with DCMU present the electron transport from H_2O to SiMo is not coupled to phosphorylation, Giaquinta et al. (1974c). When DCMU is absent the additional SiMo reduction is coupled to ATP formation with a P/e$_2$ ratio of about 0.4. This P/e$_2$ ratio is corrected as described in the legend. The P/e$_2$ ratio near 0.4 is consistent with SiMo accepting electrons near plastoquinone such that just the photosystem II- dependent phosphorylation is activated (Gould and Izawa, 1973).

The uptake or accumulation of NH_4^+ linked to electron transport from H_2O to SiMo (+DCMU) is shown in Table VI. Amine uptake can be used as a measure of internal H^+ accumulation (Crofts, 1968). With DCMU present there is very little NH_4^+ uptake as compared to the case without DCMU. This pattern is consistent with current views that internal H^+ accumulation is an essential part of membrane energization for phosphorylation.

The intriguing feature of these data is that the conformational change measured by DABS binding requires that protons from water oxidation (or some other hydrogen donor) be available within the membrane; conditions coincident with energization for phosphorylation. The trivial explanation that SiMo may be acting as an uncoupler and dissipating the proton gradient is not reasonable for several reasons: (1) $H_2O \rightarrow$ SiMo (-DCMU) is coupled to ATP formation, it is only the $H_2O \rightarrow$ SiMo (+DCMU) that is not coupled, (2) Silicomolybdate does not inhibit any of the other phosphorylatin reactions,

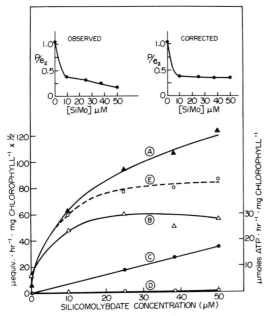

Fig. 2. Efficiency of phosphorylation during silicomolybdate-mediated electron transport in the absence and presence of DCMU using KCN-treated chloroplasts. Inhibition of photosystem I by KCN treatment was according to previously described procedures (Ouitrakul and Izawa, 1973). The reaction mixture for the electron transport and phosphorylation assays contained in 2 ml; 0.1 M sucrose, 50 mM tricine-NaOH (pH 8.0), 3 mM $Na_2H^{32}PO_4$, 0.8 mM ADP, 0.25 mM ferricyanide (FeCy), and chloroplasts equivalent to 40 µg of chlorophyll. Silicomolybdate concentrations were as indicated and the DCMU concentration (when present) was 2.5 µM. Illumination was with heat-filtered white light of approximately 1000 ergs $cm^{-2} sec^{-1}$. Electron transport and ATP formation supported by silicomolybdate are shown by curves A and B, respectively. Curves C and D are for electron transport and ATP formation in the presence of DCMU. Curve E is drawn from points obtained by subtracting curve C from A. Note that electron flow from water to SiMo in the absence of DCMU has a K_m of 10 µM for SiMo whereas the K_m for SiMo in the presence of DCMU in the 35-40 µM. These data are suggestive of two sites of SiMo reduction with markedly different affinities for the electron acceptor.

i.e. PMS or diaminodurene mediated cyclic, the dark acid-base or post-illumination phosphorylation (Giaquinta and Dilley, 1975). Therefore, we propose that in the $H_2O \rightarrow SiMo$ (+DCMU) case the water protons are directly deposited to the external media rather than released within the membrane as must occur in the normal situation. If so we must conclude that electron flow between Q, the primary acceptor of photosystem II, and plastoquinone is somehow necessary to allow protons from water oxidation to be accumulated within the membrane.

RELATION TO MEMBRANE STRUCTURE

Availability of H^+ from water oxidation within the membrane is

System	moles NH_4^+ per mg chl
1. $H_2O \rightarrow$ no acceptor	0.49
2. $H_2O \rightarrow$ SiMo	3.19
3. $H_2O \rightarrow$ no acceptor + DCMU	0.0-0.1
4. $H_2O \rightarrow$ SiMo + DCMU	0.0-0.1

Table VI. NH_4^+ uptake during electron transfer from water to silicomolybdate in the presence and absence of DCMU. The reaction mixture contained in 3 ml: 0.1 M choline chloride, 10 mM Tris-tricine (pH 7.0), 0.33 mM NH_4Cl, and chloroplasts equivalent to 66 μg of chlorophyll. Concentrations of silicomolybdate (SiMo) and DCMU were 33 and 5 μM, respectively. Illumination was with heat-filtered red light (Corning filter 2304) of 200 Kergs cm^{-2} sec^{-1} intensity. NH_4^+ uptake was determined with a Beckman Cationic electrode as the decrease in NH_4^+ concentration in the suspending medium. See Giaquinta et al. (1975) for further details.

required for the conformational change. Protons delivered inside the membrane via the plastoquinone shuttle apparently are not able to reach the site(s) necessary to bring about the conformational change. This is suggested by the fact that I⁻ methylviologen electron transport drives H^+ uptake, apparently via the plastoquinone shuttle, supports ATP formation (Izawa and Ort, 1974) with a P/e_2 of near 0.5, yet does not support the DABS- or iodoacetate-detected conformational change (Table IV). Perhaps there exists a type of compartmentation within the membrane such that protons released by water oxidation have access to regions inaccessible to protons taken across the membrane by the plastoquinone shuttle or by the PMS cyclic system.

It is not clear at this time what relationship the conformational change may have to the mechanism of Site II ATP formation. There seems to be no connection between Site I phosphorylation and the conformational change. It is unlikely that the actual phosphorylation mechanism on the coupling factor differs significantly between Sites I and II. There could well be site-specific differences in the events associated with the delivery of energy to the coupling factor. These studies imply different ways that protons interact with the membrane depending upon their origin. Because proton electrochemical potential gradients are believed to be intimately involved in the energy transduction process, it follows that different membrane-proton interactions could involve different ways of transducing the energy associated with the proton gradients developed in the two different modes.

There is recent evidence from this laboratory concerning the direct utilization of protons within the membrane for driving ATP formation before they equilibrate with the inner osmotic space

(Ort and Dilley, 1976; Ort, Dilley and Good, 1976). This fits well with the concept that protons are possibly compartmentalized within the membrane, depending on their source.

Direct deposition of protons within the membrane seemingly would require association sites such as fixed negative charges. It is known that H^+ accumulation leads to Mg^{++} efflux and in many studies K^+ efflux as well (Dilley and Vernon, 1965). Those results and other studies (Dilley and Rothstein, 1967; Walz, Goldstein and Avron, 1974) clearly establish that chloroplast membranes have abundant fixed negative charges that act as buffering groups for the accumulated H^+, with the mobile Mg^{++} or K^+ ions exchanging out as the protons enter. Protonation of polyelectrolytes generally causes changes in conformation due to alterations of the electrostatic forces within the molecule. It is reasonable to view the internal proton accumulation from water oxidation as a sufficient cause or driving force for membrane component conformational changes.

Two aspects of the conformational change dealt with in these studies remain puzzling. One is the nature of the interaction between electron transport in the Q to plastoquinone region and the water oxidation system that appears to control the directionality of water proton deposition. The other is the mechanistic role, if there is one, of the photosystem II-linked conformational change in energy transduction.

It is possible that cytochrome b559 is involved in the interaction that seems to control the directionality of proton release, although the evidence supporting this is very indirect. Giaquinta et al. (1974b) presented data showing that cyt b559 has its redox potential driven to the low potential form in the presence of DABS and photosystem II electron transport, but not when DABS is presented in darkness or with DCMU present. Further work is necessary to elucidate whether this effect is indicative of cyt b559 being involved in the conformational change in a critical way or whether these effects are incidental.

SUMMARY

Spinach chloroplast thyalkoid membranes undergo an apparent conformational change linked to photosystem II electron transport. The conformational change is measured by the increased binding of chemical modification reagents. The characteristics of the conformational change indicate that it is caused by protons, released from photosystem II water oxidation, being bound to fixed charge groups of membrane proteins.

Protons from water oxidation can be released directly into the outer phase, rather than be accumulated inside the membrane. The condition giving this effect is to not permit electrons to flow between photosystem II and plastoquinone, but to have them go from the photosystem II reaction center directly to silicomolybdate, an acceptor that is reduced by photosystem II even in the presence of the inhibitor DCMU. Under these conditions the membranes do not show the conformational change nor can ATP synthesis be driven by the photosystem II electron transport. It is not known whether the conformational change is an essential aspect of the photosystem II ATP synthesis mechanism or whether it is merely an indicator of H^+ accumulation linked to photosystem II.

ACKNOWLEDGEMENTS

This work was supported in part by NIH Grant # GM19595, N.S.F. Grant # GB30998, and an N.I.H. Research Career Development Award to R.A.D.

REFERENCES

Boyer, P.D., Cross, R.L. & Momsen, W. (1973) *Proc. Nat. Acad. Sci. U.S. 70:* 2837.

Crofts, A.R. (1967) *J. Biol. Chem. 242:* 3352.

Dilley, R.A. (1969) In *Progress in Photosynthetic Research,* Vol. III (Metzner, H. Ed.) p. 1354.

Dilley, R.A. & Giaquinta, R.T. (1975) In *Current Topics in Membranes and Transport,* Vol. 7 (Bronner and Kleinzeller, Eds.) p. 49 New York: Academic Press.

Dilley, R.A. & Rothstein, A. (1967) *Biochim. Biophys. Acta 135:* 427.

Dilley, R.A. & Vernon, L.P. (1965) *Arch. Biochem. Biophys. 111:* 365.

Giaquinta, R.T. & Dilley, R.A. (1973) *Biochem. Biophys. Res. Comm. 52:* 1410.

Giaquinta, R.T. & Dilley, R.A. (1974) *Proceedings 3rd International Congress on Photosynthesis* (Avron, M. ed). p. 883.

Giaquinta, R.T., Dilley, R.A., Selman, B.R. & Anderson, B.J. (1974a) *Arch. Biochem. Biophys. 162:* 200.

Giaquinta, R.T., Dilley, R.A., Anderson, B.J. & Horton, P. (1974b) *Bioenergetics 6:* 167.

Giaquinta, R.T., Dilley, R.A., Crane, F.L. & Barr, R. (1974c) *Biochem. Biophys. Res. Comm. 59:* 985.

Giaquinta, R.T. & Dilley, R.A. (1975) *Biochim. Biophys. Acta 387:* 288.

Giaquinta, R.T., Dilley, R.A. & Ort, D.R. (1975) *Biochemistry 14:* 4392.

Gould, J.M. & Izawa, S. (1973) *Biochim. Biophys. Acta. 314:* 211.

Green, D.E. & Ji, S. (1972) *J. Bioenergetics 3:* 159.

Izawa, S. & Ort, D.R. (1974) *Biochim. Biophys. Acta 357:* 127.

Lockau, W. & Selman, B.R. (1976) *Z. Naturforsch. 31c:* 48.

Mitchell, P. (1966) *Biol. Rev. 41:* 445.

Means, G. & Feeney, R. (1971) *Chem. Mod. of Proteins.* Holden-Day

Quitrakul, R. & Izawa, S. (1973) *Biochim. Biophys. Acta 305:* 105.

Ort, D.R. & Dilley, R.A. (1976) *Biochim. Biophys. Acta. In Press.*

Ort, D.R., Dilley, R.A. & Good, N.E. (1976) *Biochim. Biophys. Acta. In Press.*

Ort, D.R. & Izawa, S. (1973) *Plant Physiol. 52:* 600.

Timasheff, S.N. & Gorbunoff, M.J. (1967) *Ann. Rev. Biochim. 36:* 13.

Trebst, A. (1974) *Ann. Rev. Plant Physiol.* *25:* 423.
Walz, D., Goldstein, L. & Avron, M. (1974) *Eur. J. Biochem.* *47:* 403.

ENERGETICS AND CONTROL OF TRANSPORT IN NEUROSPORA

Clifford L. Slayman

Yale School of Medicine

INTRODUCTION

Water transport per se has been very little studied in fungi. Physiologically significant transpiration has been shown to occur in several species of basidiomycetes, and is probably significant for aerial mycelium in many species. Guttation may also occur in some species, particularly Phycomyces blakesleeanus, where formed water droplets have been reported between the sporangiophore wall and the enclosing lipoid cuticle (Burnett 1968). Any hydrostatic pressure, arising from the hyperosmotic state of fungal cytoplasm --relative to most liquid environments--has been widely assumed to contribute to the conspicuous cytoplasmic streaming of fungi, and particularly to hyphal extension at the growing apices (Rothstein, 1964; Bartnicki-Garcia 1973; Jennings, Thornton, Galpin and Coggins, 1974). A few experiments, such as those of Park and Robinson (1966) on Aspergillus niger and Fusarium oxysporum have shown that hyperosmotic solutions collapse apical pressure (judged by the amount of cytoplasmic extrusion occurring upon rapid lysis of the apices) and simultaneously interrupt growth, apparently without halting the accumulation of precursor molecules for wall extension. Certain fungal hyphae also display a remarkable ability to adapt to anisotonic solutions, as judged by a)rehydration of plasmolyzed hyphae (Neurospora crassa) after 5-20 min. in playmolyzing concentrations of sucrose (Slayman and Slayman, unpublished observations); b) resumption of growth within a few minutes, in the experiments of Park and Robinson (1966); and c) the ready adaptation of both terrestrial and marine fungi (excluding phycomycetes) to growth in environments of almost any osmolarity (Jones and Jennings 1964; Jennings 1973). In the case of one species, Dendryphiella salina, this adaptability appears to reside both in a strong transport discrimination between potassium and sodium and in the ability to compensate osmotic challenges by adjusting the intracellular levels of free carbohydrates, particularly mannitol and arabitol (Allaway and Jennings 1970, 1971; Jennings and Austin 1973).

The notion that water transport in biological systems is strictly passive--occurring in response to a chemical activity gradient, to a pressure gradient, or (because of solvent-solute interaction) to the flow of a specific solute--might now be called the Central Dogma of Osmoregulation. None of the experiments on fungi contradict this dogma, so that in searching for mechanisms, we should most profitably focus on the mechanisms of solute transport; and it is in this area that studies on Neurospora have proved particularly useful, both with respect to fungal physiology and with respect to general models and mechanisms of transport, applicable to all organisms. Without question, the single feature which most clearly distinguishes transport in Neurospora--and in other fungi, in bacteria and plants, and in organelles such as mitochondria and chloroplasts--from transport through the plasma membranes of animal cells is the relative independence of transport and stabilization of osmotic pressure. For most microorganisms and plant cells, encasing walls able to withstand turgor pressures of 10 atm. or more replace the sodium pumps of animal cells; and, presumably, for intracellular membraneous organelles the colloid osmotic pressure of organellar sap is roughly balanced by that of the surrounding cytoplasm. Consequently, organelles and walled cells seem to have developed (or retained) H^+-pumping, rather than Na^+-pumping, as the dominant transport process, and appear also to have developed membrane potentials (electrical gradients), rather than solute concentration gradients, as the primary means for transferring energy among different transport systems.

THE CHEMIOSMOTIC MODEL

Our present view of the organization of transport in the plasma membrane of Neurospora can be summarized in the following model (Slayman and Gradmann 1975). A large portion--perhaps 25% (Slayman, Long and Lu, 1973)-- of the total metabolic turnover of ATP is used to drive a large steady-state efflux of H^+ ions, which serves both to stabilize intracellular pH and (via the return current) to generate a membrane potential near -200 mV (cell interior negative). The major fraction of return current normally flows through a system in which K^+ and H^+ entry are coupled to Na^+ exit (K^+ exit, under steady-state conditions), while a smaller fraction is carried by efflux of organic acid anions or by H^+-coupled uptake or organic solutes. Following a period of special nutrient deficit, e.g., carbon or nitrogen starvation, a major fraction of the return current can be carried through individual nutrient transport systems (e.g., glucose uptake coupled to H^+ influx), in which circumstance the membrane becomes sharply depolarized.

Evidence for the first part of this model, linking the generation of membrane potential to ATP splitting, comes from a detailed
comparison of the time-courses of ATP depletion and membrane depolarization during the onset of respiratory blockage (Slayman, Lu
and Shane, 1970; Slayman, Long and Lu, 1973). Treatment of
Neurospora hyphae with potassium cyanide (1-25 mM, 25°C) causes,
after a 3-5 sec delay, depolarization at a maximal velocity greater
than 30 mV/sec and a time-constant of 6 sec., so that the membrane
potential goes from ca. -200 mV to -50 mV in 10-20 sec. Depolarization is rapidly reversed on washout of cyanide, displaying a
prominant post-inhibitory overshoot if washout is very rapid. Cessation of electron transfer through the mitochondrial respiratory
chair occurs 2- to 3-fold faster with cyanide than does depolarization, which rules out any direct coupling between respiration
and generation of membrane potential. More importantly, depletion
of ATP (to 10% of the control value) occurs exponentially, with a
time-constant identical to that for depolarization, but without
the delay characteristic of the membrane response. Direct plotting
of membrane potential versus ATP for the first minute of cyanide
inhibition yields rectangular hyperbolae, in which the saturating
regions come from the initial period of inhibition, when the membrane response lags behind the ATP response. Fits of the Michaelis-
Menten equation to such plots yield a voltage asymptote greater
than (-) 300 mV and a $K_{1/2}$ for APT of ~ 2 mM at 25°C. Both parameters increase at lower temperatures. The picture emerging from
these physiological results has recently received biochemical support in the finding of Scarborough (1976) that inverted vesicles
of Neurospora plasma membrane, containing an enzymatically identified ATPase, can accumulate thiocyanate (assumed to be a permeant
anion) when provided with ATP.

That pumping of hydrogen ions is the major process linking ATP
hydrolysis to membrane potential has necessarily been inferred
from indirect evidence, since unidirectional fluxes of H^+ cannot
be measured. Ions other than H^+ do not seem to be involved (Slayman 1970). Intracellular levels of K^+, Na^+, or NH_4^+, which might
be pumped outward, have little effect on the membrane potential
of Neurospora; and switching from Na^+/Na^+ exchange to net K^+ uptake
(in cells preloaded with Na^+) has similarly little effect. The
extracellular concentrations of Cl^-, NO_3^-, HCO_3^-, $SO_4^=$, phosphate,
and dimethylglutarate, which might be pumped inward, also have
little or no effect on the membrane potential. On the positive
side, net efflux of H^+, measured with a pH-glass electrode in the
extracellular solution, does correlate qualitatively with the
membrane potential under a variety of circumstances (Slayman 1970;
Slayman and Slayman 1975). However, since the normal membrane
resistance in Neurospora is the order of 10,000 $ohm.cm^2$ (Slayman
1965b), a current of ~200 $pmoles/cm^2.sec$ would be required to
generate a membrane potential of -200 mV; net H^+ efflux of this
magnitude has been observed, but only with cells maintained at pH
9. At normal pH (5.8) maximal net H^+ flux amounts to only about
20% of that required (Slayman and Slayman, unpublished observation)

which implies that 80% of the pumped H^+ efflux must be carried back inward, presumably passively.

That most of this inward H^+ flux flows normally via the potassium transport system is indicated by three lines of evidence: 1) the maximal rate of K^+/Na^+ exchange during accumulation of potassium, as well as the maximal rate of steady-state K^+/K^+ exchange, is in excess of 100 pmoles/cm^2.sec; and these are the only known transmembrane fluxes which approach the calculated H^+ flux in magnitude (Slayman and Tatum 1965; Slayman and Slayman 1968). Both exchanges are energy dependent, and the former (at least) is stopped by depolarization without metabolic blockage (see below). 3) Finally, C.W. Slayman (1970) has described a K^+-transport mutant of Neurospora which has a 3-fold elevated $K_{1/2}$ for extracellular potassium; the mutant displays membrane potentials 10% larger than those of the parent wild-type under comparable conditions, suggestive of an elevated membrane resistance (Slayman and Slayman 1965).

The final feature of our general model for transport is Neurospora--the operation of specific current-carrying transport systems for organic solutes, and particularly for limiting nutrients--has received support from several groups of recent experiments. Coupling of H^+-influx to the uptake of glucose and its analogues was the first example of this to be demonstrated in Neurospora (Slayman and Slayman 1974). The demonstration took advantage of the fact that the organism transports glucose via a low-affinity, energy-independent carrier (Glu I) when growing in glucose rich media, but derepresses the synthesis of a high-affinity, energy-dependent system (Glu II) during 1-2 hours of carbon starvation (Schneider and Wiley 1971). The electrophysiological consequence of adding glucose, or its non-metabolized analogues 3-0-methylglucose (3OMG) and 2-deoxyglucose (2DOG) to carbon-starved cells is illustrated in Fig. 1. On the left are shown an early test with 1 mM glucose (upper trace), and a later test with 2 mM 3OMG (middle) on separate hyphae in the same preparation. The bottom trace is an example of the small effect of glucose on normal, unstarved cells. At the right are shown successive brief trials with glucose and 2DOG (2 mM) on a single hypha from another preparation. All such curves from carbon-starved cells are characterized by rapid depolarization, ranging from 60 to 190 mV (110 to 150 mV, here), followed by a spontaneous partial repolarization. The initial depolarization occurs without significant depletion of intracellular ATP, and is accompanied by net influx of H^+ and sugar in a 1:1 ratio (at least for the non-metabolized sugars) (Slayman & Slayman, 1975). The depolarization and both transport functions display saturation kinetics, with the same apparent $K_{1/2}$ value (glucose:30-50 μm, 3OMG: 80-110 μm, 2DOG: not determined) (Schneider and Wiley, 1971; Slayman and Slayman, 1975; Slayman, Slayman and Hansen, 1977). Maximal H^+ influx through Glu II amounts to 25 pmoles/cm^2.sec, insufficient to give the observed magnitude of depolarization across the normal membrane resistance (Slayman, 1965b), but recent experiments by Dr. U.P..Hansen on the

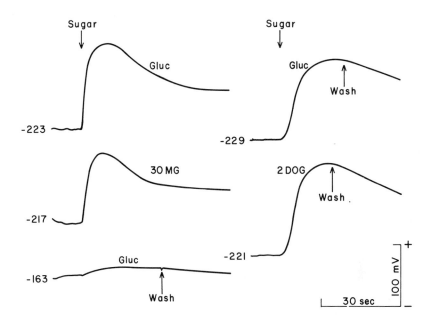

Fig. 1. *Depolarization of carbon-starved hyphae by glucose and its analogues. Mycelial cultures grown overnight (25°C) on cellophane supported by 2% agar in minimal medium (Vogel 1956) with 2% sucrose (Slayman 1965a). Patches of cellophane, with cells attached, were cut from the cultures, washed in distilled water, preincubated in minimal medium without sugar for 3-5 hours (carbon-starvation), then transferred to a recording chamber on the stage of a compound microscope. Voltage measured with KCl-filled microcapillary electrodes, in cells bathed in 0.29x minimal medium + 2.3 mM CaCl$_2$. At the down arrows 1 mM glucose, 2 mM 3-0-methylglucose (3)MG, or 2 mM 2-deoxyglucose (2DOG) was injected. Control record, lower left hand, is from an unstarved hypha. All three left-hand records are from separate preparations. Right-hand records are successive tests from a single hypha, the 2DOG test coming 10 min. after the glucose test. Speed of depolarization limited by solution exchange in the chamber. Partial repolarization in the upper 2 left-hand traces occurred in maintained presence of sugar. Number to the left of each trace is the control membrane potential.*

current-voltage relationship for the H^+/glucose cotransport mechanism indicate that carbon starvation elevates the resting membrane resistance by 2- to 5-fold. Quantitative aspects of the equivalent electrical circuit involved remain to be worked out.

Another charge-carrying transport system evoked in Neurospora by nutrient limitation is that for ammonium ions. As is shown in Fig. 2, within 2 minutes after transferring a culture to nitrogen-free medium (containing 18 mM K^+, 7 mM Na^+, and 3 mM Ca^{++}), 100 μM NH_4^+ produced a detectable depolarization, and that depolarization in-

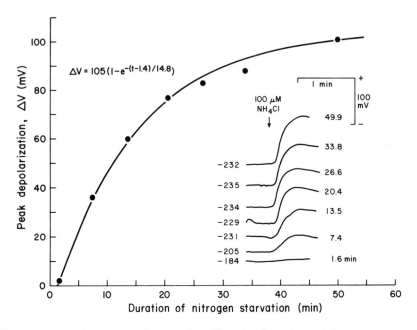

Fig. 2. *Development of ammonium depolarization with progressive nitrogen starvation. Cells grown and handled as in Fig. 1, but nitrogen-starved in 0.29x minimal medium with ammonium nitrate omitted, but 1% glucose present. Inset: Voltage traces for each test with 100 μM NH4Cl, flow of which was begun at the down arrow. Apparent delay in response is due to fluid dead space. Numbers to the left of each trace are the resting membrane potentials; numbers to the right are the times at which NH4Cl was tested (in minutes, after beginning of nitrogen starvation). Washout commenced at the end of each trace. Main plot: Amplitude of each deflection (inset) plotted against test time. Curve fitted by least squares on a semilogarithmic plot.*

creased, along a simple exponential curve with a time-constant of 15 minutes, to a maximum of more than 100 mV. At higher concentrations of ammonium, the time-course of depolarization was more complicated, showing two distinct components (Fig. 3), so that a plot of peak depolarization versus the NH_4^+ concentration (over the interval 3 μM to 100 μM) revealed two saturating ranges, having apparent $K_{1/2}$ values of 7 μM and 8.5 mM. Whether the two components of the depolarizing response to NH_4^+ represent two separate processes in the membrane or represent dual behavior of a single transport system is not yet clear. Whichever may be the case, both components are remarkably specific for NH_4^+ and respond little, if at all, to lipid soluble cations such as triphenylmethylphosphonium ion, to potassium, to neutral amino acids, or even to methylammonium ions (listed to top in Fig. 4). [Fig. 4 also demonstrates the

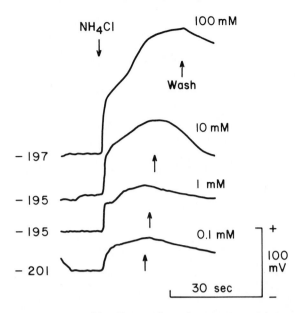

Fig. 3. Biphasic curves of ammonium depolarization. Cells managed as in Fig. 1, but ammonium-starved by 3-1/2 hours incubation in 0.29x minimal medium with KNO_3 substituted for NH_4NO_3; sugar also omitted, so that starvation was incomplete (compare lower trace with upper trace in Fig. 2). Left-hand numbers, control membrane potentials; right-hand numbers, test concentrations of NH_4Cl. Experiment of Dr. N.A. Walker.

response to 10 mM NH_4Cl, of a hypha which had been nitrogen starved; and, as a control, the response of a normal hypha to metabolic

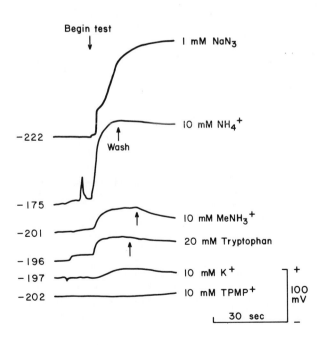

Fig. 4. Effects of various ions on nitrogen-starved _Neurospora_. Cells handled as in Fig. 2. Conventions as in Fig. 3. Tests with $MeNH_3Cl$, tryptophane, and triphenylmethylphosphonium bromide ($TPMP^+Br^-$) were made on the same preparation; other tests all on separate preparations. Upper curves shows a typical effect of the respiratory inhibitor NaN_3, for comparison with the ammonium effect; voltage calibration bar for this trace should read 110 mV. Experiment in collab-

oration with Dr. N. A. Walker.

blockage by 1 mM sodium azide.] Although we have not yet made the appropriate flux measurements, the simplest interpretation of the NH_4^+ results (or at least, of the depolarization by micromolar concentrations) is that nitrogen starvation causes derepression of synthesis of a specific NH_4^+ carrier.

Depolarization by ammonium ions also clarifies an unexpected observation made several years ago during studies on competitive inhibition of net potassium transport in Neurospora. Sodium and rubidium acted as simple competitive inhibitors of potassium uptake; but ammonium, in addition to functioning as a competitive inhibitor, also caused a finite delay of potassium accumulation. Four K^+-uptake curves, at different ammonium concentrations, are plotted in Fig. 5. [The experiment was conducted in the following

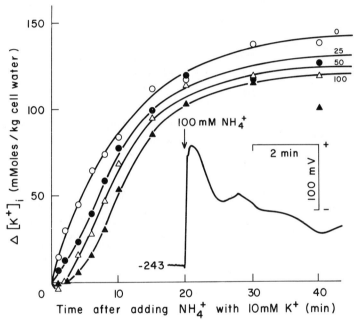

Fig. 5. Influence of ammonium ions on net uptake of potassium. Cells grown in liquid shaking suspensions at 25ºC; standard minimal medium, but with KC1 replaced by Na C1, and 50 µM KC1 added back (Slayman and Slayman, 1968). After 15-16 hours of growth, cells were harvested, washed in distilled water, and preincubated 20-40 min. in a pH 5.8 buffer containing 20 mM dimethylglutarate, 25 mM Na (OH), and 1% glucose. Then KC1 (10 mM) and 0, 25, 50 or 100 mM NH_4C1 were added. At intervals plotted, cells were harvested, rinsed, dried, weighed, and extracted for flame analysis. Each plotted point is the average for two simultaneous trials. Experiment of Dr. C. W. Slayman. Inset: Voltage record from a mature hypha grown as described for Fig. 1, but incubated for 1 hr. in nitrogen-free potassium phosphate buffer. 100 mM NH_4C1 added at the down arrow. Spontaneous repolarization occurred over a 4-min. interval. Conventions as in Fig. 3.

manner. Potassium-depleted cells were prepared by overnight growth on low potassium; after 15 hours the cells were harvested, washed, and resuspended in a pH 5.8 buffer containing 25 mM Na^+. After 20 minutes of incubation (shaking, 25^oC), a solution of ammonium and potassium chlorides was added, and aliquots were taken at appropriate intervals for cellular K^+ analysis.] At higher ammonium concentrations (50, 100 mM), a small net efflux of potassium was actually observed prior to uptake; and at all concentrations the apparent maximal rate of uptake (maximal slope in the curves of Fig. 5) were delayed by 6-8 min. A comparable voltage curve (for 100 mM NH_4^+) is shown in the inset to Fig. 5, revealing one important feature of the NH_4^+ effect that was not mentioned above, <u>viz.</u>, a spontaneous partial repolarization in the maintained presence of ammonium ions. At least two phases are evident in the repolarization curve: one with a time-constant of ~30 sec, which accounts for about 60% of the total repolarization; and the second with a linear (but noisy) recovery for several minutes. We cannot yet draw an exact correspondence between various features of the voltage curve and the inhibition/restoration of K^+ uptake. But both the rough time-courses shown in Fig. 5 and the additional fact that ammonium concentrations below 10 mM had little effect on net flux of potassium suggest that the second (slower) component of depolarization/repolarization, rather than the first (faster) component, must be related to the K^+-flux results.

The results that I have reviewed up to this point fit into the general scheme for transport in prokaryotic organisms and walled eukaryots that was first formalized by Mitchell (1963), and which might be termed the "Generalized Chemiosmotic Hypothesis." As has already been mentioned, it differs from current views of transport in animal cells mainly in giving primacy to hydrogen ions rather than to sodium ions, and in the quantitative significance of membrane potential--rather than ionic concentration gradients--as a driving force.

CONTROL PHENOMENA

However, a number of previously unassembled observations on transport and electrical phenomena in <u>Neurospora</u> indicate that at least one overriding control mechanism must be superimposed upon the basic transport processes. A hint of this is obtained in the data of Figs. 1 and 5; specifically, in the fact that glucose- or ammonium-induced depolarization is followed shortly by spontaneous partial repolarization. Such a tendency to "correct" the membrane potential toward the prior steady-state has been observed repeatedly in the electrophysiological studies on <u>Neurospora</u>. Changes in the ionic composition of the medium (Slayman 1965a), addition or removal of respiratory inhibitors (Slayman 1965b), and temperature shifts all produce similar wave-forms of voltage response.

Several examples of this overshoot-and-correction behavior of membrane potential, during rapid warming or cooling, are shown in Fig. 6. Typical responses to lowered temperature are shown in the

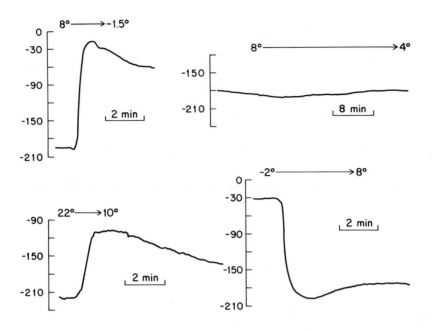

Fig. 6. *Demonstration of the biphasic response of membrane potential to temperature shifts. Cells handled as in Fig. 1, but without carbon starvation; recordings made in a pH 5.8 buffer containing 20 mM dimethylglutarate, 25 mM K(OH), 1 mM CaCl₂, and 1% glucose Temperature shifts completed in the intervals indicated by the lengths of the horizontal arrows. Temperature measured with a microthermocouple, 100 μm tip diameter, placed 20 μm from the hypha. All records from the same preparation; left-hand records from the same hypha.*

left-hand column; these two traces are from the same hypha, and they point to one amusing consequence of the correction process: that successive adaptation to small downshifts of temperature can leave hyphae in a condition where cooling of a few degrees depolarizes nearly as much (from near -200 mV to near -30mV) as cooling from room temperature to 0°C does with unadapted hyphae. The two traces of the right-hand column (Fig. 6) demonstrate the additional point that the magnitude of observable voltage correction is reduced by slowing the temperature shift, a result which probably derives from a decreased <u>rate</u> of the initial voltage shift.

Now, it is obvious that such overshoot-and-correction behavior need not, in the mathematical sense, require the operation of a feedback control mechanism. All that is required is for the environmental change (i.e., temperature shift, addition of an inhibitor) to produce two independent chemical reactions which impinge on the membrane with opposite signs and different time-constants; and that this occur <u>for almost every kind of environmental change</u>!

We might suppose, for example, that a temperature downshift rapidly inhibits the H^+-efflux pump, and causes a slow decrease of membrane leakage conductance. The former change would produce rapid depolarization; the latter, a delayed repolarization. [This type of explanation for the temperature curves in Fig. 6 is being explored, but has not yet yielded definitive experimental results.] Or, a transport process may be feedback-regulated intrinsically, without being part of a larger scheme of metabolic control. A likely example of this is the phenomenon of "transinhibition" which has been described for methionine transport in Neurospora (Pall 1971). Sulfur-starved cells (cells in which the synthesis of a methionine transport system has been derepressed) take up tracer-labelled methionine at a maximal velocity of 2-4 mmoles/kg cell water.min ($K_{1/2}$ = 20-25 µM) when first presented with methionine. If, however, they are first loaded with cold methionine for 15 min. and are then washed and tested with tracer-labelled methionine, the unidirectional influx is inhibited by 60% or more. Transinhibition of methionine transport occurs with other amino acids, but only those (such as ethionine, norleucine, and cysteine) which are structural analogues of methionine. The simplest interpretation of these findings is that intracellular methionine and its analogues bind to, and inhibit cycling of, the methionine carrier.

But more elaborate control mechanism also operate in Neurospora, as is attested by the complex effects of metabolic inhibitors illustrated in Fig. 7. All cells used for this figure were carbon-starved 3 hours or longer; when they were treated with (A) 1 mM cyanide, or (B) 30 µM carbonylcyanide-m-chlorophenyl hydrazone (CCCP, an uncoupler of oxidative phosphorylation). Membrane potential (V_m), intracellular ATP ($[ATP]_i$), and glucose influx via Glu II (J_{in}) were then followed for 10 min. Upon cyanide inhibition, all three parameters fell rapidly and together for 15-20 sec; thereafter, V_m recovered toward 80% of the control value, but J_{in} and $[ATP]_i$ continued downward to a quasi-stable level at about 15% of the control values. With CCCP, on the other hand, membrane potential and glucose influx fell for the first 15 sec, while the ATP concentration fell slowly; but after about 90 sec $[ATP]_i$ and V_m recovered in parallel toward 50-60% of the control values, while J_{in} stabilized at 5%. Since CCCP should depolarize plasma membranes by inducing H^+-permeability (LeBlanc, 1971) before getting to the mitochondria, the relationships before 20 sec. in Fig. 7B were expected, and--indeed--could be taken as further evidences for a current-driven glucose influx. Very unexpected, however, was the total dissociation of glucose flux from membrane potential at longer times and for both inhibitors. Also unexpected was the obvious dissociation of membrane potential from the ATP level after 30-60 sec in cyanide. Evidently, Neurospora possesses one or more mechanisms for restoring its membrane potential, in the face of a metabolic threat, whether or not the ATP level can be kept high. Such mechanisms seem to operate in part by suppressing solute fluxes that would normally depolarize the membrane. These

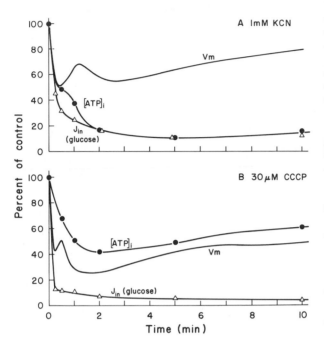

Fig. 7. Dissociation of the effects of metabolic inhibitors on membrane potential (V_m), on intracellular ATP levels ($[ATP]_i$), and on glucose influx via Glu II (J_{in}). Cells for glucose flux measurements were grown in liquid shaking cultures for 15 hours, 25°C, in minimal medium + 2% sucrose. They were then harvested, rinsed, resuspended in sugar-free minimal medium, incubated for 3 hours; and then given 1) KCN or CCCP, followed by 2) 1 mM glycose (^{14}C-labelled, given at the intervals plotted). For each time point, five samples were harvested at 30 sec intervals and assayed by scintillation counting. Each flux was estimated as the slope, at zero-time, of the smooth curve drawn through all 5 points. For the ATP measurements, cells were handled in the same manner as just described, except that after starvation inhibitor was added, and the cells were harvested into petroleum ether-dry ice at the intervals plotted. ATP was assayed with firefly luciferase (Slayman 1973). For voltage measurements, cells were handled as in Fig. 1, but tested with CN or CCCP instead of sugar. All data shown are averages for duplicate or triplicate determinations. CCCP: carbonylcyanide m-chlorophenyl-hydrazone. Control valves - A: V_m equals -227 mV, $[ATP]_i$ equals 2.7 mmoles/kg cell water (mM), J_{in} equals 12.0 mM/min: B: Vm equals -224 mV, $[ATP]_i$ equals 3.0 mM, J_{in} equals 9.4 mM/min. Experiments in collaboration with Dr. U. -P. Hansen and Dr. C. W. Slayman.

two statements are approximately equivalent, electrically, to the idea that membrane potential is stabilized by control of membrane leakage current.

A somewhat more stark manifestation of control processes imping-ing on the plasma membrane of Neurospora is the triggering of overt oscillations in potential by metabolic downshifts, in the respiratory mutant known as <u>poky f</u> (NSX f a). [This strain, which is grossly deficient in cytochromes b and aa₃, has derepressed synthesis of a terminal oxidase which branches from the respiratory chain at the flavoprotein level, thus bypassing phosphorylating sites II and III, but allowing site I phosphorylation and substrate-level phosphorylation in glycolisis and the TCA cycle to function. The "alternate" oxidase is insensitive to cyanide, so that the net effect of maximal cyanide treatment in <u>poky f</u> is a metabolic downshift of ~55%, compared with 98% in wild-type <u>Neurospora</u>.] In <u>poky f</u> the ATP level responds to cyanide by dipping to about 45% of the control value in the first 20 sec, and then recovering to a quasi-steady level of 75-80% after 90 sec (25°C). Extensive metabolic studies (Slayman, Rees, Orchard and Slayman, 1975) led to postulation of a control system that conserves ATP by regulating ATP <u>consumption</u> in concert with ATP generation. The response of membrane potential to cyanide inhibition in <u>poky f</u> appears much more complex, comprising a series of damped <u>oscillations</u>, of which one sample is shown by the top pair of traces in Fig. 8.

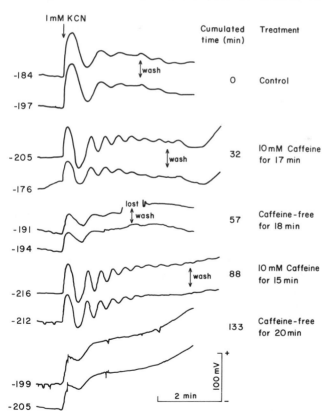

Fig. 8. Cyanide-induced oscilla-tions of membrane potential in poky f Neurospora; modification with caffeine. Cells handled as in Fig. 6. Each pair of curves is traced from records of two hyphae record-ed simultaneously in the same pre-paration. Succes-sive traces made at the time inter-vals indicated in the third column. Treatment describ-ed in the fourth column. Other conventions as in Fig. 3.

81

As is customary, these oscillations began with a depolarizing swing of about 100 mV, followed by nearly complete repolarization (often, hyperpolarization), and then by damped waves for 2 more cycles. Gradmann and Slayman (1975) presented a detailed analysis of inhibitor-triggered voltage oscillations in poky f, from which they concluded a) that the disparity between time-courses of ATP and membrane potential is genuine and not a consequence--for example--of population versus single-cell sampling; b) that associated changes of (slope) resistance are small and are probably a consequence, rather than a cause, of the voltage changes; and c) that the primary electrogenic H^+ pump must oscillate to cause the voltage swings.

Unfortunately, the latter conclusion has one rather awkward consequence. When a quantitative comparison is made between membrane potential and ATP levels (Figs. 14 and 15, Gradmann and Slayman, 1975), the H^+ pump is required to accelerate during periods of relative ATP depletion, which would be contrary to the postulated general metabolic control mechanism in poky f. In the context of general metabolic regulation, the cyanide-induced oscillations would more sensibly arise from control of depolarizing transport systems or membrane leakage currents, which could decelerate during periods of ATP depletion. The main fact which weighed critically against this interpretation in the earlier analysis was the small size of membrane resistance changes during the voltage oscillations. But current-voltage curve measurements (I-V curves) on both the H^+/glucose cotransport system (Hansen and Slayman, 1977; Slayman, Slayman and Hansen, 1977) and the primary H^+ pump (Slayman and Gradmann, 1975) raise the distinct possibility that the major transport systems in the plasma membrane of Neurospora may normally function as quasi-current sources, so that large changes in current (or energy drain) can in fact occur with only small changes in membrane resistance. Obviously, therefore, the origin of voltage oscillations in poky f needs to be reexamined with the aid of detailed I-V curve studies and measurements of the major ionic fluxes (H^+, K^+, and Na^+).

In order to understand the control process which is responsible for voltage oscillations in poky f Neurospora, the non-membrane elements involved, as well as the membrane elements, need to be determined. Some intriguing circumstantial evidence suggests that in Neurospora, as in many other types of cells, cyclic 3',5'adenosine monophosphate (cAMP) may play an important role. [The operation of a cAMP system for control of growth morphology in the organism has been established by Scott and Solomon (1975), and its possible involvement in a "stringent" response to nutrient restriction has also been hinted (DeCarlo and Somberg, 1974).] Certain extrinsic agents, such as quinidine and histamine (Scott and Solomon, 1975) depress intracellular cAMP in Neurospora and also depress the cyanide-induced oscillations in poky f. Other agents, chiefly caffeine, elevate intracellular cAMP and accentuate the voltage oscillations by increasing their frequency and persistence (see the descending sequence of records in Fig. 8). The electrical records require cautious interpretation, however, since caffeine alone evokes transient depolarization (not shown) and removal of caffeine has a prolonged depolarizing action (compare the control traces with the caffeine-free traces at 57 and 133 min). We do not yet have direct measurements of cAMP under the conditions of these experiments.

One final question which should be considered is whether the membrane potential, _per se_, is monitored and controlled; in other words, does the membrane potential lie within the controlling feedback loop(s), or merely aside the loop, acting as a passive voltmeter? Of the various putative control phenomena which I have described, only the cyanide-induced oscillations have yet yielded information on this point. The probable answer is that the membrane potential does not lie within the feedback loop, since imposed depolarizing displacements of ca. 130 mV have no influence on the phase or frequency of the oscillations (Gradmann and Slayman, 1975). Furthermore, once the oscillations have damped out, they cannot be reelicited by hyperpolarizing or depolarizing currents. One variation of an experiment on this point is shown in Fig. 9.

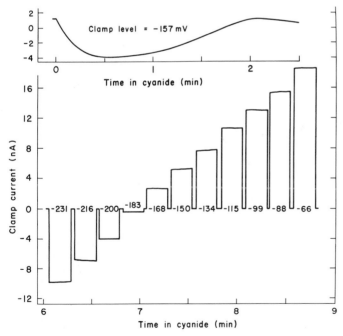

Fig. 9. Failure of membrane voltage steps to reactivate the postulated control mechanism after damping of cyanide-induced oscillations. Poky f strain. Cells handled as in Figure 6. Membrane potential near the voltage-recording electrode was clamped by passing current from another electrode 45 μm away. Clamping current recorded by a current-to-voltage converter, serving as a virtual ground. Resting membrane potential -177 mV. Upper trace: current required to hold the potential at -157 mV for first 2.5 min of cyanide inhibition; lower trace: current required to sustain voltage steps indicated, beginning 6 min after introduction of cyanide. Note absence of systematic drift within the individual pulses. Experiment of Dr. D. Gradmann.

For this experiment oscillations were induced in the usual manner by cyanide, but a point-clamp of the membrane potential near the voltage-recording electrode was imposed by passing current from another intracellular electrode (cable space constant for the hypha was about 500 μm). The current required for clamping was monitored during cyanide administration, and was observed to oscillate as expected (the first two minutes only are shown in the top trace of Fig. 9). After 6 minutes, when no further current swings were visible, the membrane potential was clamped, for 12-15 sec. intervals, at 11 voltage levels lying between -231 mV and -66 mV (resting potential = -180 mV). The current records during these steps were essentially flat, showing no systematic creep which might have indicated a control reaction to reestablish the resting membrane potential.

It is tempting to make some generalizations, which should apply not only to the cyanide-induced oscillations in poky f, but perhaps also to the other examples of "controlled" behavior which were mentioned (especially, Figs. 5, 6, and 7): 1) The primary function of these control processes is to conserve energy by limiting ATP dissipation in response to any restriction of ATP generation. 2) Transport must be regulated in concert with other energy-dissipating reactions both because a failure to do so would displace the cytoplasmic composition far from normal and because transport normally accounts for 25% of the total energy consumption in Neurospora. 3) Since most transport processes across the Neurospora plasma membrane are electrogenic, metabolic regulation is inevitably reflected in the behavior of membrane potential, without the potential per se being regulated. 4) It follows from simple considerations of electric circuits that whenever both the ATP levels and the membrane potential are returned to near normal (by a control mechanism), but at a reduced rate of energy turnover, both the secondary (voltage dissipating) and the primary (voltage generating) processes must be regulated.

SUMMARY AND CONCLUSIONS

Transport systems in Neurospora, freed from the necessity of osmotic regulation, have developed membrane potential as the critical mediary in energy transfer. Potentials near -200 mV are developed by a primary ATP-dependent H^+ pump, and are in turn dissipated for the rapid or concentrative transport of K^+, NH_4^+, sugars, and probably also amino acids and anions. The basic transport phenomena in the organism fit into Mitchell's generalized "Chemiosmotic" hypothesis.

Evidence is now emerging for the existence of one or more control processes superimposed on the basic transport mechanisms and manifest in several phenomena: in spontaneous adaptation of membrane potential to depolarization by transported substrates or metabolic inhibitors; in the differential dissociation of transport fluxes, membrane potential, and ATP levels under different conditions; and in overt oscillations of membrane potential induced in certain circumstances by metabolic inhibitors. The primary function of these control reactions is probably to regulate energy dissipation, in concert with energy generation by the organism.

ACKNOWLEDGEMENTS

The author is indebted to several collaborators, whose material contributions have been noted in the figure legends: Dr. N.A. Walker (University of Sydney), Dr. D. Gradmann (University of Tübingen), Dr. Carolyn Slayman (Yale University), and especially Dr. U. -P. Hansen (University of Kiel), who has most forcefully pressed the notion of transport modulation by control systems. The work has been supported by RCD Award GM-20164 and Research Grant GM-15858 from the U.S. National Institute of General Medical Sciences.

REFERENCES

Allaway, A.E. & Jennings, D.H. (1970) *New Phytol. 69:* 581.
Allaway, A.E. & Jennings, D.H. (1971) *New Phytol. 70:* 511.
Bartnicki-Garcia, S. (1973) *Soc. Gen. Microbiol. Symp. 23:* 245.
Burnett, J.H. (1968) *Fundamentals of Mycology,* pp. 244. New York: St. Martin's Press.
DeCarlo, R.R. & Somberg, E.W. (1974) *Arch. Biochem. Biophys. 265:* 201.
Gradmann, D. & Slayman, C.L. (1975) *J. Membrane Biol. 23:* 181.
Hansen, U.-P. & Slayman, C.L. (1977) In *Coupled Transport Phenomena in Cells and Tissues,* (Hoffman J.F. & Schultz, S.G. eds) New York: Raven Press. In press.
Jennings, D.J. (1973) In *Ion Transport in Plants,* (Anderson, W.P. ed.) p. 323. New York: Academic Press.
Jennings, D.H. & Austin, S. (1973) *J. Gen. Microbiol. 75:* 287.
Jennings, D.H., Thornton, J.D., Galpin, M.F.J., & Coggins, C.R. (1974) *Soc. Exp. Biol. Symp. 28:* 139.
Jones, E.B.G. & Jennings, D.H. (1964) *Trans. Brit. Mycol. Soc. 47:* 619.
LeBlanc, O.H., Jr. (1971) *J. Membrane Biol. 4:* 227.
Mitchell, P. (1963) *Biochem. Soc. Symp. 22:* 142.
Pall, M.L. (1971) *Biochim. Biophys. Acta 233:* 201.
Park, D. & Robinson, P.M. (1966) *Ann. Bot. 30:* 425.
Rothstein, A. (1964) In *The Cellular Functions of Membrane Transport* (Hoffman, J.F. ed.) p. 23 Englewood Cliffs; Prentice Hall.
Scarborough, G.A. (1976) *Proc. Natl. Acad. Sci., U.S.A. 73:* 1485.
Schneider, R.P. & Wiley, W.R. (1971) *J. Bacteriol. 106:* 479.
Scott, W.A. & Solomon, B. (1975) *J. Bacteriol. 122:* 454.
Slayman, C.L. (1965a) *J. Gen. Physiol. 49:* 69.
Slayman, C.L. (1965b) *J. Gen. Physiol. 49:* 93.
Slayman, C.L. (1970) *Amer. Zool. 10:* 377.
Slayman, C.L. (1973) *J. Bacteriol. 114:* 752.
Slayman, C.L. & Gradmann, D. (1975) *Biophys. J. 15:* 968.
Slayman, C.L., Long, W.S., & Lu, C.Y.-H. (1973) *J. Membrane Biol.*

14: 305.

Slayman, C.L., Ly, C.Y.-H. & Shane, L. (1970) *Nature 226:* 274.

Slayman, C.L. & Slayman, C.W. (1965) Abstr. 23rd Internat. Cong. Physiol. Sci. (Tokyo), Item 129.

Slayman, C.L. & Slayman, C.W. (1968) *J. Gen. Physiol. 52:* 424.

Slayman, C.L. & Slayman, C.W. (1974) *Proc. Nat. Acad. Sci. 71:* 1935.

Slayman, C.L., Slayman, C.W. & Hansen, U.-P. (1977) In *Transmembrane Ionic Exchanges in Plants,* (Internat. Wkshp. ed. by A. Monnier & M. Thiellier) C.N.R.S., Rouen/Paris, July 1976. In Press.

Slayman, C.W. (1970) *Biochim. Biophys. Acta 211:* 502.

Slayman, C.W., Rees, D.C., Orchard, P.P., & Slayman, C.L. (1975) *J. Biol. Chem. 250:* 396.

Slayman, C.W. & Slayman, C.L. (1975) In *Molecular Aspects of Membrane Phenomena,* (Kaback, H.R. et al., eds) P.233. Berlin: Springer-Verlag.

Slayman, C.W. & Tatum, E.L. (1965) *Biochim. Biophys. Acta 102:* 149.

Vogel, H.J. (1956) *Microbial Gen. Bull. 13:* 42.

WATER AND SOLUTE REGULATION IN INVERTEBRATES:
CHAIRMAN'S INTRODUCTORY REMARKS

Arthur M. Jungreis

The University of Tennessee

Invertebrates represent the bulk of all animal fauna and occupy
all possible biological niches. Living as they do under both aerobic and naturally anaerobic conditions, in fresh, brackish and
sea water, in salt lakes, on land, in air and in Xeric habitats,
it comes as little surprise that the mechanisms evolved by these
organisms to cope with their incredibly diverse habitats are indeed numerous. From phytophagous insects whose environments are
virtually devoid of Na^+, but which contain an over-abundance of
K^+, Mg^{++} and Ca^{++}, to marine organisms, which must osmoregulate in
environments containing excessive quantities of Na^+ in the absence
of K^+ and the presence of scant quantities of Mg^{++} and Ca^{++}, sur-
vival has caused the evolution of adaptive responses by epithelia
which are unique to specific groups of invertebrates and which
differ completely from those regulatory mechanisms possessed by
plants or chordates.
    The responses of invertebrate cells to changes in the environ-
ment fall into two broad categories, namely osmo-regulating and
osmo-conforming. Coupled with these osmotic regulatory responses
is the near universal ability of invertebrates to regulate volume,
be it intracellular or extracellular. The plethora of adaptive
mechanisms exhibited by invertebrates would be of little value to
the physiologist interested in transport were it not for the phy-
letically unique features possessed by numerous representatives of
virtually all Classes and Orders. Features possessed by most invertebrates include a tubular body plan with tubes frequently of
sufficient size as to permit their excision and use in a chamber
either in the natural state or as flat sheets of epithelia. These
tubes of epithelia are normally composed of only one or two cell
types, and have advantages over vertebrate epithelia both in being
only one cell thick and in having minimal associated musculatures.

87

Tissue specializations characteristic of vertebrates is frequently lacking or less pronounced. For example, integument of acoelomate and pseudocoelomate organisms retain many transport functions normally ascribed exclusively to the gut of coelomate organisms. And lastly, the relative independence from the rigidly defined stenothermal range characteristic of homeotherms facilitates the study of homeostatic control mechanisms involved in solute and water regulation.

One measure of the maturity of a field is the willingness of its proponents to discuss the significance of their observations within the context of adaptive mechanisms for the representative species. Transport workers interested in invertebrates are fast approaching this ideal. However, many workers continue to apologize for studying invertebrates, even to fellow invertebrate physiologists! This sorry state can in good measure be rectified by careful perusal of the ensuing papers on invertebrate epithelia. These papers will deal with volume control by the Crayfish antennal gland, extra-renal salt and water movement by the larval salt gland of brine shrimp, midgut of a Malacostracan prawn, gut of a Nematode, and mantle tissue of fresh water bivalves, and the use of the larval Lepidopteran midgut as a model system for the study of ion transport.

COMPARATIVE ASPECTS OF INVERTEBRATE EPITHELIAL TRANSPORT

Arthur M. Jungreis

University of Tennessee

INTRODUCTION

In insects, fluids, whose osmotic pressures range from hypo-osmotic to hyper-osmotic, are secreted by a variety of epithelia. However, unlike the control mechanisms possessed by most vertebrate tissues, the primary solute involved in effecting water movement across epithelia in insects from a wide range of Orders is K^+ rather than Na^+. A comparative analysis of K^+ transport systems in these insect epithelia has not heretofore been attempted. It is therefore the intent of the author to examine selected aspects of K^+ transport by insect epithelia to learn whether their transport properties are dissimilar or similar to those of Na^+ transporting vertebrate epithelia.

Transport of alkali metal cations into and across epithelia is normally catalyzed by sodium-potassium-dependent adenosine triphosphate phosphohydrolase (Na^+-K^+-ATPase, E.C. 3.6.1.4) (Skou, 1957, 1964). This observation led Keynes (1969) to propose that all epithelial cells possess Na^+-K^+-ATPases. One of the seemingly universal features of Na^+-K^+-ATPases is their sensitivity of inactivation to G-Strophanthin (ouabain) (Glynn, 1957).

A variety of insect epithelia have been reported to possess cation transport mechanisms which are insensitive to ouabain at concentrations up to 1 x 10^{3}M, a concentration several orders of magnitude higher than that needed to fully inhibit cation transport across most mammalian cells (see Dunham and Hoffman, 1971; Vaughan and Cook, 1972; Gardner and Frantz, 1974). Five different insect epithelia have been reported to possess ouabain insensitive K^+ transport systems. Of these, four: integuments of Hyalophora cecropia (Jungreis, 1973; In preparation) and Manduca sexta (Jungreis, In preparation), labial glands of Antheraea pernyi and A. polyphemus

(Kafatos, 1968), midguts of H. cecropia (Haskell, Clemons and
Harvey, 1965; Jungreis and Vaughan, 1976, 1977; Vaughan and Jungreis,
1976, 1977), Danaus plexippus (Jungreis and Vaughan, 1976; Vaughan
and Jungreis, 1977), A. pernyi (Wood, 1972) and M. sexta (Jungreis
and Vaughan, 1976, 1977; Vaughan and Jungreis, 1976, 1977; Blankemeyer, 1977), and Malpighian tubules of Carausius morosus (Pilcher,
1970) were studied in species whose hemolymphs contain little Na^+
(1-3 mM) and high levels of K^+ (22-74 mM) (Table 1). The remaining two tissues: the rectum of Schistocerca gregaria (Irvine and
Phillips, 1971) and the Malpighian tubules of Calliphora erythrocephala (Berridge, 1968), Glossina moritans (Gee, 1976), Locusta
migratoria (Anstee and Bell, 1975), Rhodnius prolixus (Maddrell,
1969), and S. gregaria (Peacock, Bowler, and Anstee, 1972; Maddrell,
Personal Communication) were studied in species whose hemolymphs
contain lower levels of K^+ (6-37 mM) and levels of Na^+ comparable
to those present in blood of vertebrates (97-241 mM) (Table 1).

EVALUATING OUABAIN INSENSITIVITY

A number of questions can be raised regarding the reported ouabain insensitivity of these tissues. Firstly, were the assays
carried out under conditions conducive for the measurement of ouabain inhibition? Secondly, what alternate explanations can be
derived to account for the observed insensitivity? Thirdly, is the
presence in hemolymph of high K^+ and low Na^+ sufficient to define
whether an epithelium is likely to be insensitive to ouabain? And
lastly, can one predict the nature of epithelial sensitivity towards ouabain in insects which have low K^+ and high Na^+ levels in
hemolymph?
Inspection of assay conditions under which the various epithelia
were studied initially revealed only that some were studied in vivo,
some in vitro in hemolymph, and some in vitro in saline (Tables 2,
3). More careful examination revealed a distinct trend, wherein
with the exception of S. gregaria - which contained only 5 mM K^+ -
all the respective incubation media in which rectum, midgut, integument and labial gland were studied contain K^+ at concentrations in
excess of 30 mM (Table 2). Although K^+ at low levels will stimulate Na^+-K^+-ATPases (Rivera, 1975; Jungreis and Vaughan, 1976;
Vaughan and Jungreis, 1977), in its absence, ouabain inhibition of
Na^+-K^+-ATPase can not occur, while in the presence of excess K^+,
inhibition by ouabain is blocked (Glynn, 1957; Gardner and Frantz,
1974; Jungreis and Vaughan, 1976; Vaughan and Jungreis, 1977). Insect tissues known to be sensitive to ouabain have a near maximal
stimulation of their Na^+-K^+-ATPase activity when the concentration
of K^+ in only 5 mM (Rivera, 1975; Jungreis and Vaughan, 1976; Vaughan and Jungreis, 1977), although microsomes derived from Malpighian tubules of L. migratoria were found to be sensitive to ouabain
when the concentration of K^+ in the assay solution was 20 mM. For
example, when the concentration of K^+ is 7.5 mM, Na^+-K^+-ATPase in
neuronal tissue from M. sexta is 50% inhibited by 5×10^{-6}M ouabain
(Jungreis and Vaughan, 1976; Vaughan and Jungreis, 1977), a value

Table 1. Concentrations of Na⁺, K⁺ and Cl⁻ in hemolymphs of insects reported to possess epithelia insensitive to ouabain.

SPECIES	STAGE	mM K⁺	Na⁺	Cl⁻	REFERENCES
Antheraea polyphemus	Adult	54	3	21	2
Antheraea pernyi	Adult	38	3	20	8
Calliphora erythrocephala	Larva	37	148	--	1
Carausius morosus	Adult	18-28	8-15	--	3,12
Danaus plexippus	Adult	74	9	53	7
	Larva	33	7	46	7
Glossina morsitans	Adult	8-12	120-140	100-120	4
Hyalophora cecropia	Adult	22	2	31	7
	Larva	24	2	9	7
Locusta migratoria migratoriodes	Adult-Nymph	35	155	--	3
	Larva	101-282	138-241	--	6
	Larva-fed	29	97	--	5
	Larva-starved	13	108	--	5
	Adult-fed	18	109	--	5
	Adult-starved	11	103	--	5
Manduca sexta	Larva	38	1	29	7
Oncopeltus fasciatus	Adult	80	80	--	9
Rhodnius prolixus	Adult	4-6	158	--	11
Schistocerca gregaria	Adult-starved	11	108	115	10
	Larva	21	187	--	3

(1) Boné, 1944; (2) Carrington & Tenney, 1959; (3) Duchateau, Florkin & Leclercq, 1973; (4) Gee, 1975; (5) Hoyle, 1954; (6) Hoyle, 1956; (7) Jungreis, Jatlow & Wyatt, 1973; (8) Kafatos, 1968; (9) Mullen, 1957; (10) Phillips, 1964; (11) Ramsay, 1953; (12) Wood, 1957.

Table 2. *The concentrations of Na⁺ and K⁺ in salines or hemoly-mphs used to bathe various insect epithelia - other than Malpighian tubules - when sensitivity towards ouabain was being determined.*

SPECIES	K^+(mM)	Na^+(mM)	REFERENCES
Antheraea pernyi	32-38	0-3	7,10
Antheraea polyphemus	54	3	2
Danaus plexippus	7.5	110	6,9
Hyalophora cecropia	32	0	1,3,4,5,10
	7.5	110	6,9
Manduca sexta	7.5	110	6,9
Schistocerca gregaria	5	81	8

(1) Blankemeyer, 1977; (2) Carrington & Tenney, 1959; (3) Harvey & Nedergaard, 1964; (4) Harvey, Wood, Quatrale & Jungreis, 1975; (5) Haskell, Clemons & Harvey, 1965; (6) Jungreis & Vaughan, 1977; (7) Kafatos, 1968; (8) Treherne, 1959; (9) Vaughan & Jungreis, 1977; (10) Wood, 1972.

Table 3. *The concentrations of Na⁺ and K⁺ in salines or hemoly-mphs used to bathe Malpighian tubules when sensitivity towards ouabain was being determined.*

SPECIES	K^+(mM)	Na^+(mM)	REFERENCES
Calliphora erythro-cephala	140	0	2
	56	84	2
	0	140	2
Carausius morosus	18	15	7
Glossina morsitans	10	140	3
Locusta migratoria	8.6	143.5	1,5
Rhodnius prolixus	8.6	129	4
Schistocerca gregaria	8.6	143.5	5
	25	127	5
	20	100	6

(1) Anstee & Bell, 1975; (2) Berridge, 1968; (3) Gee, 1976; (4) Maddrell, 1969; (5) Maddrell, Personal Communication; (6) Peacock, Bowler & Anstee, 1972; (7) Pilcher, 1970.

identical to that calculated by Anstee and Bell (1975) for $\underline{L}$. migratoria Malpighian tubules (in the presence of 20 mM K^+). When the concentration of K^+ is increased to 25 mM, even if the concentration of ouabain is increased 100 fold to 5 x 10^{-4}M, Na^+-K^+-APTase activity is now only 30% inhibited. Thus, studies attempting to document insensitivity of integument, midgut and labial gland toward ouabain are all invalid because the concentration of K^+ in the respective incubation media was excessive. This is not to say that some or all of these tissues are not ouabain insensitive (as I shall now demonstrate), but only that initial experimental conditions were not conducive for the expression of this property by these epithelia.

The rectum of $\underline{S}$. gregaria was assayed under proper conditions and indeed appears to be refractory to ouabain. In the case of Malpighian tubules, with the exception of $\underline{C}$. erythrocephala - which was assayed in salines which would have reversed or prevented ouabain inhibition - tubules of $\underline{C}$. morosus, $\underline{G}$. morsitans, $\underline{R}$. prolixus, $\underline{S}$. gregaria and $\underline{L}$. migratoria were assayed under conditions which would have permitted inhibition by ouabain. Of these, only tubules in $\underline{S}$. gregaria and $\underline{L}$. migratoria are reported to be sensitive to ouabain (Peacock, Bowler and Anstee, 1972; Anstee and Bell, 1975). These results are subject to considerable criticism, since the assay solution was 20 mM in K^+ when tubules from these species were studied, and K^+ stimulated Na^+-K^+-ATPase activity was determined by measuring the rate at which inorganic phosphate was liberated, even though these authors employed 4 mM phosphate buffer in their assay. In addition, Maddrell measuring rates of fluid secretion by intact tubules from these species in vitro was unable to verify their results (Maddrell, Personal Communication). Clearly, the nature of ouabain sensitivity in Malpighian tubules of $\underline{L}$. migratoria and $\underline{S}$. gregaria requires further investigation.

The presence of high levels of K^+ during assay is only one of many reasons why insect epithelia appear refractory to ouabain. A second consideration is the affinity of the specific Na^+-K^+-ATPases toward ouabain, namely how well do the respective Na^+-K^+-ATPases bind ouabain? Neuronal tissues from Monarch butterflies, Danaus plexippus, possess Na^+-K^+-ATPases which have extremely low affinities toward ouabain. Thus, at its maximal solubility in aqueous solution (3 x 10^{-2}M), ouabain is unable to inhibit more than 40% of the (7.5 mM) K^+ stimulated Na^+-K^+-ATPase activity (Jungreis and Vaughan, 1976, 1977; Vaughan and Jungreis, 1976, 1977). A third consideration is whether the respective epithelia do indeed possess Na^+-K^+-ATPases. Keynes' (1969) work would lead one to conclude in the affirmative, yet recent work on ouabain sensitivity of Lepidopteran tissues revealed that midguts of larval and adult stages of $\underline{H}$. cecropia, $\underline{M}$. sexta and $\underline{D}$. plexippus failed to selectively bind ouabain (intact tissues, cell suspensions, and tissue homogenates were measured), and lacked K^+ stimulated ouabain inhibitable ATPase (Jungreis and Vaughan, 1976, 1977; Vaughan and Jungreis, 1976, 1977). Thus the larval Lepidopteran midgut must transport K^+ (and Na^+, see Harvey and Zerahn, 1971) by mechanisms that do not depend upon ouabain sensitive K^+ stimulated Na^+-K^+-ATPases.

OUABAIN INSENSITIVITY AND HIGH HEMOLYMPH Na^+

The questions, "Do integument, labial gland and Malpighian tubules from some or all of the species mentioned earlier possess ouabain inhibitable Na^+-K^+-ATPases?", remain unresolved experimentally. It is the opinion of this author that many of these epithelia also lack Na^+-K^+-ATPases. This view is drawn from several lines of evidence, namely a) these epithelia transport cations electrogenically and preferentially secrete K^+ to the virtual exclusion of Na^+ (Berridge, 1968; Maddrell, 1971; Maddrell and Phillips, 1976), and b) there is virtually no sodium in hemolymph of the respective species. When Na^+ is present at high levels, its presence can serve to reduce net Na^+ uptake across the midgut from the gut contents, in turn reducing the need to excrete Na^+ via the Malpighian tubules. Insects which utilize Na^+ in this fashion have minimal capacities to excrete Na^+ via the Malpighian tubules. This is corroborated by several lines of evidence. In H. cecropia, the concentration of Na^+ varies according to that in the diet (Jungreis, Jatlow and Wyatt, 1973; Harvey, Wood, Quatrale and Jungreis, 1975; Jungreis, 1977) and the Malpighian tubules in this species are unable to excrete appreciable quantities of Na^+ (Irvine, 1969). The concentration of Na^+ in hemolymph of the locust, Locusta migratoria migratorioirdes, reflects that of the diet, with virtually no Na^+ but significant K^+ secretion by Malpighian tubule (see review by Maddrell, 1971) under conditions of starvation (Hoyle, 1954) (see Table 1). In essence, Na^+ may very well represent a passive component in hemolymph of these species, whose presence permits energy conservation via reduction in the need for Na^+ excretion, inasmuch as Na^+ would normally passively re-enter the hemolymph via the gut contents in response to the concentration differential.

DISCUSSION

Insect epithelia are reported to be insensitive to ouabain during Na^+-independent active transport of K^+. Rectum and Malpighian tubule were initially assayed correctly and do indeed appear to be insensitive to ouabain. Integument, labial gland and midgut were initially assayed under conditions which would have blocked detection of an effect of ouabain had it been present. Of these three latter tissues, only midgut has recently been re-examined under the proper conditions of assay. Again it was found to be insensitive to ouabain by virtue of an absence of both plasma membrane ouabain binding and K^+ stimulated Na^+-K^+-ATPases.

The insensitivity toward ouabain reported for Malpighian tubule and rectum is puzzling. The cation distributions in hemolymph (Table 1) would lead one to agree with Keynes (1969) that Na^+-K^+-

94

ATPases are universally present, yet these tissues actively transport K^+ in the absence of Na^+, but not vice versa. Levels of Na^+ in hemolymph comparable to those observed in vertebrate blood, rather than the K^+ concentrations in the range 6-37 mM tend to implicate Na^+-K^+-ATPases in cation regulation. However, these levels of Na^+ may only reflect a widespread mechanism of energy conservation employed by insects, wherein the rate of active Na^+ excretion (and thus energy utilization) by the Malpighian tubules is reduced when the rate of Na^+ entry into hemolymph via the midgut is restricted. Such a restriction in Na^+ movement will occur if - coupled with either low rates of fluid secretion or passive elimination of Na^+ by the Malpighian tubules - the concentrations of Na^+ in hemolymph and midgut contents are equal.

Regulation of cation movements in insects which secrete Na^+ preferentially over K^+ is likely to be different in insects which do the converse. In insects that actively secrete Na^+ but not K^+, epithelia probably possess ouabain sensitive Na^+-K^+-ATPases. Further, the concentration of K^+ in their hemolymph would probably reflect an admixture of the maximal concentration that will permit maintenance of bioelectric potentials across electrically excitable tissues, and the maximal concentration that will minimize both the net uptake of K^+ across the midgut and the net secretion by the Malpighian tubule-hindgut-rectum complex.

SUMMARY

Five insect epithelia are reported to be insensitive to ouabain. Of these tissues, only rectum and Malpighian tubules from a variety of species were assayed correctly so as to validate this conclusion. The remaining epithelia: Lepidopteran midgut, Lepidopteran labial gland, and Lepidopteran integument were all initially assayed improperly under conditions which disrupt and block ouabain inhibition. Midgut has recently been re-examined and found insensitive to ouabain. However, the basis of insensitivity by this tissue is the absence of K^+ stimulated Na^+-K^+-ATPase activity. Other mechanisms which could be responsible for the observed insensitivities of integument, Malpighian tubule and labial gland are also discussed.

ACKNOWLEDGEMENTS

Supported by National Science Foundation Grant PCM75-23456 and Biomedical Sciences Support Grant #RR-07088.

REFERENCES

Anstee, J.H. & Bell, D.M. (1975) *J. Insect Physiol.* 21: 1779.
Berridge, M.J. (1968) *J. Exp. Biol.* 48: 159.
Blankemeyer, J.T. (1977) Ph.D. Thesis, Temple University.
Bone, G.J. (1944) *Ann. Soc. Belg. Med. Trop.* 24: 229.

Carrington, C.B. & Tenney, S.M. (1959) *J. Insect. Physiol. 3:* 402.
Duchateau, Gh., Florin, M. & Leclercq, J. (1953) *Arch. Int. Physiol. Biochim. 66:* 573.
Dunham, P.B. & Hoffman, J.F. (1971) *J. Gen. Physiol. 58:* 94.
Gardner, J.D. & Frantz, C. (1974) *J. Memb. Biol. 16:* 43.
Gee, J.D. (1975) *J. Exp. Biol. 63:* 381.
Gee, J.D. (1976) *J. Exp. Biol. 64:* 357.
Glynn, I.M. (1957) *J. Physiol., Lond., 136:* 148.
Harvey, W.R. & Nedergaard, S. (1964) *Proc. Natn. Acad. Sci., U.S.A. 51:* 757.
Harvey, W.R., Wood, J.L., Quatrale, R.P. & Jungreis, A.M. (1975) *J. Exp. Biol. 63:* 321.
Harvey, W.R. & Zerahn, K. (1971) *J. Exp. Biol. 54:* 269.
Haskell, J.A., Clemons, R.D. & Harvey, W.R. (1965) *J. Cell Comp. Physiol. 65:* 45.
Hoyle, G. (1954) *J. Exp. Biol. 31:* 260.
Hoyle, G. (1956) *Nature, Lond., 178:* 1236.
Irvine, H.B. (1969) *Am. J. Physiol. 217:* 1520.
Irvine, H.G. & Phillips, J.E. (1971) *J. Insect Physiol. 17:* 381.
Jungreis, A.M. (1973) *Am. Zoologist 13:* 270A.
Jungreis, A.M. (1977) In: *Mechanisms of Dormancy and Developmental Arrest* (Clutter, M.E. Ed.) New York: Academic Press. In Press.
Jungreis, A.M., Jatlow, P. & Wyatt, G.R. (1973) *J. Insect Physiol. 19:* 225.
Jungreis, A.M. & Vaughan, G.L. (1976) *The Physiologist 19:* 449A.
Jungreis, A.M. & Vaughan, G.L. (1977) *J. Insect Physiol.* In press.
Kafatos, F.C. (1968) *J. Exp. Biol. 48:* 435.
Keynes, R.D. (1969) *Q. Rev. Biophys. 2:* 177.
Maddrell, S.H.P. (1971) *Adv. Insect Physiol. 8:* 199.
Maddrell, S.H.P. & Phillips, J.E. (1976) *In Perspectives in Experimental Biology Vol. 1. Zoology* (Davies, S. Ed.) p. 179, Oxford: Pergamon Press.
Mullen, J.A. (1957) *Nature, Lond. 180:* 813.
Peacock, A.J., Bowler, K. & Anstee, J.H. (1972) *Experientia 28:* 901-902.
Pilcher, D.E.M. (1970) *J. Exp. Biol. 53:* 465.
Phillips, J.E. (1964) *J. Exp. Biol. 41:* 39.
Ramsay, J.A. (1953) *J. Exp. Biol. 32:* 183.
Rivera, M.E. (1975) *Comp. Biochem. Physiol. 52B:* 227.
Skou, J.C. (1957) *Biochim. Biophys. Acta 23:* 394.
Skou, J.C. (1964) *Prog. Biophys. 14:* 131.
Treherne, J.E. (1959) *J. Exp. Biol. 36:* 533.
Vaughan, G.L. & Cook, J.S. (1972) *Proc. Natn. Acad. Sci., U.S.A. 69:* 2627.
Vaughan, G.L. & Jungreis, A.M. (1976) *Am. Zoologist 16:* 268A.
Vaughan, G.L. & Jungreis, A.M. (1977) *J. Insect Physiol.* In press.
Wood, D.W. (1957) *J. Physiol. Lond., 138:* 119.
Wood, J.L. (1972) *Ph.D. Thesis, Cambridge, University.*

SOLUTE AND WATER MOVEMENT IN THE ROUNDWORM
ASCARIS SUUM (NEMATODA)

C.G. Beames, Jr., J.M. Merz and M.J. Donahue

Oklahoma State University

INTRODUCTION

Ascaris lumbricoides and a physiological variety (var. suum)
are large parasitic roundworms which infect man and swine. The
adult worms live in the upper small intestine of their host and
feed on the intestinal contents. The worm's cuticle is permeable
to water, certain ions and some chemicals, but its intestine is
the principle route for the uptake of nutrients (Fairbairn, 1957,
Lee, 1965, and Brand, 1966). The intestine also plays a role in
osmoregulation and excretion (Lee, 1965, Harpur, 1969, and Harpur
and Popkin, 1973). There is evidence that the worm's hypodermis
is involved in osmoregulation and possibly with excretion (Lee,
1965). Our discussion of solute and water movement is limited
however to the work that has been done with the intestine of Ascaris.

CYTOLOGY OF THE INTESTINE

The intestine is essentially a straight tube which is often divided into three regions, the anterior region, the mid region or
mid-gut and the posterior region. In a large adult female worm
(avg. wt. 5 gms) the organ will be approximately 25 cm in length
and will vary in diameter from a narrow inlet-about 1.0 millimeter
in diameter at the pharynx-to a wide termination about 3.5 milli-
meters in diameter-at the proctodaeum. The intestinal wall (Fig. 1)
is composed of a single layer of high columnar cells resting on a
thick basement membrane (basal lamella). There are no muscular or
connective tissue components in the traditional sense. Microvilli
cover the luminal surface of these cells and suggest, among other
things, an absorption function (Kessel, et al., 1961, and Sheffield,
1964). The cells are very slender, being approximately 5 to 7
microns in width in comparison to their 70 micron height. The

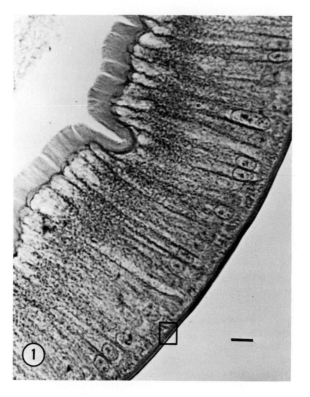

Fig. 1. A longitudinal section of the intestinal wall viewed by light microscopy. The lumen of the intestine appears at the upper left and shows some detritus. The pseudocoelomic cavity appears at the lower right. The intestinal wall consists of palisade-like high columnar cells, each cell extending from the thick basal lamina to the lumen. The luminal surface of these cells is crowned by microvilli. A small black rectangle in the lower part of the micrograph frames the basal lamina which runs from its upper right to its lower left corners. Figure 2 illustrates this framed view at higher magnification; X523.5; bar equals 10 μ. (Peczon, et al., 1975).

junctions of their lateral walls are straight and unusually visible by light microscopy. The basement membrane is relatively tough, and sheets of it that are 0.5 cm^2 or larger can be isolated free of cells and, in fact, of cell membranes (Figs. 2-3).

The intestine is easily removed from the worm, and it is an excellent tissue for _in vitro_ study of transmembrane and transepithelial transport. Until recently, however, little attention has been given to the physiology of this organ.

Differences in the size and contents of the cells of the anterior, mid and posterior regions of the intestine together with cytochemical and biochemical determinations of secretory activity and the distribution of digestive enzymes, suggest that the anterior region is primarily secretory while the mid and posterior regions are involved more with absorption (Lee, 1965). Determinations of the rate of movement of 3-0-methylglucose across _in vitro_ sac preparations of the three regions support the cytochemical and biochemical determinations (Schanbacher and Beames, 1973).

Fig. 2. An electron micrograph of a mechanically isolated basement membrane. The orientation and field match that of the black rectangle of Figure 1. Fragmented membranes from the basal in-foldings of the intestinal cells remain adhered to the luminal side. The basement membrane is primarily a fine feltwork forming a thick and a thin sublamina. The thick sublamina shows inhomogeneous zones with coarser materials woven into the finer feltwork. A thin basal sublamina borders the pseudocoelomic cavity and is devoid of the coarser materials; X193,500; bar equals 0.1 μ. (Peczon, et al., 1975).

LIPIDS

Fatty Acids - Harpur (1969) presents data which indicate that in Ascaris volative fatty acid end-products of carbohydrate catabolism leave the worm via the feces. His determinations also suggest that there is selective reabsorption of acids which are excreted into the lumen of the intestine.

Long chain fatty acids move across sac preparations of the mid region of the intestine when they are presented to the luminal

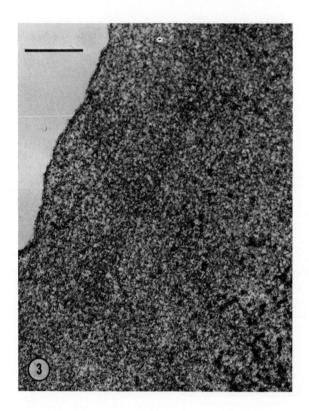

Fig. 3. An electron micrograph of the luminal surface of the basement membrane showing the surface freed of membranous debris after sonication. The fineness of the primary fletwork is illustrated plus an increasing inhomogeneity of structure toward the pseudocoelomic side (lower right); X140,250; bar equals 0.1 μ. (Peczon, et al., 1975).

surfaces as a fatty acid-albumin complex or when they are complexed with bile salts in micellar solution (Beames and King, 1972). The rate of movement is increased when the fatty acid is complexed with a bile salt and when glucose is in the incubation medium. In addition, the rate of movement is a function of the concentration of fatty acid in the luminal solution.

Glycerolipids and Sterols - Sac preparations have been used for in vitro determinations of the rates of movement across the intestine of Ascaris of triglycerides (glycerol tripalmitate) and monoglycerides (rac-glycerol 1-oleate) (Beames, et al., 1974b), and of cholesterol and β-sitosterol (Beames, et al., 1974a). Triglycerides are not hydrolyzed in the luminal solution of these preparations, and their rate of movement into and across the intestine is only a small fraction of that observed with free fatty acids. On the other hand, monoglycerides are taken up by the intestine at a rate similar to that for free fatty acids. Monoglycerides are hydrolyzed in or at the cell surface, and the released free fatty acids are either subsequently incorporated into tissue glycerolipids and/or moved across the sac preparations.

Movement of sterols across the intestine is quite slow when

contrasted to that of free fatty acids or monoglycerides. However, the rate of cholesterol transport is approximately twice that observed for β-sitosterol. The cells of the intestine accumulate cholesterol but not β-sitosterol, and this specificity for choles- terol probably accounts for the differences in observed rates of movement. There is very little or no esterification of either sterol by cells of the intestine.

AMINO ACIDS

Little information is available regarding movement of amino acids across the intestine of Ascaris. Read (1966) reports that strips of intestine accumulate ^{14}C-labeled histidine, methionine, glycine and valine at nonlinear rates with respect to concentration, and there is competitive inhibition between amino acids. He concludes that the intestine accumulates amino acids by some mediated process. However, this work is all qualitative. The anthelmintic mebendazole [methyl-5, (6) benzoly-2-benzimadozole carbamate] inhibits the up- take of amino acids by intact worms in vitro (Van den Bossche and De Nollin, 1973). In view of the fact that the intestine is the initial and a major site of action of mebendazole (Borgers and De Nollin, 1975), the above observation provides indirect evidence of the movement of amino acids across the intestine.

SUGARS

Membrane Bound Enzymes - Palma et al., (1970) report that mal- tase and sucrase activity is associated with the luminal side of sac preparations of the intestine. Their results suggest that the activity of the enzymes is independent of the transport of mono- saccharides. The above disaccharidases as well as palatinase, trehalase, and an enzyme which hydrolyzes 5'-adenosine monophosphate (at neutral pH) are associated with the intestinal brush border (Gentner, et al., 1971 and Van den Bossche and Borgers, 1973). Nonspecific acid phosphatase(s) is present in the intestine, and some 80 percent of the activity is confined to the brush border (Borgers, et al., 1970; and Borgers and Van den Bossche, 1972). The disaccharidases undoubtedly play a role in the assimulation of carbyhydrates by the worm. It is reasonable to assume that the phosphatases play a similar role. Many investigators have shown a functional relation of surface enzymes to transport with cestodes and acanthocephala, and this work is discussed well in the review of Pappas and Read (1975).

Movement of sugars - Absorption of sugars by the intestine of Ascaris is the subject of a number of papers. Sanhuzia, et al., (1968) report the movement of several sugars across Ascaris mid- gut in vitro. They show that glucose and fructose are absorbed rapidly from the intestinal lumen and appear in the pseudocoelomic fluid. The rate of absorption of glucose is in part controlled by a phloridzin sensitive process and upon the presence of sodium

101

ions in the lumen. With the conditions they employ, little or no absorption of galactose or 3-0-methylglucose is observed. Their results indicate that specific processes for glucose transport exist on the luminal surface of the epithelial cells, but this and other hexoses are not moved across the intestine against a concentration gradient.

Castro and Fairbairn (1969) compare cuticular and intestinal absorption of glucose by Ascaris. They report that isolated ribbons of the intestine absorb glucose freely against a concentration gradient, but there is no significant uptake of the sugar by the cuticle. With the conditions they employ, the intestine does not accumulate 3-0-methylglucose against a concentration gradient.

Beames (1971) reports a series of determinations of the effect of various gases (95% N_2-5% CO_2, 95% O_2-5% CO_2, 95% air-5% CO_2, and 99+% N_2) and the presence of glucose upon the movement in vitro of 3-0-methylglucose, galactose and fructose across sac preparations of the intestine of Ascaris. 3-0-methylglucose moves from the luminal fluid to the pseudocoelomic fluid when glucose is available as a substrate and 95% N_2-5% CO_2 is the exchangeable gas. Little transport of this molecule is observed when glucose is omitted or when one of the gases other than 95% N_2-5% CO_2 is employed in the system. Fructose moves from the luminal fluid to the pseudocoelomic fluid, and the movement is enhanced by the addition of glucose to the pseudocoelomic fluid. Galactose does not move to any significant extent.

Movement of 3-0-methylglucose across sac preparations of the intestine is studied in more detail by Schanbacher (1974). He reports that the endogenous carbohydrate of the intestine is rapidly depleted in vitro, and this depletion is prevented by an exogenous source of glucose. Measurements are presented that indicate that the intestine utilizes its endogenous carbohydrates to move 3-0-methylglucose across in vitro sac preparations. The addition of glycogen or trehalose to the pseudocoelomic side of the intestine does not increase the movement of 3-0-methylglucose. Measurements of the effect of pH, temperature, and substrate concentration indicate that the system for moving 3-0-methylglucose across the intestine is a saturable carrier-mediated process that has an optimal pH of 6.5. The Q_{10} values for the increase in movement of 3-0-methylglucose with each 5 C rise in temperature from 25 C to 45 C range from 2.85 to 3.17. When these values are expressed in an Arrhenius plot of the ln flux versus 1/T, a straight line results with a slope of -10033. The energy of activation from this slope is 19938 calories/mole of 3-0-methylglucose that moves across the intestine. This value is in the range of many mediated processes in biological systems. Saturation of the system can only be demonstrated by correcting for a diffusion component. Kinetic studies of the movement of 3-0-methylglucose indicate that the transport system has a K_m=22.7 mM 3-0-methylglucose and a V_{max}=4.19 µmoles 3-0-methylglucose/cm^2/hour. Results with competitive sugar studies indicate that the sugar transport system has specificity and recognizes slight structural changes in the sugar molecules. D-glucose and D-fructose show the greatest inhibitory effect (80% inhibition) of 3-O-methylglucose movement at high inhibitor concentrations (I/S = 5). D-galactose at a high concentration (I/S = 5) causes a slight reduction (28% inhibition) in the movement of 3-O-methylglucose. Iodoacetamide and sodium fluoride drastically reduce the movement of 3-O-methylglucose across the intestine. This suggests that the energy for the transport process is obtained from the tissues catabolism of carbohydrates. Maximum movement of 3-O-methylglucose is dependent upon the presence of sodium ions, and the system is insensitive to ouabain at concentrations of 1 X 10^{-3}M.

FLUID TRANSPORT

Harpur and Popkin (1973) present results of a series of determinations of fluid transport across the intestine. Everted and noneverted sac preparations are employed, and fluid movement is measured by weight difference. Under isosomotic conditions the movement is from the luminal to the pseudocoelomic side of the intestine. With the everted sac preparation, the rate of movement of water is increased nearly 16 times by the addition of glucose to the luminal solution. Glucose is itself moved from the luminal to pseudocoelomic solution against a concentration gradient, and no measurable difference in the osmolality of the two solutions develops during the process. Although glucose causes a large flux of fluid from the luminal to the pseudocoelomic side of the sacs, the net flux of Na$^+$ remains the same. The addition of ouabain (1X10^{-5}M) to the pseudocoelomic solution produces an increase in the fluid movement. Photomicrographs (Figs. 4 and 5) show that the intercellular spaces dilate during fluid transport.

MOVEMENT OF SOLUTES AND THE SHORT CIRCUIT CURRENT

An electrical potential (15-30 mV) develops across isolated ribbons of the intestine of Ascaris when they are positioned as a membrane separating two compartments of a Ussing chamber (Merz and Beames, 1975). The polarity is pseudocoelomic negative with respect to the luminal surface and the potential can be adjusted to zero with a short circuit current (SSC) of 20-30 μA. Addition of glucose (5-10 mM) to the pseudocoelomic side of the intestine increases the time that the system is able to maintain the potential. A rapid decrease in the SSC follows the addition of glucose to the luminal solution as is shown in Fig. 6. Further decreases in the SSC follow increases in the concentration of glucose in the luminal solution until the concentration reaches approximately 20 mM. Above 20 mM there is no additional decrease in the SSC which suggests that the system can be saturated. Similar results are produced with fructose, 3-O-methylglucose and galactose as is shown in Fig. 7. The total decrease in the SSC is quite small, however, with galactose.

Measurement of the transmural flux of Na$^+$ and Cl$^-$ indicates that neither ion is solely responsible for the SSC, although both

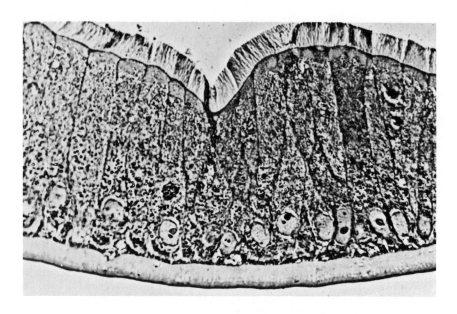

Fig. 4. Section from everted sac of <u>Ascaris</u> intestine, not transporting water (Phase contrast X750). (Harpur and Popkin, 1973).

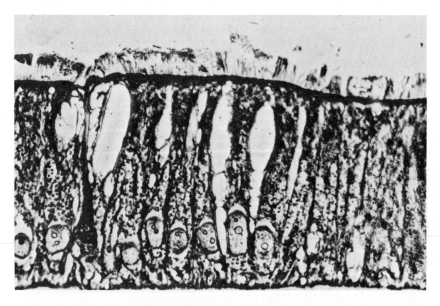

Fig. 5. Section from everted sac of <u>Ascaris</u> intestine taken after it was transporting water for ½ hour (Phase contrast X750). (Harpur and Popkin, 1973).

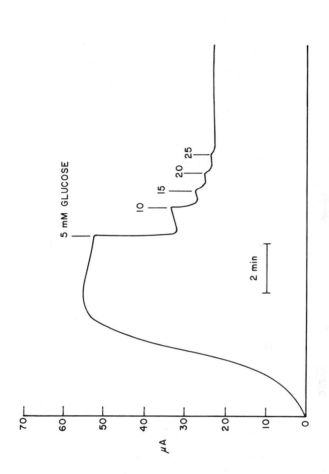

may make up a part of the observed value (Merz and Beames, 1976). Unidirectional flux measurements indicate that movement of Na^+ across the plasma membrane is nearly equal to the Na^+ diffusional component (shunt pathway) across the intestine. The relative permeability of the intestine to ions, as measured by diffusional potentials, indicates that it is as permeable to K^+ as it is to Na^+, but it is less permeable to Cl^-. Diffusional potentials across the isolated basement membrane (basal lamella) of the intestine indicate that it is more permeable to Cl^- than it is to either Na^+

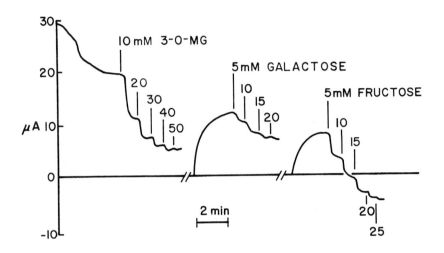

Fig. 7. Change in the short circuit current across the intestine of Ascaris in response to the addition of various sugars to the luminal solution. Gas phase: 95% N_2 - 5% CO_2. Temperature 38 C.

or K^+. Collectively, the results suggest that Na^+ is a major current conducting ion for the system. As with other biological systems, the observed potential across the intestine appears to be a function of the permeability characteristics of the membranes, junctional complexes and ion transport processes of the system.

Measurements of the uptake of glucose and Na^+ by the epithelium, under conditions which presumably block the metabolism of the tissue, indicate a stoicheometric movement of the sugar and cation (Merz and Beames, 1976). The change in the SSC under such conditions is equivalent to the uptake of Na^+.

A model which illustrates our speculation on the movement of the charged particles and glucose into and across the intestine of Ascaris is presented in Fig. 8. Sodium ions are actively transported into the lateral intercellular space followed by chloride follows. The spontaneous potential across the intestine is due to the net segregation of the charged particles (Na^+ and Cl^-) across the two main diffusional barriers, the "tight junctions" and basement membrane. Glucose uptake is a carrier mediated mechanism which involves the cotransport of sodium ion.

SUMMARY

In general the mechanisms for absorption of lipids by the intestine of Ascaris are the same as the mechanisms for absorption of lipids by the mammalian intestine.

The worms intestine probably accumulates amino acids by some

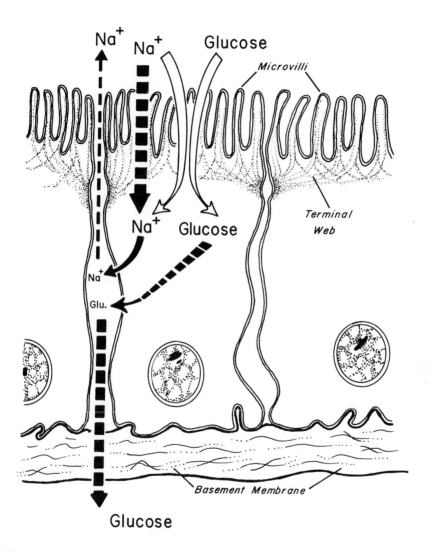

Fig. 8. Model for the movement of charged particles and glucose across the intestine of Ascaris. Broken lines represent diffusion, open lines represent coupled transport and solid lines represent active transport. Movement of ions and glucose is described in detail in the text.

mediated process(es). There is only one brief report which deals directly with the subject, however, and more information is needed.

Results of determinations of sugar transport of the intestine of Ascaris indicate that a carrier mediated mechanism operates to move specific sugars into and across the organ. The "carrier sites" are located in the apical region of the cells. The process requires energy that is derived from carbohydrate catabolism of the epithelium, and it is capable of moving sugars against a concentration gradient. The process is sensitive to phloridzin, and it is dependent upon sodium ions (K^+ may substitute) in the luminal solution. There is a net flux of fluid and sodium ions from the luminal to the pseudocoelomic side of the intestine during the transport of glucose (the only sugar studied in this regard), and the bulk of this flow appears to occur via the intercellular spaces of the epithelium. In the above respects the absorption of sugars by the worm's intestine is similar to the absorption of sugars by the intestine of vertebrates. There are some very interesting and significant differences, however, in the absorption of sugars by the intestine of Ascaris and the intestine of vertebrates. Galactose is not transported effectively by the worm's intestine and ouabain does not appear to inhibit the transport of sugars by the intestine of Ascaris.

An electrical potential exists across the intestine that is pseudocoelomic negative with respect to the lumen. The polarity of this potential is the reverse of that of the intestine of most vertebrates, and it is significantly larger. The explanation for this observation appears to be that the worm's intestine behaves in a manner similar to that described for the rabbit gall bladder (Machen and Diamond, 1969) where sodium ions diffuse back into the luminal solution from the intercellular spaces faster than chloride ions.

ACKNOWLEDGEMENTS

The authors wish to thank Ms. Dee Collins and Mary West for help in preparing the manuscript. Supported by NIH Grant AI 12783 to C.G.B.

REFERENCES

Beames, C.G., Jr. (1971) *J. Parasit. 57:* 97.
Beames, C.G., Jr. & King, G.A. (1972) In *Comparative Biochemistry of Parasites* (Van den Bossche, H. ed.) N.Y. Academic Press.
Beames, C.G., Jr., Bailey, H.H., Rock, C.O. & Schanbacher, L.M. (1974a) *Comp. Biochem. Physiol. 47A:* 881.
Beames, C.G., Jr., Bailey, H.H., Schanbacher, L.M. & Rock, C.O. (1974b) *Comp. Biochem. Physiol. 47A:* 889.
Borgers, M. & De Nollin, S. (1975) *J. Parasit. 61:* 110.
Borgers, M. & Van den Bossche, H. (1972) In *Comparative Biochemistry of Parasites* (H. Van den Bossche, ed). N.Y.: Academic Press.

Borgers, M., Van den Bossche, H. & Schaper, J. (1970) *J. Histochem. Cytochem. 18:* 519.

Brand, von T. (1966) *Biochemistry of Parasites,* N.Y.: Academic Press.

Castro, G.A. & Fairbairn, D. (1969) *J. Parasit. 55:* 13.

Fairbairn, D. (1957) *Expt. Parasit. 6:* 491.

Gentner, H., Savage, W.R. & Castro, G.A. (1972) *J. Parasit. 58:* 247.

Harpur, R.P. (1969) *Comp. Biochem. Physiol. 28:* 865.

Harpur, R.P. & Popkin, J.S. (1973) *Can. J. Physiol. Pharm. 51:* 79.

Kessel, R.G., Prestage, J.J., Sekhon, S.S., Smalley, R.L. & Beames, H.W. (1961) *Trans. Am. Micro. Soc. 80:* 103.

Lee, D.L. (1965) *The Physiology of Nematodes.* Oliver and Boyd, Edinburgh.

Merz, J.M. & Beames,C.G., Jr. (1975) Abs. 212, 50th Ann. Meeting Am. Soc. Parasitol.

Merz, J.M. & Beames, C.G., Jr. (1976) *The Physiologist 19:* 293.

Palma, R., Orrego-Malte, H. & Salinas, A. (1970) *Comp. Biochem. Physiol. 34:* 405.

Pappas, P.W. & Read, C.P. (1975) *Expt. Parasitol. 37:* 469.

Read, C.P. (1966) In *Biology of Parasites* (E.J.L. Soulsby, ed.). N.Y.: Academic Press.

Sanhuzia, P., Orrego, H., Palma, R., Parsons, D.S., Salinas, A. & Oberhauser, E. (1968) *Nature 219:* 1062.

Schanbacher, L.M. (1974) Ph.D. Thesis, Oklahoma State University, Stillwater, Oklahoma.

Schanbacher, L.M. & Beames, C.G., Jr. (1973) *J. Parasit. 59:* 215.

Sheffield, H.G. (1964) *J. Parasit. 50:* 365.

Van den Bossche, H. & De Nollin, S. (1973) *Int. J. Parasit. 3:* 401.

Van den Bossche, H. & Borgers, M. (1973) *Int. J. Parasit. 3:* 59.

Machen, T.E. & Diamond, J.M. (1969) *J. Membrane Biol. 1:* 194.

Peczon, I, Venable, R., Beames, C.M. & Hudson, A. (1975) *Biochem 14:* 4069.

SOLUTE AND WATER MOVEMENT IN
FRESHWATER BIVALVE MOLLUSKS
(Pelecypoda; Unionidae; Corbiculidae; Margaritiferidae)

Thomas H. Dietz

Louisiana State University

INTRODUCTION

All bivalve molluscs living in freshwater maintain blood osmotic pressure at concentrations above that of the medium in which
they live. However, the freshwater mussels maintain blood osmolalities which are lower than that found in any other aquatic organism
(Krogh, 1939; Robertson, 1964; Prosser, 1973). Nevertheless, the
freshwater bivalves have the same osmo- and iono-regulatory problems as do other organisms whose blood osmolalities exceed the environment. Water continuously enters and must be excreted whereas
ions must be absorbed from the medium against a substantial electrochemical gradient.
    Evidence that bivalves could absorb ions from dilute media was
first presented by Krogh (1939), who noted that salt depleted
Anodonta would accumulate Cl^- from 1mM NaCl solutions. Krogh suggested Cl^- was transported by an exchange system since Cl^- was
absorbed from NH_4Cl and $CaCl_2$ solutions with a net loss of cations,
while Na^+ was absorbed from dilute $NaHCO_3$ solutions. Apart from
Krogh's early studies, there is little information available on
ionic and osmotic regulation in freshwater bivalves (Chaisemartin
et al., 1968; Chaisemartin, 1969; Schoffeniels and Gilles, 1972;
Dietz and Branton, 1975).

ION LEVELS IN BIVALVES

The concentrations of major blood ions in representatives of three of the four families of freshwater bivalves found in the northern hemisphere are presented in Table 1. The blood is about 15-20 mM Na^+, 0.5 mM K^+ and is a saturated or supersaturates solution with respect to Ca^{++} (Potts, 1954; Burton, 1976). Inorganic phosphate is present at low concentrations (0.1-0.2 mM) and is thought to be effective in preventing crystallization of $CaCO_3$ (Burton, 1976). However, when blood is exposed to air, precipitation of some calcium and protein-carbonate complexes is noted (Potts, 1954). An alkaline blood pH is characteristic of bivalves and other molluscs (Wilbur, 1964; Bedford, 1973; Burton, 1976). The blood composition of Corbicula manilensis is significantly different from other bivalves in having NaCl as the predominant salt and little HCO_3^-.

REGULATION OF IONS

Ion balance in unfed mussels can be maintained only when salt is accumulated to offset diffusive and renal losses. Routes of ion accumulation would be across epithelia, including the gut. Drinking rates of 0.1-1.0 ml/g dry tissue - hr were measured in L. subrostrata by uptake of inulin from the bath. Uptake of $SO_4^=$ under comparable conditons is negligible. While drinking could account for 10-100% of the accumulated NaCl, it seems unlikely that this is the preferred route since it would aggrevate problems of water balance (Dietz, unpublished).

To demonstrate active transport of an ion it is necessary to show movement against electrical and concentration gradients. The in vivo transepithelial electrical potential (TEP) for L. subrostrata is shown in Figure 1. In pond water (0.5 mM NaCl, 0.4 mM $CaCl_2$, 0.2 mM $NaHCO_3$, 0.05 mM KCl) the TEP is -10 to -15 mV blood negative to the medium (Dietz and Branton, 1975). The TEP is Na^+ independent but Ca^{++} dependent. These data are in agreement with data reported for an in vitro clam mantle preparation (Kirschner, Sorenson and Kriebel, 1960; Istin and Kirschner, 1968). Kirschner and coworkers noted this TEP was a Ca^{++} diffusion potential and not due to active ion transport.

Table 1. *Blood ion composition in freshwater mussels.*

Species	Total Solute mOsm/l	Concentration mM/1 ($\bar{x}\pm$SEM,N)						REF
		Na	K	Ca	Cl	HCO$_3$	pH	
Unionidae								
Ligumia subrostrata	47±1 (12)	20.6±0.7 (14)	0.6±0.1 (5)	3.6±0.3 (11)	12.5±1.0 (11)	11.5±0.5 (6)	7.927±0.062 (5)	1
Carunculina texasensis	45±0 (10)	15.4±0.6 (10)	0.5±0.1 (5)	4.7±0.2 (10)	11.4±0.5 (10)	11.1±0.3 (5)	7.623±0.016 (3)	2
Anodonta grandis	55±1 (6)	19.5±0.3 (6)	0.5±0 (6)	5.8±0.3 (6)	16.1±0.5 (6)	11.2±1.0 (6)	7.356±0.006 (6)	2
Anodonta cygnea	42±0 (20)	15.6±0.3 (4)	0.5±0 (5)	8.4±0.4 (17)	11.7±0.3 (14)	14.6±0.8 (14)		3
Margaritiferidae								
Margaritifera hembeli	39±0 (8)	14.6±0.3 (8)	0.3±0 (8)	5.2±0.1 (8)	9.3±0.2 (8)	11.9±0.2 (8)	8.120±0.038 (8)	2
Margaritifera margaritifera		14.4±1.7 (27)	0.5±0.1 (24)	7.8±1.0 (28)	11.4±1.9 (22)	-----	-----	4
Corbiculidae								
Corbicula manilensis	69±1 (10)	28.9±1.0 (10)	0.9±0.1 (5)	12.8±0.7 (10)	24.7±0.3 (10)	2.9±0.2 (10)	7.505±0.068 (10)	2

1. Murphy and Dietz, 1976
2. Dietz (unpublished)
3. Potts, 1954
4. Chaisemartin, 1968

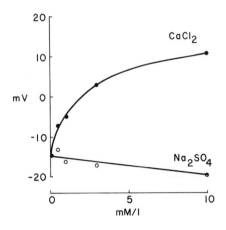

Fig. 1. In vivo transepithelial potential as a function of $CaCl_2$ or Na_2SO_4 concentration in L. subrostrata (Adapted from Dietz and Branton, 1975).

Knowledge of the electrical and chemical gradients between the animal and its environment permits calculation of passive flux ratios for diffusive ion movements using the Ussing (1949) flux ratio equation:

$$J_i/J_0 = (C_0/C_i) \exp (FE/RT)$$

where J = unidirectional flux and C = concentration, F = Faraday's constant, E = the TEP, R = the gas constant and T = absolute temperature. For L. subrostrata in pond water the predicted flux ratio for both Na^+ and Cl^- is about 0.06 (Dietz and Branton, 1975).

Unidirectional fluxes for Na^+ and Cl^- were determined for mussels acclimated in pond water. The influxes (J_i) were measured by the disappearance of isotope from the bathing medium (Dietz and Branton, 1975). Net fluxes (J_{net}) were determined from ion concentration changes in the bathing solution and the efflux (J_0) was calculated by the difference ($J_i - J_{net}$). Measured influxes and effluxes were equivalent indicating that the animals were essentially in a steady state (Table 2). Since the observed J_0's contain a renal component the epithelial flux ratios (J_i/J_0) would be greater than 1. This observed flux ratio does not agree with the predicted flux ratio, therefore, both Na^+ and Cl^- must be actively transported. Although rates of ion transport in the unionid species are similar, those measured in C. manilensis are significantly higher. This difference probably reflects the more recent immigration of C. manilensis from brackish into freshwater (Sinclair, 1971). However, transport rates for C. Manilensis reported here are an order of magnitude less than those reported for M. margaritifera (Chaisemartin et al., 1968; Chaisemartin, 1969).

INULIN CLEARANCE

Inulin clearance rates in bivalves are unusually high for freshwater animals and range between 0.1 and 0.4 ml/g dry tissue - hr (Picken, 1938; Potts, 1954b; Martin et al., 1958; Kirschner, 1967;

Table 2. Unidirectional Na and Cl fluxes in freshwater mussels acclimated to pondwater.

Species	ueq/g dry tissue - hr				Reference
	J_i^{Na}	J_o^{Na}	J_i^{Cl}	J_o^{Cl}	
Ligumia *subrostrata*	1.13±0.16 (8)	1.32±0.25 (8)	1.48±0.36 (13)	1.65±0.49 (13)	Dietz & Branton, 1975
Carunculina *texasensis*	1.36±0.10 (10)	1.14±0.14 (10)	1.18±0.08 (15)	1.56±0.25 (15)	*
Corbicula *manilensis*	10.52±1.02 (10)	9.48±1.61 (10)	6.19±1.08 (15)	6.81±1.01 (15)	*

*Unpublished

Chaisemartin et al., 1970: Murphy and Dietz, 1976). This level of inulin filtration indicates a daily water filtration rate of about 100% of the total tissue weight. However, most filtered water is apparently reabsorbed (Little, 1965). To check for possible convection salt movements the fluxes of Na^+ and Cl^- were determined before and after the bathing medium was made isosmotic using 50 mM mannitol (J_i^{Na} before 0.96 ± 0.20 µeq/g dry-hr, after 80 ± 27% of control, N = 8; J_i^{Cl} before 1.36 ± 0.21 µeq/g dry-hr, after 75 ± 10% of control, N = 10). Although the fluxes tend to be slightly depressed the reduction is not significant and convective salt movement is minimal (Dietz, unpublished).

KINETICS OF Na AND Cl TRANSPORT

Unidirectional influxes were determined for Na^+ and Cl^- in pond water acclimated L. subrostrata exposed to a range of NaCl concentrations between 0.1-2.0 mM (Figure 2). Transport systems for both Na^+ and Cl^- are saturable. The capacity (V_{max}) for Na^+ transport is about 2 µeq/g dry tissue - hr and the V_{max} for Cl^- is 1 µeq/g dry tissue - hr. The affinity (K_s) of the transport systems for both Na^+ and Cl^- is between 0.1 and 0.15 mM. These transport rates and affinities are similar in magnitude to a variety of freshwater animals (Shaw, 1963; Chaisemartin, 1969; Alvarado and Moody, 1970; Dietz and Alvarado, 1970; Kirschner, 1973; Dietz and Alvarado, 1974; Dietz, 1974a).

A classic technique for stimulating ion transport is to subject the animals to distilled water (Krogh, 1939). Prolonged exposure of L. subrostrata to deionized water leads to a rapid decline in blood NaCl (Figure 3). However, there is a simultaneous increase of Ca^{++} and HCO_3^- which tends to minimize the decline in total blood solute (Murphy and Dietz, 1976). When salt depleted mussels are returned to 0.5 mM Na_2SO_4 or 1 mM choline chloride solutions

115

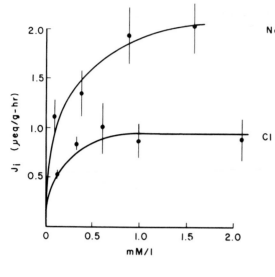

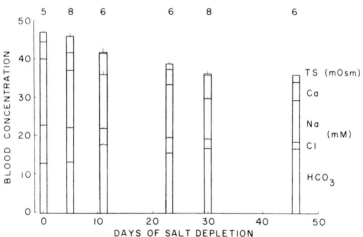

Fig. 3. Blood solute composition in L. subrostrata subjected to deionized water. The column height represents total solute (TS) and the specific ion concentrations are represented by the subdivision. Vertical lines represent one standard error of the mean and the number of animals is given above each bar (Murphy and Dietz, 1976).

there is a significant net uptake of Na^+ or Cl^-, respectively (Table 3). The elevated J_{net} is due to both an increased J_i and a decrease in J_0. The reduced J_0 is probably due to enhanced renal reabsorption of salts and lower diffusive losses across the epithelia. Preliminary experiments have indicated the increase in

Table 3. Effect of salt depletion on unidirectional fluxes of Na from Na_2SO_4 and Cl from Choline Cl.

Species	Ion	N	ueq/g dry tissue-hr		Reference
			J_i	J_o	
Ligumia subrostrata	Na	5	2.24±0.22	0.56±0.15	Murphy and Dietz, 1976
	Cl	6	1.82±0.25	0.59±0.14	Dietz and Branton, 1975
Carunculina texasensis	Na	5	3.37±0.25	0.50±0.06	unpublished
	Cl	9	2.33±0.40	1.35±0.37	unpublished

NaCl transport is due to an elevated V_{max} with no change in affinity (Branton, unpublished). It is noteworthy that Na^+ and Cl^- transport are independent, apparently Na^+ is transported in exchange for an endogenous cation (NH_4^+ or H^+ while Cl is exchanged for HCO_3^- or OH^-. Ligumia subrostrata is ammonotelic (Dietz, 1974b) and has high concentrations of blood HCO_3^- for ion exchange. However, these exchange mechanisms have not been studied.

EFFECTS OF PHARMACOLOGICAL AGENTS

Although ion transport in mussels can be stimulated it is remarkably refractory to inhibition by a number of pharmacological agents (Table 4). The following substances had no effect on chloride fluxes when dissolved in the bathing medium: Furosemide, Amiloride, 4-acetamino-4'-isothiocyano stilbene-2, 2' disulfonic acid. Tiocyanate caused a 40% reduction in J_i but only at a high concentration. Furthermore, "stilbene" injection had no effect on either J_i or J_o. Acetazolamide injection had no effect on J_i but resulted in a 300% increase in J_o. This effect may be due to inhibition of renal Cl^- reabsorption.

Preliminary experiments of Na^+ fluxes indicate acetazolamide injection was without effect on J_i or J_o. Amiloride (0.5 mM) in the bathing medium reduced J_i to 20% of controls with no apparent effect on J_o. These data point out that important differences in Na^+ and Cl^- transport properties exist between vertebrates and invertebrates (Epstein et al., 1973; Kirschner, 1973; Dietz, 1974a; Garcia-Romeu and Ehrenfeld, 1975; Alvarado et al., 1975).

CONCLUSION

These data indicate freshwater mussels possess well developed transport systems for maintaining ion balance while living in dilute solutions. In bivalves the rates of transport and the affin-

Table 4. Effect of various drugs on Cl fluxes in Unionids.

Drug	Conc.	Location	N	Flux as % of Control J_i^{Cl}	J_o^{Cl}
SCN	1mM	Bath	7	60[*]	90
Furosemide	1mM	Bath	8	98	170[*]
Amiloride	0.1mM	Bath	5	71	76
Stilbene	0.15mM	Bath	6	67	89
Stilbene	56μg/g dry	inj	5	89	128
Acetazolamide	76μg/g dry	inj	9	114	336[**]

[*]$P < 0.05$
[**]$P < 0.01$

ity of the transport mechanism toward Na^+ and Cl^- are indistinguishable from those of other freshwater animals. However, the mussels are relatively insensitive to selected drugs. Nevertheless, the bases for maintaining unusually low concentrations of blood solutes is not due to a deficiency of ion transport capacity. It is possible that the low blood solute levels may be due to the limited capacity of the bivalve kidney to excrete water.

ACKNOWLEDGEMENTS

I wish to thank Hoechst Pharmaceuticals for the generous gift of Furosemide, Carl Booker for able technical assistance and Ms. Avia Dimattia for typing the manuscript. M. Vidrine identified and J.R. Womack provided many of the mussels. Supported by NSF Grant BMS75-05483.

REFERENCES

Alvarado, R.H. & Moody, A. (1970) Am. J. Physiol. 218: 1510.
Alvarado, R.H., Dietz, T.H. & Mullen, T.L. (1975) Am. J. Phsiol. 229: 869.
Bedford, J.J. (1973) Arch. Int. Physiol. Biochem. 81: 819.
Burton, R.F. (1976) In Perspectives in Experimental Biology (P. S. Davis, ed) Pergamon, Oxford. 1: 7.
Chaisemartin, C. (1968) C.R. Soc. Biol. (Paris) 162: 1193.
Chaisemartin, C. (1969). C.R. Soc. Biol. (Paris) 163: 2422.
Chaisemartin, C., Martin, P.N. & Bernard, M. (1968) C.R. Soc. Biol. (Paris) 162: 523.
Chaisemartin, C., Martin, P.M. & Bernard, M. (1970) C.R. Soc.

Biol. (Paris) 164: 877.

Dietz, T.H. (1974a) *Comp. Biochem. Physiol. 49A:* 251.

Dietz, T.H. (1974b) *Biol. Bull. 147:* 560.

Dietz, T.H. & Alvarado, R.H. (1970) *Biol. Bull. 138:* 247.

Dietz, T.H. & Alvarado, R.H. (1974) *Am. J. Physiol. 218:* 764.

Dietz, T.H. & Branton, W.D. (1975) *J. Comp. Physiol. 104:* 19.

Epstein, F.H., Maetz, J. & Renfis, F. De (1973) *Amer. J. Physiol. 224:* 1295.

Garcia-Romeu, F. & Ehrenfeld, J. (1975) *Am. J. Physiol. 228:* 845.

Istin, M. & Kirschner, L.B. (1967) *Ann. Rev. Physiol. 29:* 169.

Kirschner, L.B. (1973) In *Transport mechanisms in Epithelia* (H. H. Ussing & N.A. Thorn, eds) p. 447 New York. Academic Press.

Kirschner, L.B., Sorenson, A.L. & Kriebel, M. (1960) *Science 131:* 735.

Kirschner, L.B., Greenwald, L. & Kerstetter, T.H. (1973) *Am. J. Physiol. 224:* 832.

Krogh, A. (1939) *Osmotic regulation in Aquatic Animals:* Cambridge Univ. Press.

Little, C. (1965) *J. Exp. Biol. 43:* 39.

Martin, A.W., Harrison, F.M., Huston, M.H. & Stewart, D.M. (1958) *J. Exp. Biol. 35:* 260.

Murphy, W.A. & Dietz, T.H. (1976) *J. Comp. Physiol. 108:* 233.

Picken, L.E.R. (1937) *J. Exp. Biol. 14:* 20.

Potts, W.T.W. (1954a) *J. Exp. Biol. 31:* 376.

Potts, W.T.W. (1954b) *J. Exp. Biol. 31:* 614.

Prosser, C.L. (1973) *Comparative Animal Physiology:* Saunders, Philadelphia.

Robertson, J.D. (1964) In *Physiology of Mollusca* (K.M. Wilbur & C.M. Yonge, eds). 1: 283.

Schoffeniels, E. and Gilles, R. (1972) In *Chemical Zoology* (M. Florkin & B.T. Scheer, eds). 7: 393.

Shaw, J. (1963) *Viewpoints Biol. 2:* 163.

Sinclair, R.M. (1971) *Sterkiana 43:* 11.

Ussing, H.H. (1949) *Acta Physiol. Scand. 19:* 43.

ASPECTS OF FLUID MOVEMENT IN THE
CRAYFISH ANTENNAL GLAND

J. A. Riegel

University of London

INTRODUCTION

The antennal gland of the crayfish appears to conform to the functional pattern expected of a filtration: reabsorption kidney (Kiegel & Cook, 1975). Primary urine is isosmotic and is formed in the coelomosac as a hemolymph (blood) filtrate. As is shown in Fig. 1, fluid is resorbed in most parts of the antennal gland. About two-thirds of the water and inorganic solutes are reabsorbed during passage of the primary urine from the coelomosac to the most distal portion of the tubule. The markedly dilute urine of the crayfish appears to be due primarily to hyperosmotic reabsorption of solute in the urinary bladder.

ROLE OF FORMED BODIES IN WATER METABOLISM

Much of the evidence for the current concept of urine formation in the crayfish has been based upon micropuncture studies (Peters, 1935; Riegel, 1963, 1965, 1966a,b, 1968; Cook, 1973). These studies revealed the presence within most parts of the antennal gland of vesicular elements or formed bodies (Riegel, 1966a,b, 1968). Formed bodies appear to serve as vehicles of solute secretion, since foreign materials (e.g., various dyes including phenol red) are sequestered within them. However, formed bodies also contain proteolytic and possibly other hydrolytic enzymes. It thus seems probable that formed bodies are merely the extracellular manifestation of lysosomes (Riegel, 1970b).

Attempts to isolate and analyse formed bodies led to the discovery that they possess a marked ability to attract water. This water-attracting ability was the basis of a model of transepithelial water movement, for which details were recently published

121

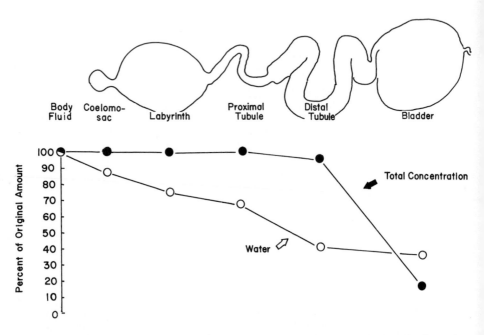

Fig. 1. Changes in the osmotic pressure (total concentration) and water content (reciprocal of the inulin urine: hemolymph concentration ratio) of primary urine as it passes through the crayfish antennal gland.

(Riegel, 1970b). The proposed model envisaged the extrusion of formed bodies into spaces within or adjacent to cells. The formed bodies attract water because of hydrolytic reactions which occur within them. The membrane surrounding formed bodies appears to be impermeable to solutes. Thus, movement of water into them creates a local concentration effect toward which more water is attracted. As depicted in Fig. 2, water and solutes tend to move out of the spaces in which the formed bodies are swelling due to hydrostatic pressure. The direction in which the fluid flows and both the quantity and quality of the flowing fluid depends upon the resistance to fluid movement offered by passages giving access to the spaces in which the formed bodies swell.

UNIQUE FEATURES OF THE FORMED BODY MODEL

The major difference between the formed body model and those proposed by Curran (1960) and associates, and Diamond (1967) and associates is that the present model virtually demands that fluid move in extracellular spaces rather than through the cytoplasm of cells. Indeed, most recent studies favor extracellular spaces as

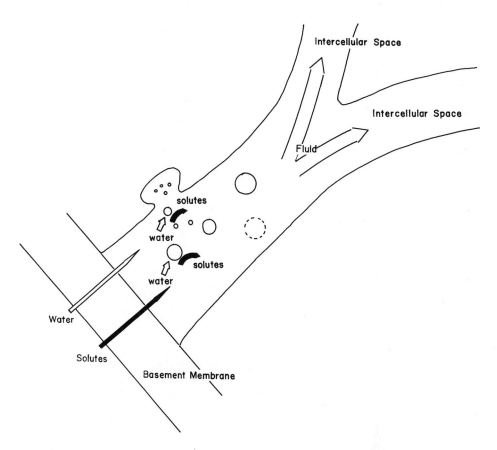

Fig. 2. Resume of the formed-body model of fluid movement. Formed bodies are extruded into the intercellular spaces where they swell by uptake of water. Solutes in the intercellular space tend to be concentrated by the water uptake by formed bodies and the exclusion of solutes by formed bodies. Water and permeable solutes flow into the intercellular spaces. This flow is due to an osmotic pressure differential along one of the passages giving access to the intercellular spaces. Net transepithelial fluid flow is due to hydro-static pressure within the intercellular spaces. The direction taken by this net flow is along the path of least resistance lead-ing out of the intercellular space. In this model, it is assumed that the basement membrane is responsible for setting up an osmotic pressure differential in the intercellular space. It is further assumed that the intercellular spaces and apical junctions of the cells provide less resistance to fluid flow than does the basement membrane.

the major route of water and (especially) solute movement across epithelia.

A further consequence of the formed-body model is that it is no longer necessary to think of secreting and filtering epithelia in terms of selective permeability. Inherent in the model is the possibility that so-called 'secretory' epithelia exist whose permeability characteristics are similar to those of 'filtering' epithelia. In fact at least one such filtering 'secretory' epithelium has been described, namely, the Malpighian tubule of the pill millipede, Glomeris marginata Latreille (Farqutharson, 1974). Carrying this argument a step further leads to the possibility that real differences do not exist between secretory and filtering epithelia, at least in terms of the mechanisms underlying fluid movement. That is, all transepithelial fluid movement may be underlain by the swelling of formed bodies, but in some this process may be assisted by externally-applied (i.e. capillary or hemocoelar) hydrostatic pressure.

PROCEDURES FOR QUANTITATING WATER MOVEMENT BY FORMED BODIES

The present paper summarizes the results of attempts to quantify the contribution by formed bodies to the observed rates of fluid movement. All studies were carried out in vitro. Measurements were made of the rate of fluid movement across Millipore filters having a pore diameter of 0.1 μm. The membranes separated droplets of Ringer and coelomosac fluid. The general procedure can be summarized as follows: the size of the droplet of coelomosac fluid was reduced to a few nanolitres by lowering the pressure on the side of the membrane containing the Ringer droplet. The Millipore filter was then punctured, thereby equalizing the pressure on both of its sides; the growth in volume of the coelomosac-fluid droplet was measured over a timed period. The details of the apparatus used and the method of making the measurement will be described in a future publication. Under conditions where fluid flow could not have been due to an osmotic pressure gradient (fluid flow across membranes of 0.1 μm pore diameter) the rate of fluid flow was relatively rapid (Fig. 3). Furthermore, for coelomosac droplets of equal size, the flow rate increased with increased area of contact between droplet and membrane.

In the intact animal at least two factors not present in the in vitro experiments tend to restrict fluid flow: firstly, the flow will be opposed by the osmotic pressure of hemolymph colloids. Secondly, pores in the filtering surface very likely are much smaller than 0.1 μm. Consequently, measurements were made of fluid flow rate into coleomosac-fluid droplets under circumstances where such flow was opposed by colloid osmotic pressures as high 80 cm H_2O. As shown in Fig. 4, fluid flow into coelomosac-fluid droplets can occur at an appreciable rate even when such flow is opposed by a large colloid osmotic pressure.

Initial studies of fluid flow versus colloid osmotic pressure were made using Pellican ultrafiltration membranes which were impermeable to molecules whose molecular weights exceeded about

124

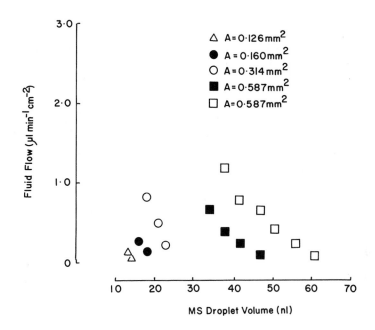

Fig. 3. The influence of area on fluid flow across Millipore filters into droplets of coelomosac fluid. The initial size of the coelomosac-fluid droplets was a few nanolitres, and the average pore diameter of the filters was 0.1 μm.

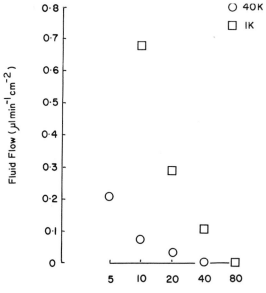

Fig. 4. The rate of fluid flow into droplets of coelomosac fluid under conditions where flow was opposed by varying the colloid osmotic pressure. Fluid flow proceeded across Pellican ultrafiltration membranes which were permeable to compounds up to 1000 molecular weight (1K, open squares) or 40,000 molecular weight (40K, open circles).

125

1000 (i.e., 1K membranes). These studies were repeated using ultrafiltration membranes having a molecular weight cut-off between 25,000 and 40,000 molecular weight. This molecular-weight range approximates that of the permeability of the epithelium of the antennal gland filtering surface.

In a second series of studies fluid flow into droplets of coelomosac fluid was compared with flow into droplets of colloid Ringer (7.5 cm H2O) and coelomosac fluid from which the formed bodies had been removed. As shown in Fig. 5, the rate of flow into droplets containing formed bodies exceeds that caused primarily by fluid flow in response to concentrated colloids (Ringer and filtered coelomosac fluid).

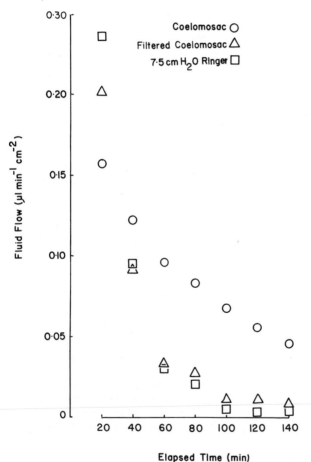

Fig. 5. The rates of fluid flow into droplets of coelomosac fluid, coelomosac fluid from which formed bodies were removed by filtration and Ringer which contained colloid (7.5 cm H2). In each case fluid flow was opposed by a colloid osmotic pressure of 10 cm H2O. Fluid flow proceeded through pellican ultrafiltration membranes (40K).

DISCUSSION

The results of the foregoing studies seem clear: formed bodies are indeed responsible for fluid movement into coelomosac fluid droplets in vitro. Furthermore, this movement occurs at an appreciable rate even when opposed by colloid osmotic pressures of 10 to 20 cm H_2O. Recent measurements of the colloid osmotic pressure of crayfish haemolymph averaged only 6.8 cm H_2O (Riegel & Cook, 1975). Therefore, it appears that conditions exist in the intact crayfish which are conducive to a role for formed bodies in the formation of urine.

The question remains: can extrusion of formed bodies account for primary urine formation in the crayfish? Cook (1973) has estimated the rate of primary urine formation (i.e., inulin clearance) in the crayfish. The value is 2.5 μl min^{-1} for a 20 gram animal. The fastest rate of fluid flow seen in Fig. 5 occurs under conditions which might be expected to prevail at the filtration site, namely a large formed-body volume relative to the total fluid volume, and a large surface area relative to the total fluid volume (Fig. 4). Were primary urine formation in the crayfish due entirely to swelling of formed bodies, then the filtering surface (20 g animal) would have to be about 16 cm^2 (0.157 μl min^{-1}cm-2 ÷ 2.5 μl min^{-1}). Rough estimates made of the external dimensions of a few coelomosacs average about 1 mm^2. Therefore, infolding of the walls of the coelomosacs in the two antennal glands of a 20 gramme crayfish would have to increase the filtering surface some 800 fold. The general hemocoelar hydrostacic pressure of the crayfish averages about 15 cm H_2O (Picken, 1936). Since the coelomosac of the antennal gland has an arterial blood supply, there is little doubt that arterial pressure aids the filtration process. However, in lower Crustacea, no arterial supply exists, and it is questionable whether or not hemocoelar pressure favours filtration. In the latter examples primary urine formation could be due entirely to the energy supplied by swelling formed bodies.

REFERENCES

Cook, M.A. (1973) Doctoral Thesis, University of London.
Curran, P.F. (1960) *J. Gen. Physiol.* 43: 1137.
Diamond, J.M. & Bossart, W.H. (1967) *J. Gen. Physiol.* 50: 2061.
Peters, H. (1935) *Z. Morph. Okol. Tiere* 30: 355.
Picken, L.E.R. (1936) *J. Exp. Biol.* 13: 309.
Riegel, J.A. (1963) *J. Exp. Biol.* 40: 487.
Riegel, J.A. (1965) *J. Exp. Biol.* 42: 379.
Riegel, J.A. (1966a) *J. Exp. Biol.* 44: 379.
Riegel, J.A. (1966b) *J. Exp. Biol.* 44: 387.
Riegel, J.A. (1968) *J. Exp. Biol.* 48: 587.
Riegel, J.A. (1970a) *Comp. Biochem. Physiol.* 35: 843.
Riegel, J.A. (1970b) *Comp. Biochem. Physiol.* 36: 403.
Riegel, J.A. & Cook, M.A. (1975) *Fortschr. Zool.*

INTESTINAL WATER AND ION TRANSPORT IN FRESHWATER MALACOSTRACAN PRAWNS (CRUSTACEA)

G.A. Ahearn, L.A. Maginniss, Y.K. Song, and A. Tornquist

University of Hawaii at Manoa

INTRODUCTION

There are potentially four major sites of water and ion exchange in crustaceans: gills, kidney (antennal glands), integument, and gastrointestinal tract. A considerable body of information currently exists describing the ion and water regulatory functions of crustacean gills (Koch et al., 1954; Bielawski, 1964; Croghan et al., 1965; Mantel, 1967; King & Schoffeniels, 1969), excretory organs (Shaw, 1961; Lockwood, 1961; Sutcliffe, 1961a,b; Potts & Parry, 1964; Gross & Capen, 1966; Riegel, 1970; Lockwood & Inman, 1973; Prosser, 1973), and body wall (Gross, 1957; Shaw, 1961; Bryan, 1960; Rudy, 1966, 1967; Croghan & Lockwood, 1968; Sutcliffe, 1968), whereas the physiological role of the gastrointestinal tract in water and ion relations is less well understood.

The crustacean gut has been implicated as a site of water and ion transport, particularly in animals inhabiting iso-osmotic or hyperosmotic environments. Croghan's early work on *Artemia* (Croghan, 1958a,b,c,d) suggested that NaCl and water were absorbed by the gut, leaving other salts to be voided in the feces. These initial studies on the brine shrimp were later extended by Smith (1969a,b), who critically analyzed the contributions of both the gut and gills to total organismic ion and water exchange. Dall (1965, 1967) demonstrated that marine penaeid shrimp exhibit a high drinking rate as do marine fish and that the gut of this animal appears permeable to divalent cations such as calcium. The stomach of the land crab, *Gecarcinus lateralis*, was found permeable to sodium, chloride, and water; the extent of this permeability being under tight neuroendocrine control by the thoracic ganglionic mass (Mantel, 1968).

The gastrointestinal physiology of ion and water balance in

129

freshwater crustaceans is unstudied. Although drinking rates of animals living in hypo-osmotic environments are generally low (Maetz, 1974) and provide little mineral uptake, considerable dietary input of salts may make a significant contribution to total ion balance. This report presents information concerning the characteristics of sodium, chloride and water transport by the perfused midgut of a freshwater prawn, *Macrobrachium rosenbergii*.

INCUBATION MEDIA

The ionic composition and osmotic pressure of saline media in which isolated midguts were incubated were based on analyses of hemolymph and gut content samples using a Coleman Flame Photometer (Model 21), an Aminco-Cotlove chloride titrator (American Instruments, Co.) and an Advanced Instruments osmometer. Blood samples were collected by inserting a glass pipette through the thin intersegmental membrane between the cephalothorax and first abdominal segment. Clotted blood was centrifuged for 15 min at 10,000 rpm, and the clear serum decanted and frozen until analyzed. Midgut contents from freshly acquired animals were collected in flame-sealed capillary tubes, and treated in a manner identical to blood.

Table 1 presents data concerning the ionic composition and osmotic pressure of hemolymph and midgut samples. With the exception

*Table 1. Constituents of Prawn Hemolymph and Midgut Lumen** *

Constituent	Hemolymph	Midgut Lumen
Na	219.1 ± 1.8 (20)	239.1 ± 4.3 (13)
K	8.1 ± 0.5 (10)	26.0 ± 3.1 (13)
Ca	14.5 ± 0.8 (11)	10.6 ± 1.9 (13)
Cl	192.8 ± 3.0 (20)	189.0 ± 14.6 (13)
Osmotic Pressure	445.7 ± 5.1 (20)	532.8 ± 10.9 (5)

*Values are expressed in mM for ion concentrations and mOsm/Kg for osmotic pressure and represent means ± SEM. Numbers in parentheses denote number of samples.

of K, very small differences in ion concentration were observed across the midgut of this animal. Physiological saline employed to simultaneously bathe both surfaces of the intestine under control conditions contained (in mM): Na, 221.1; K, 8.1, Ca, 12.0; Mg, 4.8; Cl, 211.3; SO_4, 25.1; PO_4, 0.4; HCO_3, 0.7; pH = 7.4; osmotic pressure, 430 mOsm/kg. Sodium-free medium was prepared by substituting KCl (or choline chloride) and $KHCO_3$ (or choline bicarbonate) for the respective sodium salts, while chloride-free saline was made by sulfate substitution. Osmotic pressure on both gut surfaces was kept constant by addition of small quantitites of mannitol. When altered luminal pH was studied, dilute HCl was added directly to the control perfusate.

NET WATER TRANSPORT

Intact intestines were mounted on epoxy-coated 18-guage stainless steel needles with surfical thread, immersed in 10 ml of aerated physiological solution maintained at 23-24°C in a lucite chamber, and perfused with this saline by means of a peristaltic pump (Buchler Instruments) at a flow rate of about 125 µl/min. Net water transport in shrimp midgut was measured using procedures developed for perfused mammalian kidney nephrons (Burg & Orloff 1968) and adapted to arthropod intestine (Ahearn & Hadley 1976a,b) where the change in perfusate concentration of a radioactively labelled non-absorbed volume marker, such as ^{131}I-albumen (E.R. Squibb, Co.) after passage through the tubular preparation, provides information regarding the extent and direction of water flow. The perfusion rate (V_0, µl/min) was calculated as

$$V_0 = {}^{131}I_{tot} / {}^{131}I_0 t$$

where $^{131}I_0$ is the specific activity of the perfusion fluid (dpm/µl); $^{131}I_{tot}$ the total amount of radioactivity in a sample of midgut fluid collected after passage through the preparation (dpm); and t the duration of the collection period (min). The rate of collection (V_f, µl/min) was estimated directly from the volume (determined gravimetrically) and duration of each collection. The net water transport rate J_v (µl cm^{-2} hr^{-1}) was calculated as

$$J_v = V_0 - V_f / A$$

where A represents the estimated mucosal surface area (cm^2) determined by considering the midgut a cylinder.
^{131}I-albumen was added to the mucosal perfusate along with unlabelled bovine serum albumen (0.5% solution; fraction V; Sigma Chem. Co.) to reduce absorption of ^{131}I-activity by the perfusion pump tubing. Under these conditions there was complete recovery of the labelled volume marker from the tubing with a negligible effect of the unlabelled albumen on solution osmotic pressure. Collections of midgut effluent were made every 5 minutes and the serosal bath was periodically sampled (250 µl) for ^{131}I-leakage. Data from leaking preparations were discarded. Samples withdrawn from the serosal bath were immediately replaced with an equal volume of appropriate unlabelled medium. Effluent and bath samples of ^{131}I were added to a toluene-based scintillation cocktail and counted in a Beckman LS-230 liquid scintillation spectrometer.
The requirement of luminal sodium for new water transport from mucosa to serosa was tested by replacing the perfusate Na with equimolar K. Under control conditions with 221 mM luminal Na, net volume flow (J_v = + 42.06 $\pm$ 4.50 µl cm^{-2} hr^{-1}; mean $\pm$ SEM)(5:31;

number of guts: number of effluent collections) was significantly greater than zero (P<0.005)(Fig. 1). Upon replacing mucosal Na with K (luminal Cl remaining unaltered), net water flux was significantly reduced (P<0.01 to + 6.30 $\pm$ 4.14 µl cm^{-2} hr^{-1} (5:31).

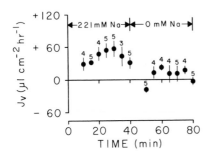

Fig. 1. *Effect of Na-free perfusate (K replacing Na) on net water transport. Control saline (221 mM Na) in serosal bath throughout exposure. Midguts were first perfused with control saline for 40 min and were subsequently exposed to Na-free luminal solution for another 50 min. A positive net water flux indicates movement from mucosa to serosa. Circles represent mean fluxes, vertical lines are $\pm$ SEM, and number signify sample size.*

This rate of water transport, in the absence of luminal Na, was not significantly different than zero (0.10<P<0.20). These results suggest that luminal sodium is necessary to support a net movement of water from lumen to blood.

Replacement of mucosal chloride by suflate (luminal Na remaining unaltered) resulted in a highly significant decrease (P<0.005) in net water flux from mucosa to serosa (Fig. 2). In this experiment control volume flow was + 45.60 $\pm$ 2.66 µl cm^{-2} hr^{-1} (5:34) and was reduced to - 8.00 $\pm$ 4.20 µl cm^{-2} hr^{-1} (5:40) under chloride-free conditions, a value not significantly different from zero (0.05<P<0.10). Therefore, both sodium and chloride are required in the

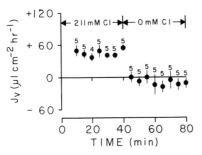

Fig. 2. *Effect of Cl-free perfusate (SO$_4$ replacing Cl) on net water trasnport. Control saline (211 mM Cl) in serosal bath throughout exposure. Incubation, intervals and symbols are as described in Fig. 1.*

lumen of the prawn midgut for the absorption of water.

The effects of increasing luminal hydrogen ion concentration on volume flow across the isolated midgut were examined by decreasing the perfusate pH from a control value of 7.4 to a test value of 6.0 (Fig. 3). Control water flux in this experiment was +62.58 $\pm$ 3.00 µl cm^{-2} hr^{-1} (5:33). At luminal pH 6.0 the net transport of water across the preparation was reduced significantly (P<0.005) to - 1.76 $\pm$ 2.76 µl cm^{-2} hr^{-1} (5:40). Thus flux at the lowered pH was not significantly different than zero (0.7<P<0.5). These data imply a potentially important inhibitory role of mucosal hydrogen ions on the magnitude of Na/Cl-dependent net water transport.

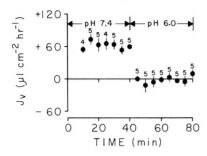

Fig. 3. Effect of luminal pH on net water transport. HCl added to perfusate to reduce pH from 7.4 (control) to 6.0 (test), while serosal bath pH maintained at 7.4 throughout incubation. Incubation intervals and symbols are as described in Fig. 1.

Control net water transport data from the above three experiments were pooled to provide more accurate assessment of J_v in the absence of transmural differences in hydrostatic pressure, osmotic pressure, or ion concentration. Under these conditions the net movement of water from the mucosa to serosa was + 50.20 ± 2.16 µl cm^{-2} hr^{-1} (15:98).

TRANSMURAL FLUXES OF SODIUM AND CHLORIDE

Ion fluxes across isolated midguts were measured in the present investigation using techniques and notation previously established in studies with vertebrate intestine (Schultz & Curran 1974). Unidirectional transmural fluxes of ^{22}Na and ^{36}Cl from mucosa to serosa (J_{ms}^{Na} and J_{ms}^{Cl}) and from serosa to mucosa (J_{sm}^{Na} and J_{sm}^{Cl}) were determined in separate experiments by placing the respective isotope in the incubation solution on one surface of the gut and measuring its rate of appearance on the opposite surface. Net transmural fluxes were calculated as the difference between the two unidirectional fluxes ($J_{net}^{Na} = J_{ms}^{Na} - J_{sm}^{Na}$ and $J_{net}^{Cl} = J_{ms}^{Cl} - J_{sm}^{Cl}$).

Preliminary measurements of ^{22}Na transmural transport across the isolated midgut of the freshwater prawn in the absence of differential transmural ion concentrations are presented in Fig. 4. The

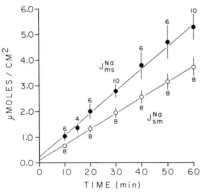

Fig. 4. Transmural fluxes of ^{22}Na using control saline on both midgut surfaces. Circles represent means, vertical lines are ± 1 SEM, and numbers refer to observations at selected time intervals.

appearance of ^{22}Na on the side opposite from that where the isotope was introduced was linear over a 60-min exposure interval for fluxes in both directions. Ion fluxes were calculated from slopes determined by regression analysis. Vertical axis intercepts were not significantly different from (P<0.05). Sodium transport from mucosa to serosa (J_{ms}^{Na}) was 5.13 $\pm$ 0.52 μmoles cm^{-2} hr^{-1} (mean $\pm$ SEM) (n = 10, number of guts measured) while transport from serosa to mucosa (J_{sm}^{Na}) was 3.63 $\pm$ 0.33 μmoles cm^{-2} hr^{-1} (n = 8), the difference (1.50 μmoles cm^{-2} hr^{-1}) being significant (P<0.05).

Transmural chloride fluxes across the shrimp midgut, resembled those of sodium in that ion movements in both directions were linear functions of time for at least 50 min with calculated slopes showing nonsignificant (P>0.05) vertical axis intercepts (Fig. 5).

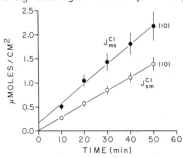

Fig. 5. *Transmural fluxes of* ^{36}Cl *using control saline on both midgut surfaces. Symbols are as described in Fig. 4.*

Based upon these initial studies, J_{ms}^{Cl} was 2.45 $\pm$ 0.36 μmoles cm^{-2} hr^{-1} (n = 10) and J_{sm}^{Cl} 1.66 $\pm$ 0.16 μmoles cm^{-2} hr^{-1} (n = 10), the net flux (0.79 μmoles cm^{-2} hr^{-1}) also being significant (P<0.05).

Transmural sodium and chloride fluxes obtained from a more complete set of studies conducted under control conditions are summarized in the first row of Table 2. Flux data presented in this

Table 2. *Transmural Fluxes of Sodium and Chloride**

Perfusate Solution**	J_{ms}^{Na}	J_{sm}^{Na}	J_{net}^{Na}	J_{ms}^{Cl}	J_{sm}^{Cl}	J_{net}^{Cl}
Control	4.83 ± 0.26 (30)	3.34 ± 0.19 (15)	1.49 ± 0.30	2.47 ± 0.20 (19)	1.76 ± 0.19 (20)	0.72 ± 0.22
Cl-free	3.39 ± 0.39 (9)	3.40 ± 0.11 (5)	-0.01 ± 0.43	---	---	---
Na-free	---	---	---	1.86 ± 0.26 (5)	1.94 ± 0.22 (6)	-0.08 ± 0.32

**Values expressed as* μmoles/cm^2 *hr. (mean* $\pm$ *SEM) and numbers in parentheses are number of guts per treatment. A positive net flux indicates transfer from mucosa to serosa.*
***Control solution present in serosal bath at all times.*

table are consistent with those obtained from somewhat smaller
sample sizes. The J_{net}^{Na} (1.49 $\pm$ 0.30 μmoles cm^{-2} hr^{-1}) was significantly greater (P<0.05) than the J_{net}^{Cl} (0.72 $\pm$ 0.22 μmoles cm^{-2} hr^{-1}),
suggesting a net transmural flux ratio of 2 Na/ 1 Cl.

Replacing luminal chloride with equimolar sulfate resulted in a
significant decrease in J_{ms}^{Na} from 4.83 $\pm$ 0.26 μmoles cm^{-2} hr^{-1} (n =
30) to 3.39 $\pm$ 0.39 μmoles cm^{-2} hr^{-1} (n = 9) (P<0.05), while J_{sm}^{Na} remained unaffected (P>0.05)(Table 2; row 2). The result of this
treatment was the total elimination of J_{net}^{Na} from 1.49 $\pm$ 0.30 (control) to - 0.01 $\pm$ 0.43 (Cl-free perfusate) μmoles cm^{-2} hr^{-1} (P>0.05).
Similarly, Na-free luminal solution (choline replacing Na) signifi-
cantly (P<0.05) lowered J_{ms}^{Cl} from 2.47 $\pm$ 0.20 μmoles cm^{-2} hr^{-1} (n =
19) to 1.86 $\pm$ 0.26 μmoles cm^{-2} hr^{-1} (n = 5) without altering J_{sm}^{Cl}
(P>0.05)(Table 2; row 3). In the absence of mucosal Na, J_{net}^{Cl}
dropped from 0.72 $\pm$ 0.22 to - 0.08 $\pm$ 0.32 μmoles cm^{-2} hr^{-1}, the
latter rate not being significantly different from zero (P>0.05).
Net transmural fluxes of both ions, therefore, appear to require
the presence of a counter ion in the perfusate and in the absence
of this counter ion the mucosa to serose fluxes of both sodium and
chloride become identical to their respective serosa to mucosa
fluxes.

INFLUXES OF SODIUM AND CHLORIDE ACROSS THE APICAL CELL MEMBRANE

Influx of ^{22}Na (J_{mc}^{Na}) and ^{36}Cl (J_{mc}^{Cl}) from the luminal solution
into the intestinal epithelium of the freshwater prawn was mea-
sured by briefly exposing the isolated preparation with saline
containing either ^{22}Na (7.5, 15, 30, and 60 sec) or ^{36}Cl (0.25,
0.50, 1.00, and 2.50 min). Immediately after this radioactivity
pulse, a 15-sec rinse of unlabelled saline followed to wash out the
bulk of luminal isotope. After this rinse the tissue was removed
from the perfusion chamber, digested in Protosol Tissue Solubilizer
(New England Nuclear Corp.), and analyzed for radioactivity using
the toluene-based cocktail and Beckman 230 scintillation counter.
Radioactivity in the rinsed tissue was considered to represent
labelled ions present in both cellular and extracellular compartments, the former increasing in isotope concentration with time,
and the latter containing a fixed isotope concentration in equilibrium with the bulk luminal solution throughout the incubation
interval.

When 211 mM Cl was present in the luminal perfusate, a significant (P<0.01) increase in tissue accumulation of ^{22}Na occurred over
the incubation interval (Fig. 6, top left). Sodium influx across
the apical cell membrane under these conditions was determined from
the slope of the tissue accumulation curve using regression analysis
and found to be 1.26 $\pm$ 0.20 μmoles cm^{-2} hr^{-1} (n = 19). The significant (P<0.01) vertical axis intercept is taken to represent ^{22}Na
present in the tissue extracellular space. In contrast, replacement of luminal chloride by sulfate (0 mM Cl) did not result in a
significant (P>0.05) increase in tissue accumulation of ^{22}Na (nonsignificant change in slope) over what was present in the extra-

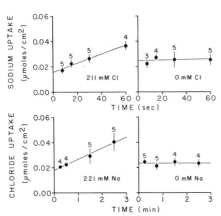

Fig. 6. Time course of tissue accumulation of ^{22}Na and ^{36}Cl across the mucosal border of the prawn midgut. Ion concentrations on figure refer to those in the perfusate, while control saline was present at all times in the serosal bath. Symbols are as described in Fig. 4. Note different time scale for uptake of each ion.

cellular compartment during a similar incubation interval (Fig. 6, top right). Luminal chloride is therefore required to facilitate the cellular entry of ^{22}Na, and in the anion's absence sodium uptake across the epithelial mucosal membrane ceases.

With a luminal concentration of 221 mM Na, J_{mc}^{Cl} was significantly greater than zero (P<0.01), as determined by regression analysis of ^{36}Cl uptake as a function of incubation time (Fig. 6, lower left). The rate of cellular chloride entry under these conditions was 0.52 + 0.15 µmoles cm^{-2} hr^{-1} (n = 18). Using a Na-free perfusate (K replacing Na), cellular uptake of ^{36}Cl ceased (P>0.05) and tissue radioactivity appeared confined to the extracellular compartment (Fig. 6, lower right). These results suggest that luminal sodium is necessary to support the unidirectional trans-membrane transport of chloride across the mucosal border of the midgut epithelium. The flux ratio for cellular entry of sodium and chloride (J_{mc}^{Na} / J_{mc}^{Cl} = 1.26/0.52 = 2.4) was similar to the net transmural flux ratio exhibited by the same ions (J_{net}^{Na} / J_{net}^{Cl} = 1.49 / 0.72 = 2.1), suggesting that the stoichiometry of ion flow across the tissue as a whole was most likely a result of rate limitations imposed by the mucosal membrane alone.

A MODEL FOR WATER AND ION TRANSPORT

Net water transport across epithelia may occur as a result of transcellular ionic, osmotic, or hydrostatic pressure gradients (Schultz & Curran 1968; House 1974). In addition, active ion transport by the tissue can generate localized increases in osmotic and hydrostatic pressures within epithelial compartments, leading to the passive flow of water from the bathing solution into this region of elevated osmolality (Curran 1960, 1962; Diamond & Bossert 1967; Diamond 1971; Patlak, Goldstein & Hoffman 1963). The overall effect of these local osmotic gradients is directional net volume flow across the tissue.

Water and ion transport across the perfused intestine of the freshwater prawn, *Macrobrachium rosenbergii*, appear to exhibit several characteristics that support the local osmosis concept of epithelial volume flow. In the absence of transmural osmotic or ionic gradients, J_v was totally dependent upon the simultaneous presence of both luminal sodium and chloride (Figs. 1,2). In their absence, or in the presence of elevated mucosal concentrations of hydrogen ion (Fig. 3), net volume flow ceased. The simultaneous requirement of both luminal sodium and chloride for J_v is clearly a result of the tight coupling between the net transmural fluxes of these two ions (Table 2), a coupling apparently originating at the entry step across the epithelial brush-border membrane (Fig. 6). The complete cessation of J_{net}^{Na} and J_{net}^{Cl}, as well as the elimination of significant J_{mc}^{Na} and J_{mc}^{Cl}, in the absence of luminal chloride or sodium, respectively, implies that virtually none of the net transport of either ion across the tissue occurs by independent pathways,

Complete dependence upon the counter ion for net Na and Cl transfer across gastrointestinal organs appears rare in animals studied to date (Table 3), although in every case at least some portion of the respective net fluxes were coupled. Ion transport in mammalian and amphibian gut share the common property of possessing both coupled and independent transfer mechanisms. Independent net ion transport in fish intestine appears less clearly developed and, if present, represents only a very small component of the ion payment across the tissue. Only in invertebrate intestine does the totality of J_{net}^{Na} and J_{net}^{Cl} appear to occur by a shared mechanism, a point which may have implications for the evolution of gastrointestinal ion transport processes.

A tentative model describing the proposed mechanism of water and ion transport in the midgut of the freshwater prawn is illustrated in Fig. 7. The model includes a shared ion transport

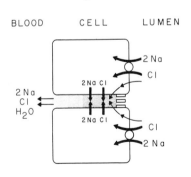

BLOOD   CELL   LUMEN

2 Na
Cl
2 Na Cl
2 Na
Cl
H₂O
2 Na Cl
Cl
2 Na

Fig. 7. Proposed model for sodium, chloride, and water transport across the prawn intestine. Bold arrows represent ion movements, while narrow arrows signify the transfer of water. A shared membrane-bound carrier process is shown in the apical cell membrane, while paucity of information precludes identification of lateral carrier mechanisms. Stippling density in the intercellular space refers to relative osmotic concentration regulated by active ion extrusion across lateral cell membranes.

carrier in the mucosal membrane which concurrently transfers sodium and chloride from perfusate to cytoplasm in the ratio 2 Na/

137

Table 3. *Comparative Gastrointestinal Net Fluxes of Sodium and Chloride*

Tissue	J_{net}^{Na}	J_{net}^{Cl}	Coupled net flux*	Independent net flux*	Reference
	(μmoles cm^{-2} hr^{-1}; mean ± SEM)				
Mammals					
Rabbit ileum	4.30 ± 0.80	2.30 ± 0.90	+	+ (Na) − (Cl)	Nellans et al. (1973)
Rat colon	8.80 ± 0.30	9.10 ± 0.90	+	+ (Na) + (Cl)	Binder and Rawlins (1973)
Bovine rumen	1.30 ± 0.20	3.60 ± 1.00	+	+ (Na) + (Cl)	Chien and Stevens (1972)
Amphibians					
Bullfrog small intestine	1.48 ± 0.26	1.18	+	+ (Na) + (Cl)	Quay and Armstrong (1969)
Fish					
In vitro sea water eel intestine	1.50	3.10	+	? (Na) + (Cl)	Hirano et al. (1976)
In vivo sea water eel intestine	24.40 ± 7.00	28.10 ± 7.40	+	? (Na) ? (Cl)	Skadhauge (1974)
Sculpin intestine	12.90 ± 0.60	11.40 ± 4.00	+	+ (Na) − (Cl)	House and Green (1965)
Molluscs					
Sea hare intestine (Aplysia californica)	---	8.39 ± 2.15	+	− (Na) − (Cl)	Gerencser et al. (unpubl.)**
Sea hare intestine (A. juliana)	---	5.68 ± 0.58	+	− (Na) − (Cl)	Gerencser et al. (1976); Gerencser (unpubl.)**
Arthropods					
Prawn midgut	1.49 ± 0.30	0.72 ± 0.22	+	− (Na) − (Cl)	Present report

* + and − signify the reported presence or absence, respectively, of coupled and independent components of net flux.

** Unpublished estimations of net flux components based on transmural potential measurements (Dr. G. A. Gerencser, University of Florida).

1 Cl (Fig. 6). Inhibitory effects of increased luminal hydrogen ion concentration on J_V (Fig. 3) may occur as a direct interaction of H^+ with the cation site on this carrier mechanism. Electron microscopy of the midgut epithelium in *M. rosenbergii* indicates numerous apical mitochondria, while the basal aspect of the cell was relatively few of these organelles (Cooke & Ahearn 1976). Ion efflux, proposed in the model, is therefore confined to the lateral membrane, near potential sources of metabolic energy input. At present it is unknown if both ions are actively transported out of the cell or whether the efflux of one ion is by active means and the other passively following the electrochemical gradient induced across the membrane. In either case the net flux of sodium and chloride into the intercellular compartment is proposed to create a local increase in osmotic pressure within this confined region, which serves to draw fluid from the perfusate creating a directional transport of water across the tissue.

This proposed model is similar in several respects to those presented for gastrointestinal epithelia in other organisms. A Na-dependent water transport mechanism, inhibited by other luminal cations (in this case K), has recently been reported for the desert scorpion ileum (Ahearn & Hadley 1976a,b). These studies did not, however, examine the potential role of luminal anions in volume flow. Using an identical experimental procedure to measure J_V as that reported in the present study on *M. rosenberaii*, these authors obtained an estimate of water transport across the scorpion ileum of $107.4 \pm 15.6 - 144.0 \pm 10.8$ μl cm^{-2} hr^{-1}, an absorption rate greater than twice that found for the freshwater prawn midgut (50.2 ± 2.16 μl cm^{-2} hr^{-1}). These transport differences may reflect adaptations to markedly dissimilar environments.

Studies examining the mechanisms of water and ion transport across intestinal preparations from amphibians and mammals have suggested that the coupled component of J_{net}^{Na} and J_{net}^{Cl} results from a shared entry process across the apical membrane followed by independent exit mechanisms out the baso-lateral border (Quay & Armstrong 1969; Nellans, Frizzell & Schultz 1973), a finding in keeping with the results for the prawn midgut. However, as indicated in Table 3, independent net transmural flux components for sodium and chloride were also found for those organisms, while apparently absent in *M. rosenbergii*.

A kinetic model for coupled NaCl influx (J_{mc}^{Na} and J_{mc}^{Cl}) across the rabbit ileum brush border suggests that a single cation and anion attach to a membrane-bound carrier mechanism forming a mobile ternary complex (Nellans, Frizzell & Schultz 1973). In this case the binding of both ions are required before the translocation process may begin. A report on a similar entry mechanism for the coupled influx of Na and Cl across the apical membrane of rabbit gallbladder epithelium has also recently been published (Frizzell, Dugas & Schultz 1975). Preliminary influx data presented for *M. rosenbergii* (Fig. 6) suggest that a ternary complex may not represent the mobile carrier form in this species, but rather a quaternary structure carrying 2 Na and 1 Cl. Similar differences between mammalian and crustacean gut have been found in the mobile carrier forms involved in translocation of Na and amino acids. In rabbit ileum, glycine (Peterson, Goldner & Curran 1970) and alanine (Curran, Schultz, Chez and Fuisz 1967) transport across the brush border membrane involve the formation of a mobile ternary complex, whereas a mobile quaternary complex composed of two sodium ions and one amino has been reported for glycine influx in the midgut of the marine shrimp, *Penaeus marginatus* (Ahearn 1976). These differences in mobile carrier structure for transport of both ions and amino acids may be the result of progressive or divergent evolution; in either case, in response to possible selection pressures such as diet or maintenance of ionic and osmotic homeostasis.

CONCLUSIONS

Net water transport (J_v) across the intestine (midgut) of the freshwater prawn, *M. rosenbergii*, depends upon luminal sodium and chloride and is inhibited by elevations in luminal hydrogen ion concentration. The ion dependency of J_v results from the operation of a coupled entry process for sodium and chloride in the apical cell membrane in combination with proposed separate extrusion mechanisms for the two ions located in the lateral cell borders. The stoichiometry of sodium and chloride mucosal influxes suggests that the coupled apical entry process requires the attachment of 1 Na and 1 Cl for translocation. No evidence was found for the presence of independent transport mechanisms for sodium and chloride in the brush border membrane as has been reported for vertebrate intestinal tissues. Hydrogen ion apparently competes with sodium for attachment to the coupled transfer process, but is unable to support a flow of water. Directional water movement across the tissue is proposed to occur as a result of local elevations in osmotic pressure in epithelial compartments (intercellular spaces) following the efflux of sodium and chloride from the cells. Differences in ion transport properties between crustacean and vertebrate intestine may have their basis in membrane-bound protein evolution.

ACKNOWLEDGEMENTS

We thank Drs. Robert C. May (Hawaii Institute of Marine Biology) and G. Alex Gerencser (University of Florida) for critically reviewing the manuscript during its preparation. We also thank Dr. Gerencser for allowing us to refer to his unpublished work on Aplysia intestine. Appreciation is extended to Mr. Thomas J. Harris for analyzing midgut content samples. This investigation was supported by the University of Hawaii Sea Grant Program under Grant no. 04-3-158-29 from the National Oceanic and Atmospheric Administration and by the National Science Foundation under Grant no. BMS 74-20663.

REFERENCES

Ahearn, G.A. (1976) *J. Physiol. 258:* 499.
Ahearn, G.A. & Hadley, N.F. (1976) *Am. Zoologist 16:* 225.
Ahearn, G.A. & Hadley, N.F. (1976) *Nature 261:* 66.
Bielawski, J. (1974) *Comp. Biochem. Physiol. 13:* 423.
Binder, H.J. & Rawlins, C.L. (1973) *Am. J. Physiol. 225:* 1232.
Bryan, G.W. (1960) *J. Exp. Biol. 37:* 83.
Burg, M.G. & Orloff, J. (1968) *J. Clin. Invest. 47:* 2016.
Chien, W. & Stevens, C.E. (1972) *Am. J. Physiol. 223:* 997.
Cooke, W.J. & Ahearn, G.A. (1976) *Am. Zoologist 16:* 225.
Croghan, P.C. (1958) *J. Exp. Biol. 35:* 213.
Croghan, P.C. (1958) *J. Exp. Biol. 35:* 219.
Croghan, P.C. (1958) *J. Exp. Biol. 35:* 234.
Croghan, P.C. (1958) *J. Exp. Biol. 35:* 243.
Croghan, P.C., Curra, R.A. & Lockwood, A.P.M. (1965) *J. Exp. Biol. 42:* 463.
Croghan, P.C. & Lockwood, A.P.M. (1968) *J. Exp. Biol. 48:* 141.
Curran, P.F. (1960) *J. Gen. Physiol. 43:* 1137.
Curran, P.F. (1962) *Nature 193:* 347.
Curran, P.F., Schultz, S.G., Chez, R.A. & Fuisz, R.E. (1967) *J. Gen. Physiol. 50:* 1261.
Dall, W. (1965) *Aust. J. Mar. Freshwater Res. 16:* 181.
Dall, W. (1967) *Comp. Biochem. Physiol. 21:* 653.
Diamond, J.M. (1971) *Fed. Proc. 30:* 6.
Diamond, J.M. & Bossert, W.H. (1967) *J. Gen. Physiol. 50:* 2061.
Frizzell, R.A., Dugas, M.C., & Schultz, S.G. (1975) *J. Gen. Physiol. 65:* 769.
Gerencser, G.A., Hong, S.K. & Malvin, G. (1976) *Fed. Proc. 35:* 464.
Gross, W.J. (1957) *Biol. Bull. 112:* 43.
Gross, W.J. & Capen, R.L. (1966) *Biol. Bull. 131:* 272.
Hirano, T., Morisawa, M., Ando, M. & Utida, S. (1976) In *Intestinal Ion Transport* (Robinson, J.W.L., Ed.), p. 301. University Park Press, Baltimore.
House, C.R. (1974) *Water transport in cells and tissues.* Edward-Arnold, London.
House, C.R. & Green, K. (1965) *J. Exp. Biol. 42:* 177.
King, E.N. & Schoffeniels, E. (1969) *Archs. Int. Physiol. Biochim. 77:* 105.
Koch, J.H., Evans, J. & Schicks, E. (1954) *Meded. Vlaamse Acad. Kl. Wet. 16:* 1.
Lockwood, A.P.M. (1961) *J. Exp. Biol. 38:* 647.
Lockwood, A.P.M. & Inman, C.B.E. (1973) *J. Exp. Biol. 58:* 149.
Maetz,J. (1974) In *Biochemical and Biophysical Perspectives in Marine Biology* (Malins, D.C. and Sargent, J.R., eds.), Vol. 1, p. 1. Academic Press, New York.
Mantel, L.H. (1967) *Comp. Biochem. Physiol. 20:* 743.
Mantel, L.H. (1968) *Am. Zoologist 8:* 433.
Nellans, H.N., Frizzell, R.A., & Schultz, S.G. (1973) *Am. J. Physiol. 225:* 467.

Patlak, C.S., Goldstein, D.A. & Hoffman, J.F. (1963) *J. Theoret. Biol. 5:* 426.

Peterson, S.C., Goldner, A.M., & Curran, P.F. (1970) *Am. J. Physiol. 219:* 1027.

Potts, W.T.W. & Parry G. (1964) *Osmotic and ionic regulation in animals.* Pergamon Press, Oxford.

Prosser, C.L. (ed.) (1973) *Comparative Animal Physiology.* W.B. Saunders, Philadelphia.

Quay, J.F. & Armstrong, W. McD. (1969) *Am. J. Physiol. 217:* 694.

Riegel, J.A. (1970) In *Chemical Zoology* (Florkin, M. and Scheer, B.T., eds.) Vol. VI, Pt. B, p. 249. Academic Press, New York.

Rudy, P.P. Jr. (1966) *Comp. Biochem. Physiol. 18:* 881.

Rudy, P.P. Jr. (1967) *Comp. Biochem. Physiol. 22:* 581.

Schultz, S.G. & Curran, P.F. (1968 In *Handbook of Physiology, Sect. 6: Alimentray Canal.* (Code, C.F., ed.), Vol. III, p. 1245. Am. Physiol. Soc., Washington, D.C.

Schultz, S.F. & Curran, P.F. (1974) In *Current topics in membranes and transport.* (Bronner, F. and Kelinzeller, A., eds.), Vol. 5, p. 225. Academic Press, New York.

Shaw, J. (1961) *J. Exp. Biol. 38:* 135.

Shaw, J. (1961) *J. Exp. Biol. 38:* 153.

Skadhauge, E. (1974) *J. Exp. Biol. 60:* 535.

Smith, P.G. (1969) *J. Exp. Biol. 51:* 727.

Smith, P.G. (1969) *J. Exp. Biol. 51:* 739.

Sutcliffe, D.W. (1961) *J. Exp. Biol. 38:* 501.

Sutcliffe, D.W. (1961) *J. Exp. Biol. 38:* 521.

Sutcliffe, D.W. (1968) *J. Exp. Biol. 48:* 359.

MOLECULAR MECHANISMS IN THE BRANCHIOPOD
LARVAL SALT GLAND (CRUSTACEA)

Frank P. Conte

Oregon State University

INTRODUCTION

Having selected concentrated brines as a habit in which to live, halophilic plants, animals and microbes are faced with having to solve several important environmental problems. In particular, halophilic animals must cope with (1) hypoxia brought about by the lowering of the partial pressure of gaseous oxygen (Fig. 1),

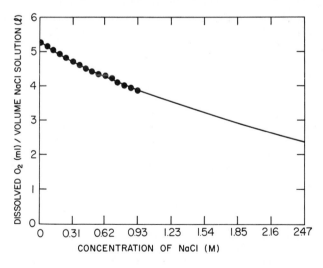

Fig. 1. Concentration of dissolved oxygen as a function of salinity (Green & Carritt 1967).

and (2) desiccation due to the decrease in available "free" water for osmosis that is also combined with the passive inward diffusion of solutes (Fig. 2). The molecular mechanism(s) involved in achiev-

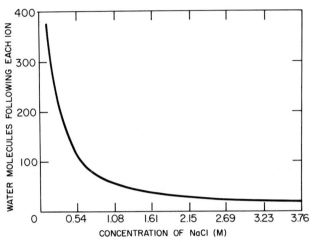

Fig. 2. Water transport per ion translocated across membrane as a function of solute concentration.

ing osmotic balance appears to be quite diverse among the halophilic biota. Halophilic bacteria for example are osmotic conformers, which neither regulate intracellular water nor electrolytes in a differential manner. Although these bacteria require a high intracellular salt concentration (> 3 M) for growth and replication, the metabolic pathways which supply energy to drive these intracellular processes are similar to those present in mesophilic bacteria. These micro-organisms have evolved a unique molecular strategy that affords protection to subcellular organelles from disaggregation when exposed to extreme ionic concentrations. It involves the alteration of the amino acid composition of the polypeptide backbone of cellular proteins to contain an extraordinarily large number of acidic amino acid residues. In this manner electrostatic repulsions do not disrupt protein conformations because the carboxyl groups are neutralized by monovalent cations (K+) preventing electrical forces from disrupting protein conformations.

Brachiopod (gill-footed) crustaceans live in the same habitat as halophilic microbes but employ a different strategy to survive in concentrated brines. Osmotic equilibrium is achieved by regulation of the concentrations of internal electrolytes. The species of brachiopod crustacea whose physiology is best understood is the brine shrimp, Artemia salina.

HYPOOSMOTIC REGULATION IN BRANCHIOPOD CRUSTACEANS (A. salina)

Ionic fluxes and osmoregulatory mechanisms in adult brine shrimp have been studied by Croghan (1958 a-d), Theut et al.,(1968) and Smith (1969a,b), who convincingly demonstrated that brine shrimp are hypoosmotic regulators (see Table 1) with the responsible organs being the gut and the middle leg segment (metepipodite). Water and electrolyte balance is achieved by having the animal swallow the external medium, absorb ions and water from the

Table 1. *Comparison of Osmotic and Ionic Concentrations, Unidirectional Ion Fluxes, and Transepithelial Electrical Potential Differences in Adult and Nauplear Brine Shrimp Adapted to 0.5 M NaCl‡*

Developmental Stage-Epithelium	Osmotic Conc. Int. (mM)	Osmotic Conc. Ext. (mM)	Sodium Conc. Int. (mM)	Sodium Conc. Ext. (mM)	Potassium Conc. Int. (mM)	Potassium Conc. Ext. (mM)	Chloride Conc. Int. (mM)	Chloride Conc. Ext. (mM)	Transepithelial Potential* Equilibrium (mv)	Transepithelial Potential* Measured (mv)	J_{in} (pmoles/sec/mg animal)	J_{out} (pmoles/sec/mg animal)	Reference
Adult- Body Integument	318	1032	172	485	10	11	153	565			Na:111 Cl: 99	Na: 86 Cl: 64	Croghan 1958b
			172	468	8	10	153	548	Na:+26 K:+ 6 Cl:-33	+23	Na:215 Cl:220		Smith 1969
			159	500									Thuet 1968
Nauplius- Cervical Gland	413	1055											Croghan 1958b
	207	922	100	467	4	16		546		+25			Mangos 1976
	262	983	140	472									Conte 1976

‡ 0.5 M NaCl approximates full strength sea water

* Sign is that of the body fluids.

Pd=diffusional water permeability = 6.9×10^{-6} cm/sec (Smith 1969)

PNa=diffusional sodium permeability = 2.8×10^{-5} cm/sec (Smith 1969)

PCl=diffusional chloride permeability = 3.1×10^{-6} cm/sec (Smith 1969)

alimentary canal, and then secrete excess ions across the gills.
Measurement of ion fluxes found gill tissue more permeable
than any other part of the external integument. Measurements of
transepithelial potential differences between the coelomic cavity
and the medium (ergo across the gill) found the external gill surface to be negatively charged. In addition, when compared to other
transporting epithelia, a low degree of osmotic water loss was
found (7.1×10^{-8}cm/sec/atm). Calculations of the equilibrium potentials for sodium, potassium and chloride indicate that movement
of sodium is passive, while outward passage of chloride to the external medium and the inward passage of potassium to the hemolymph
are both active transport processes (Smith 1969a,b).   Thuet et al.
(1968) regard the transepithelial mechanism to be one of exchange
diffusion characterized by a coupling of both unidirectional fluxes
of sodium and chloride.

     In larval brine shrimp, a limited number of studies on ionic
homeostasis and transepithelial transport mechanisms have been performed. In preliminary micro-punctures of nauplii, Croghan (1958b)
presented   evidence (Table 1) which suggested that this larval
stage was also a hypoosmoregulator. This finding was confirmed
by Conte (Conte et al.,1972) using an indirect method. He showed
that the total naupliar body (intracellular and extracellular)
sodium and potassium levels did not vary when animals were exposed
to very steep salt gradients. Collectively, these data support the
hypothesis that the larval nauplius hypoosmoregulates, and that
some type of ion transporting structure must exist to maintain the
osmotic gradient between blood and medium. The most intriguing
aspect of this regulation is that nauplii maintain ionic homeostasis in the complete absence of established excretory structures
such as gills, antennal gland, or gut. [The alimentary tract is
not completed prior to the first larval ecdysis and with regard to
ingestion and digestion of external food is non-functional. (Hootman and Conte 1974).] How are internal salt levels maintained?
The nauplius must have a structure whose function is similar to
the gill leaflet and various investigators (Dejdar, 1970; Croghan,
1958b; Conte, et al., 1972) have proposed that the cervical gland
("neck" organ) found in the cephalothorax region serves as a larval
salt gland (Conte et al.,1972; Hootman et al.,1972; Hootman & Conte
1975). Proof that the cervical gland is an ion regulating structure comes from recent measurements of the chemical and electrical
gradients that exist between the external medium, gland epithelium
and the hemolymph. Through the cooperation and guidance of Dr.
John Mangos, University of Wisconsin, Conte (unpublished observations) made exploratory micropuncture analyses of naupliar hemolymph
taken from a single nauplius. Using a specially designed funnel-
shaped glass trap held by a micromanipulator, an animal is caught
(see Fig. 3), held by the post mandibular region of the cervical
gland and transferred into a concave depression of a glass slide
to which is added a solution of mannitol equimolar to that of the
NaCl in which the nauplius had been suspended and acclimatized.
The cardiocoelomic cavity lying beneath the cervical gland is

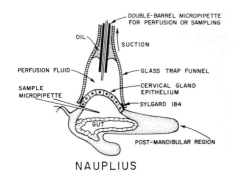

NAUPLIUS

Fig. 3. Schematic of micropuncture apparatus used to obtain hemolymph from a single nauplius.

punctured by guiding a glass capillary with a beveled tip (2-4μ) filled with dyed mineral oil between the edge of the gland and the frontal face of the labrum. As soon as the tip of the capillary enters the coelomic cavity, a small droplet of colored oil is expelled. If indeed the tip is in the cavity the oil droplet can be reaspirated and expelled freely, which is not the case if the tip is in the foregut or imbedded inside the cervical gland epithelium. Hemolymph is collected after the capillary tip is inserted into the coelomic cavity, the absence of mannitol leakage through the site of micropuncture,was insured via the following controls. If the mannitol medium contains a small trace of H^3-inulin, then monitoring the aspirated hemolymph for tritium contamination would reveal any dilution of the hemolymph caused by the medium. Thus, contaminated samples could be readily identified. Data presented in Table 1 facilitates comparisons of internal ionic concentrations and the transepithelial potential differences between larval and adult stages in development. Active chloride ion transport across the adult body epithelium and the cervical gland epithelium of the nauplius can be postulated from these data. Sodium permeability is probably associated with the downhill electrical gradient that exists between the medium and the hemolymph.

ONTOGENY OF ELECTROLYTE REGULATION

Having presented evidence that the cervical gland appears to be the ion transporting organ in the nauplius, the organ is not a functional structure until after postgastrulation (Benesch 1969), and does not exhibit excretory properties until late in the prenaupliar stage (Figs. 4,5). Therefore, two distinct modes of osmotic regulation occurring during brine shrimp embryogenesis.

The first type of osmotic adaptation occurs early in development and is related to the onset, duration and termination of dormancy. Fertilized eggs in the maternal brood sac have two routes of embryonic development, either with or without dormancy. In the nondormant route, within the brood pouch of the female development proceeds through the prenaupliar stage to the nauplius. Alterna-

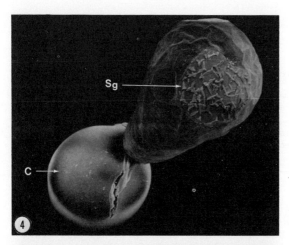

Fig. 4. Scanning electron micrograph of early prenauplius with hatching membrane covering non-functional salt gland (C - cyst, Sg - Salt gland).

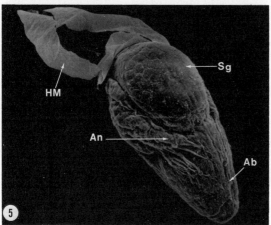

Fig. 5. Scanning electron micrograph of late prenauplius with hatching membrane removed to reveal the surface of the functional salt gland (Ab - Abdominal structure, An - second Antenna, Hm - Hatching membrane, Sg - Salt gland).

tively, the embryo may become encased in a thick shall secreted by the female during early stages of gastrulation and enter a dormant state (cyst). Interestingly, the encysting embryo while still contained in the maternal brood sac and prior to achieving the complete state of dormancy, begins storing glycogen for its energy needs in future growth (Dutrieu 1960; Clegg 1962). Trehalose, a non-reducing or disaccharide of glucose is also stored. Trehalose creates a hyperosmotic environment for the embryo, which helps it to escape from the biological restraints imposed by the chitinous shell. This mechanism develops early after the initial rehydration of the dormant cyst. It is chronicled by the appearance of the enzyme trehalase, which is indirectly responsible for the appearance of small molecular fragments, such as glycerol (Clegg 1964). Free glycerol is not metabolized. Rather, it accumulates within

148

the free-embryonic spaces until sufficient quantities have been
built up inside the cyst to create an internal osmotic gradient
greater than that of the external medium (Fig. 6). The osmotic

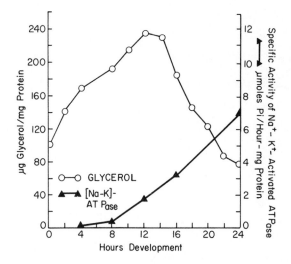

*Fig. 6. Simultaneous
measurement of internal
glycerol and Na+K-ATPase
in an asynchronous pop-
ulation incubated in
media containing 0.5 M
NaCl.*

swelling apparently generates hydrostatic pressures sufficiently
great to rupture the shell, thereby releasing the embryo. Fortunately shell rupture does not interfere with embryonic growth.

The embryo is exposed to external solutes following destruction
of the chitinous shell. External solutes now have free access to
the embryo should the trehalose-glycerol system fail to maintain
water balance, additional problems of desiccation due to steep external solute gradients will arise. And fail it does! Embryonic
membranes begin leaking substantial quantities of glycerol to the
external medium almost immediately (Fig. 7). The embryo cannot
sustain these continued losses by compensatory increases in trehalose metabolism. The embryo avoids premature death by converting
the water equilibrium process which is based upon the passively
hyperosmotic trehalose-glycerol system, to an active hypoosmotic
mechanism dependent upon ionic regulation (Conte, et al., 1976).
In the early stages of excystment (first 8 hours), before the
outer shell has ruptured, the cyst is able to imbibe water without
any net increase in sodium (Fig. 8). However, upon the beginning
of emergence (about 14-16 hours), sodium rises dramaticly coincident with a fall in internal glycerol (Figs. 6&8) and a peaking of
external media glycerol. After hatching (about 24 hours), the
embryo exhibits a pronounced regulation of internal sodium with
an increased sensitivity to external ouabain that is dependent
upon the external salinity (Ewing et al.,1972). With the onset of
sodium regulation, there is the de novo appearance of the cationic
Na+K-ATPase as evidenced by ^{14}C-leucine incorporation into sub-

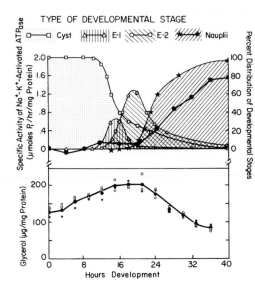

Fig. 7. Correlating of Na+ K-activated ATPase activity and glycerol levels with developmental stages. Heavy dark lines in upper figure represent Na+K-activated ATPase activity. Heavy dark lines in lower figure represent mean glycerol levels in media containing 0.25 M (o---o), 0.50 M (---), and 0.75 M (*---*) NaCl.

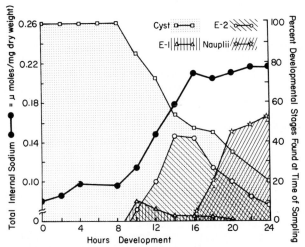

Fig. 8. Internal sodium levels of embryos during excystment of an asynchronous population in 0.5 M NaCl.

units of the enzyme (Peterson et al., 1975). The available evidence thus indicates the existence of a transition from a passive hyperosmoregulatory system dependent upon a trehalose-glycerol mechanism, to a hypoosmoregulatory system dependent upon active transport of electrolytes.

NATURE OF THE ELECTROLYTE ACTIVELY TRANSPORTED

It is usually agreed that the polarity of the transepithelial potential provides a good indication as to which ion is actively transported and which ions follow passively down the potential gradient thus created. Chloride appears to be the ionic species involved in the net outward flow of ions in the adult (Table 1). Preliminary data on the cervical gland suggests that it too is the species actively transported across the apical boundary of the epithelium. If chloride outflow is an active process, then it might well be coupled to bicarbonate transport through a specific anion-linked ATPase. An unsuccessful search for an $HCO_3^-+Cl^-$-activated ATPase in nauplii was made but only a Na^++K^+-stimulated ATPase was detected (Ewing et al.,1974).

Nauplii exhibit a salt-dependent toxicity toward acetazolamide (Ewing et al. 1972). A systematic study of naupliar toxicity toward a wide variety of sulfonamides was made (Bartel 1976). The embryo is very sensitive to sulfonamides (10^{-6} M). This sensitivity is both salt and does dependent, with the degree of toxicity apparently varying with the dissociation constants for carbonic anhydrase-sulfonamide complexes (Fig. 9). The principle action of

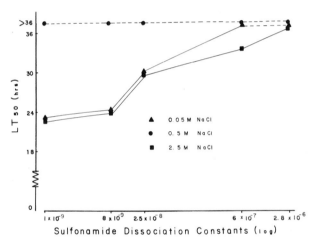

Fig. 9. *Dissociation constants of sulfonamide from erythrocyte carbonic anhydrase as related to LT_{50} at 10^{-3} M concentration.*

most sulfonamides is the inhibition of carbonic anhydrase (CA). An unsuccessful attempt to isolate and characterize a crustacean CA from the nauplius was made utilizing spectrophotometric, manometric and titrimetric techniques. Furthermore, bicarbonate transport into the cytoplasm as found by $H^{14}CO_3^-$ permeability experiments (Fig. 10) showed transport largely unaffected. We concluded that if transport required an $HCO_3^- \rightarrow Cl^-$ exchange, it does not appear dependent upon the presence of CA. Another possibility for the site of sulfonamide inhibition could lie in the respiratory-oxidative phosphorylation system. Investigation of oxygen consumption and the electron transport system in embryos exposed to the drug showed these pathways to be unaffected. Lastly, we directed our attention to the metabolic pathway of glycolysis with its attendant shunts, especially those containing Zn-metallo-enzymes. If one takes the present evidence together with earlier information

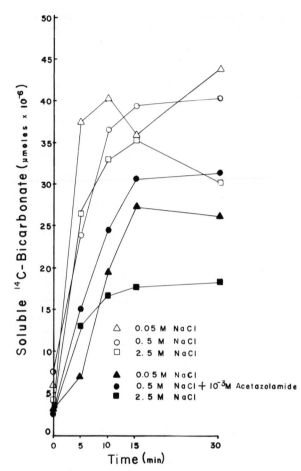

Fig. 10. Transport of ^{14}C-bicarbonate ion into cytosol of the nauplius as a function of salinity and its sulfonamide sensitivity.

Legend in figure:
△ 0.05 M NaCl
○ 0.5 M NaCl
□ 2.5 M NaCl

▲ 0.05 M NaCl
● 0.5 M NaCl + 10^{-3}M Acetazolamide
■ 2.5 M NaCl

concerning the capability of brine shrimp embryos to fix CO_2 (Clegg 1967), then a hypothesis about the nature of an anion which might substitute for Cl^- in the HCO_3^- translocation reaction becomes clearer. Suppose HCO_3^- ion uptake is coupled with dicarboxylic organic acid synthesis and its salt-regulated secretion.

C-4 DICARBOXYLIC ACID PATHWAY AS A FACULTATIVE ANAEROBIC SHUNT

Experiments were performed to determine whether nauplii excrete organic acids in response to changes in external salinity. Small numbers of nauplii were placed in sterilized artificial salines and acclimated for several hours. At selected times, nauplii and conditioned saline were poured through a filter and the filtrate was assayed for acidity. The results graphed in Figure 11 show that nauplii secrete acids, with the rate dependent upon the steepness of the salt gradient. The identity of the secreted acids was

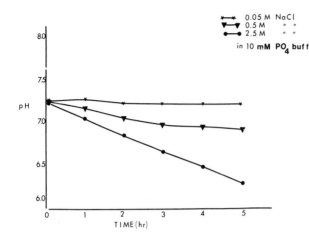

Fig. 11. Rate of acid secretion into the external medium as a function of salt gradient. Each saline contained 10mM KPO$_4$ buffer at pH 7.2, 1 mM HCO$_3$, and 3 mM CaC1$_2$.

sought. The fate of the H^{14}CO$_3^-$ incorporation reactions was care-fully followed. Earlier studies (Clegg 1967) on ^{14}CO$_2$-fixation in cysts showed that radioactivity appeared in a variety of com-pounds. Of particular interest to us was whether ^{14}CO$_2$ became in-corporated into α-keto acids. The results as shown in Figure 12 indicate that cytoplasmic ^{14}C-bicarbonate quickly becomes incor-porated into these acids. In addition, there is a build-up of labeled α-keto acids in the presence of acetazolamide (Fig. 12), which suggests that in ncuplii the drug does not inhibit the init-ial CO$_2$-fuxation reaction but blocks the pathway at another site. Recently, Clegg (1976), reinvestigating the ^{14}CO$_2$-fixation react-ion as it relates to cellular water and intermediary metabolism, found that approximately one-half of the total ^{14}CO$_2$ incorporated in the cytosol appeared in one compound, aspartic acid (>40%), with the remainder equally distributed between malate (8%), alanine (9%), and threonine (9%) as shown in Table 2. We are currently identifying the radioactive products in both the cytosol and the external medium to confirm these findings. Interestingly, Emerson (1967), studying transitions in osmotic adaptations in brachiopod crustaceans, found that the dead-end products eliminated from met-abolism were mostly free amino acids, with the most abundant species being aspartate and glutamate. We have confirmed this ob-servation (Conte et al. 1973). However, non-nitrogen containing acid compounds were not analyzed. Therefore, we propose that CO$_2$-fixation derived from exogenous bicarbonate forms oxaloacetate, which can either be transaminated into aspartate and excreted into the environment or reduced to form malate. The formation of malate by cytosolic malate dehydrogenase would provide a metabolic shunt that could substitute for the lactate pathway, since it furnishes NAD$^+$ for the continuous breakdown of glucose. Tables 3 and 4 present the experimental evidence that characterizes the cytoplas-mic carboxylase-dehydrogenase enzyme system as has been found in the naupliar cytosol. The carboxylating enzyme is phosphoenolpy-ruvate carboxykinase (GTP: oxaloacetate carboxy-lyase [transphos-

153

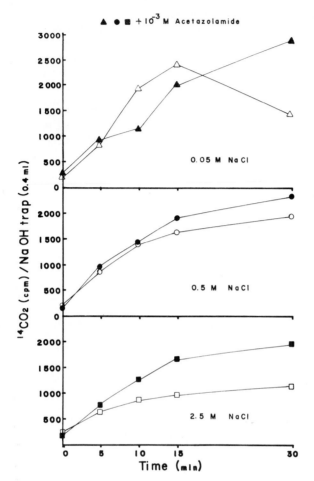

Fig. 12. The incorporation of $^{14}CO_2$ into α-keto acids of nauplii at various salinities. Open symbols are control media without inhibitor and shaded symbols are media 10^{-3} M acetazolamide.

phorylating] EC 4.1.132) and it catalyzes the carboxylation of phosphoenolpyruvate to form oxaloacetate and GTP. It is dependent upon GDP and Mn^{++} ion. The conversion of oxaloacetate to malate is catalyzed by malate dehydrogenase (L-malate; NAD oxidoreductase, EC 1.1.1.37). Purification of the two enzymes is currently in progress. Comparison of the C-4 facultative anaerobic shunt in halophilic crustaceans to Crussulacean acid metabolism in plants would be extremely interesting, since both pathways apparently have been adapted to environments that require low water usage. In addition, the need to integrate the C-4 pathway with oxidative metabolism is evident from our data presented in Table 5 which shows internal ATP concentration to be very much affected by the loss of osmotic water (Ewing et al. 1976).

Table 2. Distribution of Radioactivity in the Cold Acid-soluble Fractions of Cysts Incubated at Two Levels of Hydration.

Fraction isolated	Group Number[‡]			
	1(0.37gH$_2$O/g cyst)		2(0.73g H$_2$O/g cyst)	
	CPM/mg cysts	% of total	CPM/mg cysts	% of total
Total TCA-soluble	203.4	100.0	785.1	100.0
Amino acids:				
aspartic	84.5	41.6	285.0	36.3
threonine	17.3	8.5	78.5	10.0
alanine	21.1	10.4	69.9	8.9
glutamic	4.3	2.1	50.2	6.4
Organic acids:				
malic	16.9	8.3	74.6	9.5
citric & isocitric	4.9	2.4	30.6	3.9
fumaric	4.1	2.0	22.0	2.8
oxaloacetic	3.3	1.6	18.8	2.4
succinic & pyruvic	2.2	1.9	12.6	1.6
Nucleotides:				
UMP	1.2	0.6	19.2	2.4
CMP	0.8	0.4	10.8	1.4
Totals:	160.0	79.0	671.9	85.6

[‡]The cysts hydrations are given in parentheses next to the group numbers. The cysts were pre-hydrated at 0°C and then transferred to the vapor phase of the same NaCl solution; 3.0 M NaCl = Group 1 and 1.25 M NaCl = Group 2. Incubation took place at 23°C (Clegg 1976).

Table 3. Characterization of Larval Brine Shrimp Phosphoenolpyruvate Carboxykinase.

Incubation Mixture	Rate of Carboxylation[d] (nanomoles CO_2/min.)			Total Enzyme[c] Protein (mg)			Specific Activity of Carboxylation (nanomoles CO_2/min/mg		
	1	2	3	1	2	3	1	2	3
Complete	2.47	1.83	0.76	3.25	3.25	3.25	0.76	0.56	0.23
Omit NaH[14]CO_3	0.06	0.06	0.06	"	"	"	0.02	0.02	0.02
Omit PEP	1.15	----	0.06	"	----	"	0.35	----	0.02
Omit GDP	----	0.02	0.08	----	"	"	----	0.06	0.02
Omit PEPCK(Artemia)[a]	0.76	0.54	0.06	0.00	0.00	0.00	0.00	0.00	0.00
Omit MDH(Bovine)[b]	2.28	----	0.80	3.25	----	3.25	0.70	----	0.25
Omit NADH	----	1.45	0.72	----	3.25	"	----	0.45	0.22
Omit Mn++	----	0.06	----	----	3.25	----	----	0.02	----
Omit PEP + GDP	----	0.12	0.06	----	3.25	3.25	----	0.04	0.02
Omit PEPCK + MDH	0.85	0.07	----	0.00	0.00	----	0.00	0.00	----
Omit MDH + NADH	----	----	0.71	----	----	3.25	----	----	0.22
Omit Mn++(subst Mg++)	----	0.10	----	----	3.25	----	----	0.03	----
Omit PEP + GDP + PEPCK	----	----	0.06	----	----	0.00	----	----	0.00
Omit PEP + MDH + PEPCK	----	0.07	----	----	0.00	----	----	0.00	----
Omit NADH+MDH + PEPCK	----	----	0.06	----	----	0.00	----	----	0.00
Complete + PEPCK(0.010 ml)	----	0.89	0.34	----	0.33	0.33	----	2.74	1.05
" " (0.025 ml)	----	1.09	0.52	----	0.81	0.81	----	1.34	0.64
" " (0.050 ml)	----	1.73	0.73	----	1.63	1.63	----	1.06	0.45
" " (0.075 ml)	----	1.72	0.74	----	2.44	2.44	----	0.70	0.30
" " (0.100 ml)	2.45	1.83	0.76	3.25	3.25	3.25	0.75	0.56	0.23
" " (0.150 ml)	----	1.32	----	----	4.88	----	----	0.27	----
" " (0.200 ml)	2.25	2.45	----	6.50	6.50	----	0.35	0.38	----

The complete reaction mixture contained the following: 100 umoles imidazole(cl$^-$) buffer, pH6.6; 1.25 umoles P-enolypyruvate; 1.25 umoles GDP; 1.0 umole MnCl$_2$; 2.0 umoles GSH; 2.5 umoles NADH; 4.7 units Bovine malate dehydrogenase; 1.1 x 10^5 cpm per micromole KH[14]CO_3 and P-enolpyruvate carboxykinase (0.1 ml of 70% ammonium sulfate (A.S.) suspension) in a total volume of 1.0 ml. Placed mixture in a 25 ml stoppered Kontes metabolic reaction flask. The reaction was stopped by the addition of 1.0 ml of 2N HCl.

[a]PEPCK(Artemia) was prepared from naupliar cytosol as 45-60% A.S. precipitate and stored as a 70% A.S. suspension.

[b]Purified malate dehydrogenase isolated from pig heart and obtained from Sigma Biochemical, St. Louis, Missouri.

[c]Enzyme protein based upon Lowry assay of 60% A.S. pellet containing PEPCK(Artemia).

[d]Unit of carboxylation under these conditions: 10^{-6}mole CO_2-fixed as acid soluble(^{14}C-malate)/min./mg protein.

Table 4. Characterization of Larval Brine Shrimp Malate Dehydrogenase

Sample	A_{340} (per min)	Units[b]	Protein (ug)	Specific Activity (Units/mg protein)
Purified Bovine MDH	0.121	0.059	0.1	590
Brine Shrimp MDH; 40-60% A.S. precipitate of naupliar cytosol (Artemia)[a]	0.107	0.052	32.5	16
Brine Shrimp MDH; 2-phenoxy-ethanol extract of 24 hr nauplii homogenized in phosphate-sucrose buffer	0.133	0.064	**	**
Brine Shrimp MDH; 50% A.S. precipitate of the above extract	0.086	0.042	**	**

[a]Identical 40-60% A.S. precipitated pellet used in carboxylation reactions.

[b]One unit of malate dehydrogenase is defined as that amount of enzyme which catalyzes the oxidation of 1 micromole of NADH per minute.

**2-phenoxyethanol caused interference of the spectrophotometric measurement of protein at 280/260 mu.

Table 5. Concentration of ADP and ATP in the Cytoplasm of Nauplii at Various Salinities

Salinity (NaCl)	ADP (umoles/g wet weight)	ATP (umoles/g wet weight)	ADP/ATP
0.1	0.254±0.001	1.267±0.045	0.20
0.5	0.243±0.010	1.123±0.066	0.22
1.0	0.239	1.033	0.21
2.0	0.247	0.953	0.26
2.5	0.219±0.014	0.880±0.067	0.27
3.0	0.221	0.747	0.30

SUMMARY

Larval and adult brine shrimp (Artemia salina) are halophilic crustaceans that utilize hypoosmoregulation to maintain water balance. In both the adult and larval stages, electro-physiological measurements across the epithelium of the metepipodite and cervical gland have shown that hypoosmotic regulation occurs via a net outward movement of sodium and, possibly, chloride ions. Cation (Na^+) transport involves the development of a plasma membrane Na+ K-ATPase system and is dependent upon the electron transport system furnishing the cofactor ATP to drive Na^+ across the laterobasal cell membrane into the intercellular fluid of the coelomic cavity. Anion transport (Cl^- or HCO_3^-) is not mediated by carbonic anhydrase. Bicarbonate - aspartate exchange may be the ion translocator located at the apical boundary which is a part of a cytoplasmic PEP carboxykinase-malate dehydrogenase shunt furnishing GTP and NAD^+ for the continuous breakdown of glycogen. This pathway substitutes for the lactate pathway, since it has the decided advantage of avoiding accumulation of dead-end products.

ACKNOWLEDGEMENTS

The author wishes to thank Dr. Mary M. Becker, Mr. Steve Hand, Mr. Joe Naemura and Mrs. April Barnes for the technical assistance. Financial support was given by NSF Grant No. PCM 74-01031 A02.

REFERENCES

Bartel, L. (1976) M.S. Thesis, Oregon State University, Corvallis, Oregon.
Benesch, V.R. (1969) Zool. Jb. Anat. Bd. 86: 307.
Clegg, J. (1962) Biol. Bull, 123: 295.
Clegg, J. (1964) J. Exp. Biol. 41: 879.
Clegg, J. (1967) Comp. Biochem. Physiol. 20: 801.
Clegg, J. (1976) J. Cell Physiol. (in press).

Conte, F.P., Hootman, S.R., & Harris, P.J. (1972) *J. Comp. Physiol.* 80: 239.

Conte, F.P., Peterson, G.L. & Ewing, R.D. (1973) *J. Comp. Physiol.* 82: 277.

Conte, F.P., Droukas, P. & Ewing, R.D. (1976) Submitted to *J. Exp. Zool.*

Croghan, P.C. (1958a) *J. Exp. Biol. 35:* 213.

Croghan, P.C. (1958b) *J. Exp. Biol. 35:* 219.

Croghan, P.C. (1958c) *J. Exp. Biol. 35:* 234.

Croghan, P.C. (1958d) *J. Exp. Biol. 35:* 243.

Dejdar, E. (1930) *A. Wiss Zool. 136:* 422.

Dutrieu, J. (1960) *Archs. Zool. Exp. Gen. 99:* 1.

Emerson, D.N. (1967) *Comp. Biochem. Physiol. 20:* 245.

Ewing, R.D., Peterson, G.L. & Conte, F.P. (1972) *J. Comp. Physiol.* 80: 247.

Ewing, R.D., Peterson, G.L. & Conte, F.P. (1974) *J. Comp. Physiol.* 88: 217.

Ewing, R.D., Peterson, G.L. & Conte, F.P. (1976) Submitted to *J. Expt. Zool.*

Green, E.J. & Carritt, D.E. (1967) *J. of Marine Research 25:* 140.

Hootman, S.R., Harris, P.J. & Conte, F.P. (1972) *J. Comp. Physiol. 79:* 97.

Hootman, S.R. & Conte, F.P. (1974) *Cell. Tiss. Res. 155:* 423.

Hootman, S.R. & Conte, F.P. (1975) *J. Morph. 145:* 371.

Mangos, J.A. (1976) Personal communication

Peterson, G.L., Ewing, R.D. & Conte, F.P. (1975) *Fed. Proc. 34.*

Smith, P.G. (1969a) *J. Exp. Biol. 51:* 727.

Smith, P.G. (1969b) *J. Exp. Biol. 51:* 739.

Thuet, P., Motais, R. & Maetz, J. (1968) *Comp. Biochem. Physiol. 26:* 793.

INSECT MIDGUT AS A MODEL EPITHELIUM

James T. Blankemeyer and William R. Harvey

Temple University

INTRODUCTION

The regulation of extracellular fluid composition via transepithelial solute movement has been crucial for the evolution of
eukaryotic organisms. Despite its importance, cell biologists continue to disregard this subject, in part because literature on
this topic is large, and in part due to the absence of a single unifying model. The two principal types of analysis, electrical-struc-
tural and tracer-kinetic, each lead to sophisticated results but
seldom interact and neither has led to a convincing model. Electrical analyses have been limited by the absence of an epithelium in
which a transporting cell can be distinguished on both electrical
and structural grounds from a non-transporting cell. As a result
equivalent circuit diagrams are usually descriptions of the principal resistances of a *tissue* and do not locate active transport mechanisms within particular *cells*. The problem with tracer kinetic
analyses has been that often a small fraction of tissue ion is involved in transport (e.g. 1.4% in the frog skin; Hoshiko & Ussing,
1960). Such small transport pool sizes in turn have often forced
a circular logic in which it has been necessary to work from a
model to deduce a pool size and then to use the pool size to deduce
the model. As a result it has not been possible to locate a transport pool within a particular type of cell.

These problems have been overcome in studies on potassium transport by the insect midgut reported here and earlier studies by Harvey, Zerahn & Wood. Dr. Zerahn and I had utilized the rapid transport rate, simple structure, and simple ionic requirements of the
midgut to initiate a kinetic analysis of tracer influx which showed
that the potassium transport pool size is too small to involve all
of the cells. Wood and Harvey had applied the simplest of all 3-
compartment kinetic analyses, that of a tracer mixing as a single

161

exponential function of time, to the midgut and showed that the transport pool size is so large that it must involve all of the cells. These conflicting results created the pool size mystery. Now, Blankemeyer (1977) has been able to distinguish electrically a transporting cell from a non-transporting cell and has shown that the potassium pump is in the apical membrane of the so-called goblet cells using electrical methods alone. He has gone on to solve the pool size mystery by showing that small sizes are found in midguts isolated from leaf fed larvae or diet fed larvae under open-circuit conditions, as used by Zerahn, whereas large ones are found only in midguts from diet fed larvae under short-circuit conditions, as used by Wood.

Blankemeyer has confirmed his electrical results in detail by kinetic analyses and has synthesized many years worth of our observations and hypotheses about the midgut into a convincing model, one so simple and testable that it may become the long-sought unifying model for epithelia. The establishment of the potassium transport route means that one can now confidently isolate those parts of the transporting cells which contain the pump, using the non-transporting cells as a control, and that one can establish pump kinetic constants using the correct substrate concentrations i.e. those in the transport pool. The rapid transport rate of the midgut means that there is much pump material to isolate. The rapid inhibition of pump activity in nitrogen, its slow decay with time, and its abrupt loss developmentally provide additional tools for the analysis and resynthesis of the system which will constitute its proof. Finally, the presence of cytochrome $\underline{b_5}$ in the apical membrane of the goblet cells where spicules are uniquely present on the inner leaflet of the plasma membrane in close proximity with mitochondria probably means that the coupling of potassium transport to metabolism will soon be understood.

THE ROUTE PROBLEM

The midgut ion transport system has been reviewed recently (Harvey & Zerahn, 1972) and methods for its study have been published (Wood, 1972; Wood & Harvey, 1975; Wood, 1977a 1977b; Blankemeyer, 1977). A diagram emphasizing the properties of the midgut relevant to the transport route problem is shown as Figure 1. In excellent, short-circuited, preparations the active flux of potassium from blood side to lumen side, the so-called influx, can amount to 50 μeq cm^{-2} hr^{-1} and the passive flux toward the blood side, the efflux, can be as low as 0.5 μeq cm^{-2} hr^{-1} yielding a measured flux ratio as high as 100 to 1. Under appropriate conditions the net flux of potassium accounts almost exactly for the short-circuit current, the latter amounting to 650 μA for a one half cm^2 preparation in this idealized case. The active flux is electrogenic in that the lumen side becomes more than 140 mV positive with respect to the blood side in oxygenated solutions requiring no specific co- or counter-ion. The influx, PD, and short-circuit current all fall abruptly when the stirring gas is changed from oxygen to

162

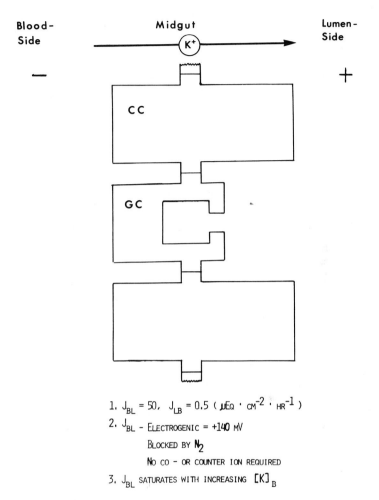

1. $J_{BL} = 50$, $J_{LB} = 0.5$ ($\mu Eq \cdot cm^{-2} \cdot hr^{-1}$)

2. J_{BL} - ELECTROGENIC = +140 MV
   BLOCKED BY N_2
   NO CO - OR COUNTER ION REQUIRED

3. J_{BL} SATURATES WITH INCREASING $[K]_B$

Fig. 1. Schematic diagram of the midgut showing its principal properties. K^+ in circle, active transport mechanism (K pump); CC, columnar cell; GC, goblet cell; J_{BL} (active) influx from blood side to lumen side; J_{LB} (passive) efflux in opposite direction.

nitrogen and promptly recover when oxygen is restored although all decay gradually with time. Finally the influx increases with increasing blood side potassium concentration and saturates at approximately 75 mM potassium.

The "route problem" is illustrated by Figure 2. Of a large number of hypothetical transport routes the three most likely ones which have not previously been eliminated are numbered 4, 5, and 6 (see Harvey & Zerahn, 1972). Route 4 is through just the goblet cells with the pool in the goblet cytoplasm and goblet cavity and the pump in the apical goblet membrane. Route 5 lies between the

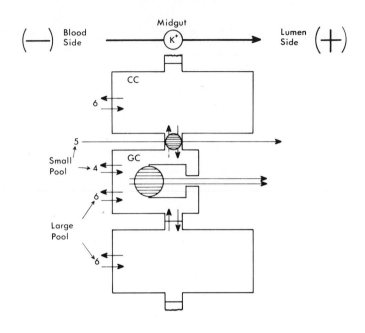

4. Mistakenly ruled out by coupling assumption.

5. Harvey and Zerahn evidence.

6. Wood and Harvey evidence.

Fig. 2. Possible transport routes indicated on schematic diagram of midgut. Route 4 is through just goblet cells with pump in apical region of goblet cell. Route 5 is between cells. Route 6 is through both columnar and goblet cells with pump in apical region of goblet cells with pump in apical region of goblet cell.

cells or along a non-mixing pathway through the cells. Both Routes 4 and 5 imply a small size for the transport pool. Route 6 is identical with Route 4 except that the columnar cells are ionically coupled to the goblet cells resulting in a large transport pool size.

Goblet cell Route 4 seemed to be ruled out by microelectrode studies (Wood et al., 1969) which implied that the columnar cells were electrically coupled to the goblet cells although direct coupling measurements had not been made. The original influx kinetic studies (Harvey & Zerahn, 1969) and subsequent work by Zerahn (1973, 1975) provided solid evidence that the pool size is small and favored the extracellular Route 5 whereas careful, steady state, studies by Wood & Harvey (1975) provided equally solid evidence that the pool size is large and favored the coupled columnar-goblet intracellular Route 6. Blankemeyer has now shown that goblet cells can be electrically distinguished from columnar cells, that the two cell types are not ordinarily coupled, and that the goblet cell Route 4 with its small pool size is the correct one. However, when

midguts from diet reared larvae are short-circuited the columnar cells become coupled to the goblet cells and then the coupled columnar-goblet cell Route 6 with its large pool size is followed.
The purpose of this paper is to review the electrical evidence for the goblet cell route, to review the kinetic analysis of pool size determinations, and to discuss Blankemeyer's largely unpublished kinetic confirmation of his electrical results.

ELECTRICAL EVIDENCE FOR GOBLET CELL ROUTE

A midgut isolated from Hyalophora cecropia is mounted as a flat sheet with the blood side (basal) upward. Provisions for measuring PDs with both extracellular and intracellular electrodes (and for passing current across the tissue) are shown in Figure 3. A photograph of a typical impalement profile is shown on the left in Figure 4. As the impalement is started both the microelectrode and

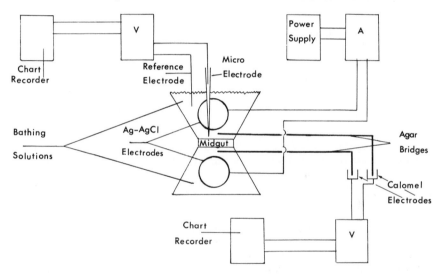

Fig. 3. Diagram of chamber used for impalement of midgut with microelectrodes, for measuring the transepithelial PD with calomel electrodes, and for passing current across the midgut with Ag-AgCl electrodes.

reference electrode are in the blood side compartment, the microelectrode showing a PD of zero. As the microelectrode is passed into the midgut tissue it shows a negative step to a PD of -27 mV with respect to the basally located reference electrode. This -27 mV PD step is designated the basal step. It is stable for many minutes, as long as an hour in favorable preparations. Upon passing the microelectrode deeper into the tissue it shows a positive step to a PD of + 105 mV with respect to the basal reference electrode, recording the transepithelial PD. The step from the -27 mV plateau to the + 105 mV transepithelial PD is designated the apical step and

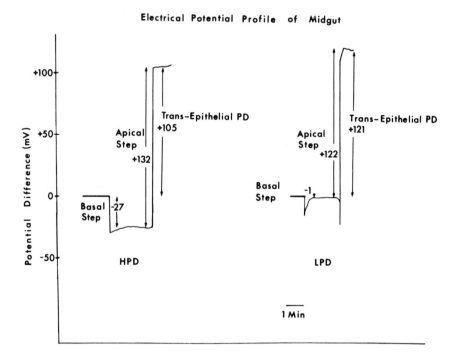

Fig. 4. *Potential profiles recorded as microelectrode is passed through midgut from blood side with reference electrode in solution on blood side. HPD profile on left and LPD profile on right. (H. cecropia, leaf)*

amounts to + 132 mV in this case. This profile is similar to the ones described by Wood et al., (1969).

However, using high resistance microelectrodes (>100 megohms; tip diameter <0.2 microns), and later even using low resistance microelectrodes (10 megohms), a second type of profile was observed and is shown on the right in Figure 4. In this profile the basal step is small, amounting to -1 mV. The profile with the -27 mV basal step was designated an High Potential Difference, HPD, profile and the one with the -1 mV basal step was designed a Low Potential Difference, LPD, profile.

With the ability to place two microelectrodes, one in a cell showing an HPD profile and the other in a cell showing an LPD profile, Blankemeyer was able to study the effects on the corresponding basal steps as he stopped the K pump in nitrogen. In Figure 5 the basal step of the LPD cell changes concurrently with the decline in the transepithelial PD in nitrogen while the basal step of the HPD cell is not affected. However, Schultz (1972) has pointed out that an ion pump cannot be located on the basis of PD profiles alone. Therefore Blankemeyer used similar techniques to measure the total cellular resistance between each microelectrode and the

166

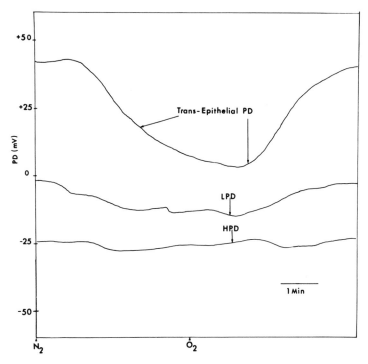

Fig. 5. Effects of stopping the K pump with nitrogen and restarting it with oxygen on the transepithelial PD (upper trace), the basal step of an LPD profile (middle trace), and the basal step of an HPD profile (lower trace). (M. sexta, diet)

external solutions and the ratio of resistance across the apical region to that across the basal region of each cell. Again, he could turn off the K pump with nitrogen and study the effects of stopping the active transport on the total resistance and on the ratios of apical to basal resistance in both HPD and LPD cells. The results were simple and clear cut. The total resistance of the LPD cell increased by 2.66 (Table 1) while the ratio of apical to basal resistance of the LPD cell increased from 4.3 to 20 as the pump was stopped in nitrogen (Table 2). At the same time the resistance and resistance ratio of the HPD cells remained virtually unchanged (Tables 1 and 2). Together these two results show that the apical resistance of the LPD cell is increased when the pump is stopped. Noting that the active potassium flux through the pump represents a large fraction of the total conductance through the midgut, Blankemeyer concluded that the conductivity of the membrane containing the pump should decrease i.e. that the resistance should increase, when the pump is stopped. Therefore, the finding that the resistance of the apical membrane of the LPD cell increased must mean that the site of active potassium transport is in that membrane.

Table 1. *Ratios of total cellular resistance in nitrogen and oxygen measured by passing current from a microelectrode within an LPD or an HPD cell to exterior solution with (second) recording electrode in cell.*

	Cellular Resistance Ratio	
	LPD	HPD
N_2/O_2	2.66	0.93

Table 2. *Apical to basal resistance ratios, measured by passing current across the epithelium with a recording microelectrode in an LPD or an HPD cell in oxygen and nitrogen.*

	Apical/Basal Resistance Ratio	
	LPD	HPD
O_2	4.3	23.5
N_2	20	30.4

He constructed an equivalent circuit diagram (Figure 6) which for the first time locates an active transport mechanism within a specific region of a cell on electrical evidence alone.

With the active potassium transport mechanism located in the apical region of the LPD cell on electrical grounds alone, Blankemeyer proceeded to show that the cell yielding the LPD profile is the goblet cell. First he plotted a frequency histogram of 227 random impalement profiles of midguts from Manduca sexta (Figure 7) and noted that they fell into three distinct populations. The most numerous are the HPD profiles which must arise in columnar cells since they are the largest and most numerous cells in the one-cell-thick epithelium. The next most numerous are the LPD profiles which must arise in goblet cells which are the next most numerous and are large. Finally, a new population of Intermediate PD profiles, IPD cells, appeared in the histogram (Figure 7). This population must arise in replacement cells since they are not seen in impalements of midguts from Hyalophora cecropia which virtually lack replacement cells.

Blankemeyer then injected dye iontophoretically into cells identified electronically as LPD cells and located the dye in goblet cells; then injected dye into HPD cells and located it in columnar cells. The localization of the dye is clear both in living preparations and in while mounts of fixed tissue. There is no doubt that the LPD cells are goblet cells and that the HPD cells are columnar cells.

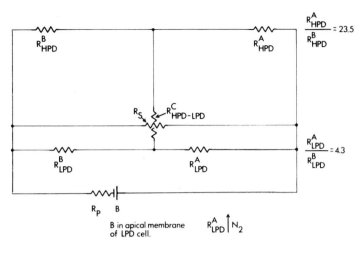

Fig. 6. Equivalent circuit diagram for the midgut in oxygen drawn so that it can be superimposed on the schematic diagrams of Figures 1 and 2 as is done in Figures 8 and 17. R refers to resistance. The superscript A refers to apical membrane, B to basal membrane, and C to the junctional membrane of the epithelial cells. The subscript HPD refers to High Potential Difference Cell, LPD to low Potential Difference Cell, and P to the potassium pump. B (Battery) signifies the PD created by the potassium pump. $R_{HPD}/R = 23.5$, $R_{PD}/R_{PD} = 4.3$. RP increases in N_2. Pump potential, B, located in apical membrane of LPD cell.

The electrical results are summarized in Figure 8 in which the equivalent circuit diagram of Figure 6 is superimposed on the schematic diagram of the midgut (Figs. 1 and 2). The transport route through just the goblet cells is shown by arrows and the pool in the goblet cytoplasm is shaded.

LACK OF COUPLING BETWEEN CELLS

The funding that goblet cells show a low PD basal step whereas columnar cells show a high PD basal step means that the conclusion by Wood et al. (1969) that these cell types are coupled must be re-examined. Recall that the goblet cell route, which we now see is the correct one, had been rejected on the grounds that all of the cells were thought to be coupled. Blankemeyer therefore studied the coupling between the midgut epithelial cells directly, using the methods popularized by Loewenstein and associates (1966) and illustrated in Figure 9. Two microelectrodes are placed in one cell and a signal is passed from one of them to ground while the ratio of voltage between the second one and ground to the voltage between a third one in an adjacent cell and ground is recorded as V1/V2. The results are summarized in Table 3. In midguts from leaf fed larvae no electrical coupling is detectable. In midguts from diet fed larvae under open-circuit conditions no coupling is detected between columnar cells (HPD to HPD cells) or between goblet and columnar cells (LPD to HPD cells). Nor is there coupling between columnar cells under short-circuit conditions. However, in

Table 3. Electrical PD coupling ratios, V_1/V_2 from HPD to HPD cells and from HPD to LPD cells. Diet-fed larvae were both M. sexta and H. cecropia; leaf fed larvae were H. cecropia.

	Open		Shorted	
	Leaf	Diet	Leaf	Diet
HPD-HPD	0	0	0	0
HPD-LPD	0	0	0	$V_1/V_2 = 0.46$

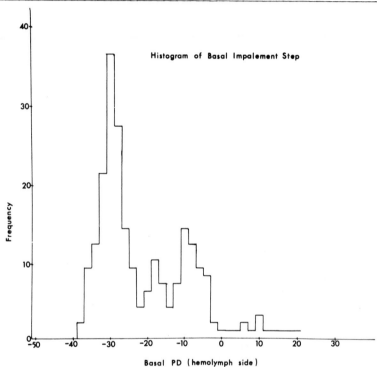

Histogram of Basal Impalement Step

Frequency

Basal PD (hemolymph side)

Fig. 7. Frequency histogram of basal PD step in 227 random impalements of isolated midgut. (M. sexta diet)

midguts from diet fed larvae the striking result is that within a few minutes after the onset of short-circuiting, columanr cells become coupled to goblet cells, finally attaining a coupling ratio of approximately 0.6 implying extensive electrical and ionic communication between these two cell types. This result suggests the solution to the pool size mystery. Harvey and Zerahn worked with leaf fed larvae and subsequently when Zerahn used diet fed larvae they were studied under open-circuit conditions. Under all of these conditions the goblet cells are not coupled to the columnar cells and the small transport pool size which they found is to be expected. However, Wood and Harvey used diet fet larvae under short-circuit conditions in which columnar cells are coupled to goblet cells and would be expected to add to the transport pool, with the result that the large transport pool size which they found is to be expected. Before reporting the results of Blankemeyer's pool

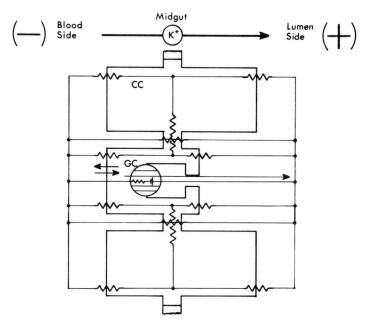

Fig. 8. *Summary of electrical results. The equivalent circuit diagram of Figure 6 has been superimposed on the "route" diagram of Figure 2. Electrical evidence alone places transport route through just the goblet cells (GC) because only the apical resistance of the goblet cells increases when the pump is stopped. This result implies that there is no coupling between the columnar cells (CC) and goblet cells (GC).*

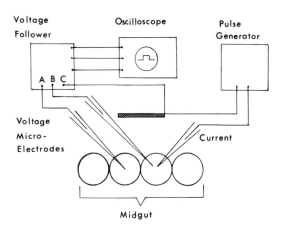

Fig. 9. *Diagram illustrating retionale for measuring electrical coupling between midgut epithelial cells as described in text.*

171

size determinations, which confirm these results, several compli-
cations and controversies regarding the non-destructive determination of pool sizes need to be clarified.

KINETIC ANALYSIS OF POOL SIZES

An entirely independent method for studying the route of ion
transport across an epithelial tissue is to measure the amount of
an ionic species that is involved in the transport route and to
use this information to deduce the route. The method is applicable
to any transport system but is especially useful for inorganic ionic
species which are not ordinarily metabolized and therefore are con-
served during transport. The amount of ionic species involved in
transport will be designated the transport pool and its size will
be designated by S. Such a pool leads to two principal types of
delay between the time when tracer is added on one side of an epithelium and the time when it appears at a steady rate on the other
side. One type is a diffusion delay, for example due to unstirred
layers, and the other type is a volume or mixing delay, for example
due to slow tracer mixing in compartments along the route.
The concept of a mixing component of a transport pool and its
measurement is illustrated in Figure 10. In the uppermost diagram
white balls are added from the left compartment to a cup and immed-

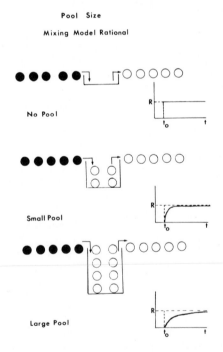

Fig. 10. Diagram illustrating
rationale for measuring size of
pool in a simple, linear mixing
model as described in text.

iately removed from the cup at the same rate, R, and placed in the right compartment. At a given time, t^0, all of the white balls in the left compartment are replaced by black balls. Black balls will appear at the constant rate, R, in the right compartment with no delay. A plot of the rate of appearance of black balls in the right compartment, R, against time will immediately jump from zero to the dotted line representing the rate previously plotted for white balls. Both time delay and pool size are zero. In the middle diagram the cup holds four balls. Now after time, t^0, when black balls replace all white balls on the left, there is a delay until the cup is entirely filled with black balls before they appear at a constant rate on the right. The plot follows a curve with a time delay before reaching the constant rate. The area between the curve and the dotted line corresponds to the amount of balls which the cup contains. This amount is called the pool size and in this case is small. Finally, in the lowermost diagram the cup is larger, the mixing time delay is longer, and the pool size is larger. If the balls were moved faster it would take less time to reach the constant rate. Intuitively, then, the size of the pool will be proportional both to the rate of moving the balls and to the time delay.

A quantitative determination of a pool size is illustrated in Figure 11. The balls now represent ions moving from compartment

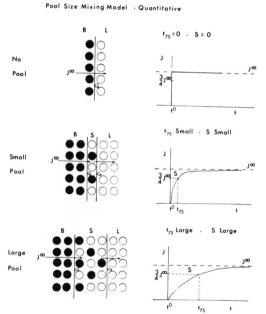

Pool Size Mixing Model · Quantitative

Fig. 11. Schematic diagram illustrating quantitatively the concept of a transport pool. In the upper diagram there are just two well mixed compartments; in the central diagram a small compartment has been added between them; and in the lower diagram a larger middle compartment has been added. The effects of the size of the middle compartment on the delay between adding isotope to B and its appearance at a constant rate in L are discussed in the text in terms of the graphs on the right. The size of the pool is shown to be zero when there is no middle compartment, to be small when the compartment is small and to be large when the compartment is large; the exact size of the pool under the simplest conditions being given by the steady state value of the flux, J^∞, multiplies by the time for 75 percent mixing, divided by 1n 4.

B to L at a constant rate, J^∞. In the upper most diagram there is
no central compartment. When tracer is added to B at time zero it
immediately appears in L at the constant rate, J^∞. The plot of
rate of appearance of label J, against time is a half rectangle,
first running parallel to the ordinate and then parallel to the
abscissa. The time for the rate of labeling to reach 3/4th of the
constant rate, t_{75}, is zero and the pool size is zero. In the
middle diagram a central compartment, S, has been added between
B and L. Now, when tracer is added to B there is a small delay
while tracer mixes in S before it appears at a constant rate in L.
The t_{75} is small and the pool is small. Finally, in the lowermost
diagram the size of the central compartment is larger, the t_{75} is
longer, and the pool size is larger. In this simplest case, there
is no diffusion delay as tracer crosses the barriers between compartments and no flux of ions back from L into S or from S to B.
One can write and solve differential equations for the process and
obtain the simple equation $S = J^\infty \cdot t_{75} / \ln 4$. One can read off
the value for J^∞ and the time when J has reached 3/4 of J^∞ and calculate the pool size using this equation. More complex cases with
changing transport rates, diffusion delays, back fluxes, and additional compartments can be analysed in much the same way (see Wood
& Harvey, 1975).

Simple influx kinetics analysis of epithelia was initiated by
Schoffeniels (1957) and applied to the frog skin by Hoshiko and
Ussing (1960). Unfortunately the amount of sodium in the frog
skin transport pool was determined by them to be a small fraction
of the total tissue sodium. Furthermore, diffusion delays in the
complex skin complicated the analysis. Finally, even though Hoshiko and Ussing determined the pool size from the data, a suspicion
crept into such analysis that the logic may be circular. A pool
size determined from an assumed model cannot be used to deduce the
model. Other methods for determining the transport pool size directly from the data were sought in a little publicized effort initiated by Zerahn and taken up in my laboratory and by others. Thus
Andersen and Zerahn (1963) plotted the amount of labeled ionic
species in the sampling compartment, L, against time (Fig. 12
"amount method"). The steady state is reached when the function
approaches an asymptote and the intercept of the asymptote with
the amount axis gives the pool size. Alternatively, Wood (1972)
plotted the rate of appearance of labeled ionic species in the
sampling compartment, L, against time (Fig. 12, "rate method") and
and calculated the pool size from the area between the steady state
rate, J^∞, and the labeling rate, J^n. Both of these methods include
both diffusion and mixing (volume) delays but methods for separating
out these components are known (Wood & Harvey, 1975). It turns out,
in the case of the midgut, that the flux data is dominated by a
single exponential component with a minor linear component which
is measurable but will be ignored in this discussion. Therefore
it may be worthwhile to digress and to compare these two methods
for calculating pool sizes from the data and to consider their
relationship with a mixing model.

174

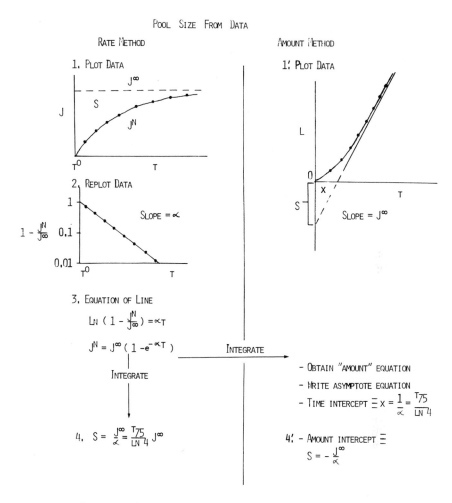

POOL SIZE FROM DATA

RATE METHOD

AMOUNT METHOD

1. PLOT DATA

1'. PLOT DATA

2. REPLOT DATA

SLOPE = α

SLOPE = J^∞

3. EQUATION OF LINE

$$\text{LN} \left(1 - \frac{J^N}{J^\infty} \right) = \alpha T$$

$$J^N = J^\infty \left(1 - e^{-\alpha T} \right)$$

INTEGRATE

INTEGRATE

- OBTAIN "AMOUNT" EQUATION
- WRITE ASYMPTOTE EQUATION
- TIME INTERCEPT $= x = \dfrac{1}{\alpha} = \dfrac{T_{75}}{\text{LN } 4}$

4. $S = \dfrac{J^\infty}{\alpha} = \dfrac{T_{75}}{\text{LN } 4} J^\infty$

4'. - AMOUNT INTERCEPT $=$

$$S = -\frac{J^\infty}{\alpha}$$

IN GENERAL, S = AREA BETWEEN STEADY STATE FLUX AND LABELING OF FLUX ∴

CAN USE TO DEDUCE MODEL.

Fig. 12. Two methods for determining the size of an influx pool, the "rate method" on the left and the "amount method" on the right are illustrated. In the simplest case when the influx is a single exponential function of time, the relationship between the two methods is illustrated. In general the size of an influx pool can be calculated directly from the data as the area between the steady state flux and the labeling of the flux, and can be used to deduce a model.

In each of the three diagrams of Figure 12 the time course of tracer appearance in L is plotted. In the "rate method" (upper

175

left plot) the flux i.e. the <u>rate</u> of eppearance, J^n, follows an upward curve which approaches a constant rate, J^∞. The pool size is the area, S, between these two functions, J^n and J^∞. In order to obtain this area by integration we need the equation fro the two functions. The constant one is simply $J = J^\infty$. To obtain the equation for the curve the labeled flux at each sampling time, J^n, is divided by J^∞, substracted from 1 and plotted against time. A straight line is obtained, the equation of which is simply $1 - J^n/J^\infty = \alpha t$. Solving for J^n we obtain the desired rate equation, $J^n = J^\infty(1-e^{-\alpha t})$. Parenthetically, the straight line immediately tells us that the mixing proceeds as a single exponential function of time with α being the mixing time constant and moreover tells us that the equation for the pool size will be identical to that given by the simplest mixing model. The integration yields the pool size $S = J^\infty/\alpha$.

The relationship of this "rate method" with Andersen and Zerahn's "amount method" is given by integration of the rate equation, $J^n = J^\infty(1-e^{-\alpha t})$. The resulting "amount equation" i.e. that describing the "amount curve" in Figure 12 is then used to obtain the equation for the asymptote. The pool size is given by its intercept on the "amount Axis" and is $S = - J^\infty/\alpha$. This equation is identical with the one for S given by the "rate method". This result means that the "rate method" and the "amount method" give exactly the same information, which they should since they use the rate and amount time courses of the same data. However, the "rate method" is exper imentally safer to use because neither J^∞ nor α are very sensitive to experimental conditions whereas the extrapolation of the asymptote in the "amount method" is particularly sensitive to such experimental conditions as the decay of the transport rate. The intercept of the asymptote on the time asix, called X in Figure 12, is equal to $1/\alpha$ and also equal to $t_{75}/\ln 4$, from which it is clear that $t_{75} = \ln 4/\alpha$ and that the equation for S from the simplest mixing model (Fig. 11) is also equaivalent to the one from the "rate" and "amount" methods (Fig. 12). Since the pool size from the data is the same as that from the simplest mixing model, arguments regarding measuring pool sizes from data vs model are irrelevant in the case of the midgut. The broader implication of this result is that one can find conditions in which the labeling of an influx proceeds as a single exponential function of time representing a simple volume delay with minimal diffusion delay. Just as enzymologists find conditions under which enzyme activity is first order with respect to concentration to simplify their analyses, so it is legitimate for physiologists to seek conditions in which the influx yields a single exponential function with time to simplify their analyses.

The point of this digression into methods of measuring pool sizes is to show that all methods in the literature, whether those of Hoshiko and Ussing (1960), Anderson & Zerahn (1963), Harvey & Zerahn (1972) or Wood & Harvey (1975) when applied to the midgut all yield single exponential functions preceded by small linear

176

delays attributable to non-instantaneous stirring in the labeling compartment or to diffusion delays in unstirred layers. By contrast, in the case of the frog skin there is an appreciable linear delay, gamma in the terminology of Wood and Harvey (1975), before a single exponential function is followed. This gamma component is largely absent in the midgut and can usually be ignored, although methods for measuring it and its contribution to the pool size are given by Wood and Harvey. Moreover, they have provided methods for dealing with a decay in the steady state flux. Finally, Wood (1977) has provided a general solution to the pool size problem.

CONFIRMATION OF GOBLET ROUTE BY KINETIC ANALYSES

Recall that Harvey and Zerahn (1969) had measured the potassium transport pool size using the "amount method" and found it to be so small that the route might lie through just the goblet cells or between the cells (Fig. 13). Zerahn had confirmed such a small pool size by isotope loading experiments (1973) and additional flux experiments (1975). On the other hand Wood and Harvey have calculated pool sizes by all of these methods (1972, 1973, 1975) and found them to be so large that the route must lie through all of the cells (Figure 14). They have confirmed their large pool sizes by isotope loading experiments (1972). There is no reason to doubt the validity of either set of determinations yet there must be an explanation for the different results.

Fortunately, in the case of potassium transport across the midgut both diffusion delays and the back flux are small enough to ignore and the complication due to decay of the active transport rate can be corrected for or minimized. Consequently, the measurement of the transport pool size by J^{∞}/α has become routine in my laboratory and the kinetic analysis of pool sizes by this non-destructive method allows repeated determinations on the same living preparation. This circumstance has enabled Blankemeyer to confirm his conclusion from electrical studies that the transport route lies through the goblet cells and has led to an explanation of the small pool vs. large pool mystery.

The pool size in midguts from diet fed larvae is shown under both open-circuit and short-circuit conditions from the same preparation in Figure 15. Under both open-circuit and short-circuit conditions (on left) the J^{∞} was estimated from the time course of the labeling of the influx, J^{n}, which tends to underestimate the pool size slightly. In hundreds of previous cases the pool size has always been found to decrease with time, as is expected since J^{∞} always decreases with time (Wood & Harvey, unpublished data). Therefore, it is with great surprise that one observes that the second pool size measured i.e. the one measured under short-circuit conditions, is three times larger than the first one measured, i.e. the one measured under open-circuit conditions. Clearly, the pool size has increased by the addition of columnar cells to the pool as predicted by Blankemeyer's electrical results.

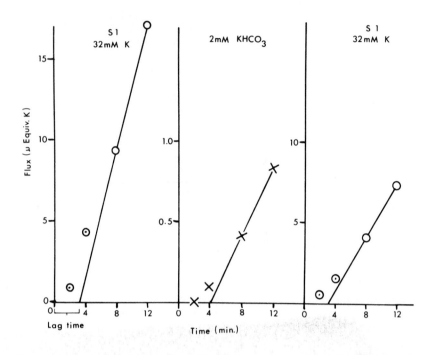

Fig. 13. "Amount" time course of influx of potassium in midgut isolated from leaf fed larva. Although the flux is changed from 119 to 6.2 and back to 49 µEq hr⁻¹ the lag time (X) remained virtually unchanged, implying that the pool size must be proportional to the flux rate. In the right hand plot when the midgut is near a steady state the pool size is about 2.5 µEq for the large spherical preparation corresponding to a small pool size of about one fourth that amount or approximately 0.6 for a flat sheet. (H. cecropia, leaf) from Harvey and Zerahn, 1969, with permission from the Journal of Experimental Biology.

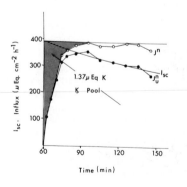

Time (min)

Fig. 14. "Rate" time course of influx of potassium in midgut isolated from diet fed larva and measured under short-circuit conditions. The size of the influx pool calculated from the influx corrected for decay in the pumping rate is large, amounting to 1.37 µEq. for the one half cm² section of midgut mounted as a flat sheet. (H. cecropia diet) From Wood and Harvey, 1975, with permission from the Journal of Experimental Biology.

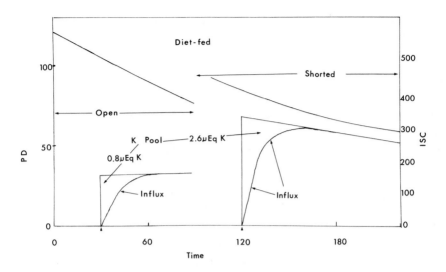

Fig. 15. *Potassium influx pool sizes in midguts isolated from diet fed larvae. In diagram on left the upper trace is the time course of the PD (left scale, mV). Isotope was added to the blood side at 30 minutes and samples were taken from the lumen side for 60 minutes. The influx was calculated and plotted as a function of time. The steady state value of the influx, J^∞, was extrapolted back to zero time and the pool size calculated as J^∞/α, α being the mixing time constant as described in the text. The pool size is small, 0.8 μEq. of potassium for the one half cm^2 section of midgut. In the diagram on the right the upper trace is the time course of the short-circuit current, I_{sc}, (right scale, μAmperes). Isotope was added to the blood side at 120 minutes and the influx followed for 90 minutes. Teh steady state value of the influx was extrapolated back to zero time and the pool size calculated as the area between that function and the influx curve. The pool size is large, 2.6 μEq of potassium for the one half cm^2 section of midgut. ($\underline{H.}$ $\underline{cecropia,}$ diet)*

The result is even more convincing when the pool size from a leaf fed larva is compared to that from a diet fed larba both under short-circuit conditions. The measurement was made at the same time after isolation and with the two J^∞s as nearly equal as possible. Again the results is conclusive (Figure 16). The pool size in the gut from the diet fed larva is four times as large as that from the leaf fed larva. Again it is clear that short-circuiting has increased the pool size by the addition of columnar cells to the pool as predicted by Blankemeyer's finding that columnar cells become coupled to goblet cells when midguts from diet fed larvae are short-circuited.

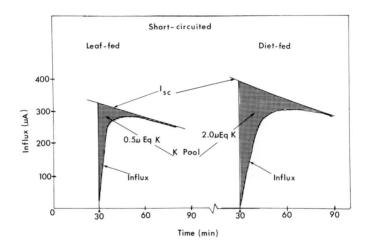

Fig. 16. *Potassium influx pool sizes in short-circuited preparations of two midguts, one isolated from a leaf fed H. cecropia larva (left) and one from a diet fed H. cecropia larva (right). The pool size was calculated in the same way as that for the short-circuited preparation at the right in Figure 15 except that the time course of the short-circuit current, I_{sc}, was taken as the measure of J^∞. The pool size in the preparation from the leaf fed larva was small, 0.5 µEq per gut whereas that from the diet fed larva was large, 2.0 µEq per gut.*

MIDGUT MODEL AND PROSPECTS

The electrical results and pool size determinations are combined in Figure 17. There is no longer any reason to include extracellular Route 5. The transport route (4) runs passively from blood side into the goblet cell and actively from the goblet cell across its apical membrane to the goblet cavity and thence passively to the lumen side. When midguts are isolated from diet fed larvae and are short-circuited an additional route (6) is followed from blood side into columnar cells, across junctional membranes to goblet cells and thence actively out of the goblet cells to goblet cavity and lumen side.

Traditional controls for studying active trasnport have been mutants lacking pumps, developmental stages lacking pumps, and pumps which have been stopped by inhibitors. Now what had been considered a vice becomes a virtue. The control cell for the transporting goblet cell is its next door neighbor, the non-transporting columnar cell. The presence of a non-transporting cell next door to a transporting cell has turned out to be the key to the localization of the pump within a specific cell type.

Now the task of isolating the pump molecules is easier. If goblet cells can be separated from columnar cells the latter can serve

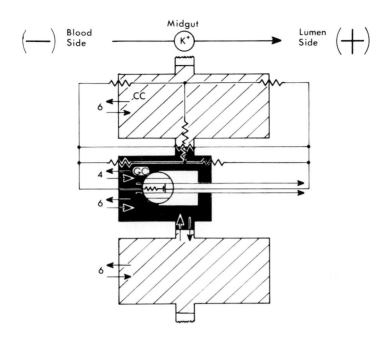

Fig. 17. Schematic diagram of midgut combining electrical results in the form of the equivalent circuit diagram from Figure 8, the identification of HPD cells as columnar cells and LPD cells as goblet cells, and the influx kinetic determinations of pool size. The potassium pump is in the apical membrane of the goblet cells. In midguts from leaf fed larvae or midguts under open circuit conditions the potassium transport route is through just the goblet cells (Route 4) with its small pool size. However, when midguts from diet fed larvae are short-circuited the columnar cells become coupled to the goblet cells and potassium now passes from columnar cells to goblet cells thereby adding the large amount of potassium in the columnar cells to the transport pool.

as controls possessing many proteins and lipids common to both cell types and therefore not unique to the transport system thereby allowing those molecules uniquely present in the goblet cell to be identified more easily. This prospect is illustrated by Figure 13 of Anderson & Harvey (1966) which shows that on the apical plasma membrane of the goblet cells, precisely where Blankemeyer has placed the potassium pump, there are spicules in close apposition to mitochondria. These spicules are not found in other portions of the goblet cell plasma membrane or on the columnar cell membranes. Moreover, Smith (1969) and others have shown that invariably insect epithelia which transport potassium possess these spicules. The isolation and characterization of the spicules and their relationship to cytochrome b_5 in the plasma membrane and to the transport

181

process are studies which make the long struggle to establish the K transport route seem worthwhile.

ACKNOWLEDGEMENT

I thank Mr. Frank Abramcheck, Miss Mary K. Hengen, Mr. Jeffery Sims for technical assistance in preparing this manuscript. This research was supported in part by a research grant (AI-09503) from the National Institute of Allergy and Infectious Diseases, U.S. Public Health Service.

REFERENCES

Anderson, B. & Zerahn, K. (1963) *Acta Physiol. Scand. 59:* 319.
Anderson, E. & Harvey, W.R. (1966) *J. Cell Biol. 31:* 107.
Blankemeyer, J.T. (1976) Ph.D. Thesis, Temple University.
Harvey, W.R. & Wood, J.L. (1972) In *Role of Membranes in Secretory Processes,* (eds. Bolis, L., Keynes, R.D. & Wilbrandt, W.) pp. 310. Amsterdam: North-Holland Press.
Harvey, W.R. & Wood, J.L. (1973) In *Transport Mechanisms in Epithelia. Alfred Benson* Symposium, V, (eds. Ussing, H.H., & Thorn, N.A. p. 342. (Munksgaard, Copenhagen & Academic Press, New York.)
Harvey, W.R. & Zerahn, K. (1969) *J. Exp. Biol. 50:* 297.
Harvey, W.R. & Zerahn, K. (1972) In *Current Topics in Membranes and Transport, Vol. 3.* (eds. Bronner, F. & Kleinzeller, A.) p. 367. New York: Academic Press.
Hoshiko, T. & Ussing, H.H. (1960) *Acta Physiol. Scand. 49:* 74.
Loewenstein, W.R. (1966) *Ann. New York Academic Sci. 137:* 441.
Schoffeniels, F. (1957) *Biochem. Biophys. Acta. 26:* 585.
Schultz, S.G. (1972) *J. Gen. Physiol. 59:* 794.
Smith, D.S. (1969) *Tissue & Cell 1:* 443.
Wood, J.L. (1972) Ph.D. Thesis, Cambridge University.
Wood, J.L. (1977a) *J. Exp. Biol. (in press)*
Wood, J.L. (1977b) *J. Exp. Biol. (in press)*
Wood, J.L., Farrand, P.S. & Harvey, W.R. (1969) *J. Exp. Biol. 50:* 169.
Wood, J.L. & Harvey, W.R. (1975) *J. Exp. Biol. 63:* 301.
Zerahn, K. (1973) In: *Transport Mechanisms in Epithelia. Alfred Benson Symposium, V,* (eds. Ussing, H.H. & Thorn, N.A.) p. 360 (Munksgaard, Copenhagen & Academic Press, New York).
Zerahn, K. (1975) *J. Exp. Biol. 63:* 295.

SOLUTE AND WATER TRANSPORT ACROSS VERTEBRATE EPITHELIA:
CHAIRMAN'S INTRODUCTORY REMARKS

Stanley G. Schultz

University of Pittsburgh

The "black box" approach to the study of solute and water movements across epithelia, in which the tissue is viewed as a single
homogeneous barrier separating two well-stirred aqueous phases, has
served us well for many years. Using this approach, the transport
properties of a wide variety of epithelia have been characterized
and considerable insight has been gained into the actions of hum-
oral, pharmacologic and in, some instances, pathogenic agents on
transepithelial solute and water transport. In addition, two
general properties of epithelia have been established beyond reasonable doubt.

First, it is clear that some solutes can be transported from
one well defined aqueous phase to the other in the absence of, or
against, electrochemical potential differences. In some instances
these "uphill" flows are energized by direct coupling to the flow
of metabolic reactions; such movements constitute irrefutable
examples of "active transport" processes (Kedem, 1961) which cannot be readily explained by "adsorption theories" or their like.
In general, it is far more difficult to establish unequivocally,
active transport across the membranes that surround nonepithelial
cells because of uncertainties with respect to transmembrane electrical potential differences and our limited knowledge of the distribution and thermodynamic activities of intracellular solutes;
the "black box" view of epithelial tissues circumvents these
difficulties.

Second, there is abundant evidence that water can be transported
across some epithelia in the absence of, or against, osmotic and/
or hydrostatic pressure differences, but that this "uphill" movement is dependent upon or coupled to solute transport and is not
directly energized by metabolic reactions (c.f. House, 1974).

However, whereas the "black box" approach is ideally suited for

183

telling us "what an epithelial tissue can do", it is obvious that this approach can provide little or no insight into "how these feats are accomplished". In order to gain insight into "mechanisms", more realistic and complex models of the epithelium must be invoked.

In 1958 Koefoed-Johnson and Ussing (1958) proposed a double-membrane model for transepithelial sodium transport by isolated frog skin which inextricably interrelated homocellular and transcellular ion transport. In 1962, Curran and MacIntosh (1962) proposed a double-membrane model for solute-coupled water transport; and in 1967 Diamond and Bossert (1967) proposed the "standing-gradient hypothesis" in which the notion of osmotic coupling of water flow to solute flow was extended to take the morphology of epithelia and, in particular, the role of the lateral intercellular spaces into account. These advances represented ingenious departures from the "black box" approach and opened a new era of research in which many efforts were focussed at relating epithelial morphology, the properties of the two limiting membranes and of the paracellular pathways, and intracellular parameters to transepithelial solute and water transport. Today, we know that many epithelia, particularly those capable of bringing about the rapid absorption of an isotonic solution, are characterized by paracellular pathways that are 'leaky' to ions, small nonelectrolytes and water. The anatomic counterparts of these pathways are the "not-so-tight junctions" and the underlying lateral intercellular spaces. Further, it seems that the properties of these pathways are not fixed but, instead, are plastic and depend upon the rate and direction of fluid transport.

There is a growing body of evidence suggesting that sodium which enters the epithelial cell, destined for transepithelial transport, is contained within a so-called "sodium transport pool". This pool comprises only a small fraction of the total sodium contained within the epithelial cells. It is not clear whether this *apparent compartmentalization* is due to the presence of multiple cell types-including some cells which are not involved in transepithelial sodium transport-or whether two or more sodium compartments are contained within individual cells responsible for transepithelial sodium transport. Finally, the relation between homocellular and transcellular sodium transport, inherent in the Koefoed-Johnson-Ussing model (1958), must be considered an open question (Schultz, 1977). Thus, it seems certain that the double-membrane models for solute (particularly sodium) and water transport will have to be modified, drastically revised, or, perhaps, discarded. In any event, they will have served their purposes brilliantly!

The contributors to this session will address some of the problems we currently face in our efforts to gain a more detailed understanding of the mechanisms responsible for sodium and water transport by vertebrate epithelia with particular emphasis on the role of paracellular pathways in isotonic fluid absorption, the morphological correlates of transepithelial solute and water transport, and the relations between intracellular ion composition and transcellular ion transport. These topics represent, at least part of, "our frontier" and hopefully the presentations will raise as many, if not more, questions than they will answer. Although we have come a long way from the "black box" we have a long way to go!

REFERENCES

Curran, P.F. & MacIntosh, J.R.   (1962)   *Nature* (*London*) *193*, 347.
Diamond, J.M. & Bossert, W.H.   (1967)   *J. Gen. Physiol. 50*, 2061.
House, C.R. (1974) *Water Transport in Cells and Tissues*, London: Edward Arnold
Kedem, O. (1961)   In *Membrane Transport and Metabolism* (Kleinzeller A. & Kotyk, A., eds) p. 87. Czech. Acad. Sci., Prague.
Koefoed-Johnson- V. & Ussing, H.H. (1958) *Acta Physiol. Scand. 42*, 298.
Schultz, S.G. (1977)   In *Molecular Specialization and Symmetry in Membrane Function*, (Solomon, A.K., Giebisch, G., Hofmann, J., Karnofsky, M. & Schultz, S.G., eds), Cambridge: Harvard Univ. Press.

THE SODIUM TRANSPORT POOL OF EPITHELIAL TISSUES

Mortimer M. Civan

University of Pennsylvania

INTRODUCTION

The concept of a Na transport pool was formulated in an effort to relate transepithelial transport to the intracellular and membrane events responsible for the transport. The pool may be defined as that amount of Na^+ which has entered the transporting cells from the bathing media, and which is *readily* available for active extrusion from the cell into the serosal or contraluminal medium. Defined in this way, the Na^+ transport pool is not necessarily identical with the intracellular content of Na^+. In fact, several lines of evidence suggest that the total intracellular content of Na^+ within epithelial and other tissues does not constitute a single homogeneous pool.

The functional heterogeneity of the intracellular Na^+ is illustrated   recent studies of the urinary bladder of the toad, a functional analogue of the mammalian distal nephron and convenient model system for investigating transepithelial active Na^+ transport. The transport properties of this tissue may be largely ascribed to a single complete surface layer of mucosal cells (DiBona et al., 1969a). Between this outer complete layer and the subjacent basal lamina is an incomplete layer of basal cells.

The most direct approach to estimating the size of the Na^+ pool is to simply measure the intracellular Na^+ content of the transporting cells. However, the mucosal cells contain only 10-20% of the total intracellular Na^+, K^+, Cl^- and water of the whole tissue; the remainder arises from the underlying submucosa and serosa (Macknight et al., 1975a). Therefore, it is first necessary to isolate the mucosal epithelium from the underlying tissue, a procedure which can be carried out remarkably simply (Gatzy and Berndt, 1968; Macknight et al., 1970; Macknight et al., 1971a; Lipton and Edelman,1971; Handler et al., 1972). At the conclusion of an ex-

187

periment, we have simply blotted the tissue dry and scraped off the epithelial cells with a glass slide. Phase and electron microscopy has established that about 95% of the epithelial cells are success- fully removed, and that <2-3% of the volume of cell scrapings arises from submucosal contamination (Macknight et al., 1971a).

In principal, the intracellular Na^+ content is calculated from measurements of: (1) the total Na^+ content of the cell scrapings, (2) the Na^+ concentration of the bathing media, and (3) the extra- cellular volume of the cell scrapings. The inulin space provides the most suitable available measure of the extracellular space in toad bladder (McIver and Macknight, 1974), and is calculated from measurements of the inulin content of the scrapings and the inulin concentration of the bathing media. When the two tissue surfaces are bathed with different solutions, ^{14}C-inulin may be added to the mucosal medium and ^3H-inulin to the serosal, permitting separate measurement of the mucosal and serosal extracellular spaces. With this technique, we have found that, under baseline conditions with both surfaces bathed with a standard Ringer's solution containing 117 mm Na^+, the intracellular Na^+ content is some 160-170 mEq/kg dry weight.

LABELLING OF THE POOL WITH RADIOACTIVE SODIUM

It was of considerable interest to determine how much of this total Na was of mucosal, and how much of serosal, origin. To ex- amine this question, paired hemibladders were mounted in chambers and bathed on both their surfaces by sodium Ringer's solution. ^{24}Na was added to the mucosal medium of one hemibladder, and to the serosal medium of the paired tissue. The total ^{24}Na content and inulin space of the cell scrapings, and the ^{24}Na concentration in the bathing media, were measured. The results have indicated that under baseline conditions, approximately 22-23% of the intracellul- ar Na^+ enters the cells from the mucosal medium, and 77-78% from the serosal medium (Macknight et al., 1975a).

These observations could be most simply interpreted within the framework of a traditional 3-compartment model consisting of the Na^+ in the two bathing media entering and leaving a single homo- geneous intracellular Na^+ pool. Within this framework, the isotope labelling experiments would indicate that the rate of Na^+ entry from the serosal medium is 3-4 times greater than that from the mucosal medium. Thus, application of a simple 3-compartment model would lead to the conclusion that a considerable amount of recy- cling of Na^+ occurs at the basolateral surface of the transporting cells, Na^+ entering from the serosal medium and being reextruded by the Na^+ pump.

Several lines of indirect evidence suggest that this interpre- tation is probably incorrect. First, tracer flux and electrical measurements, before and after adding vasopressin suggest that movement of Na^+ through the active transport transcellular pathways is unidirectional under baseline conditions (Civan, 1970). Second, addition of ouabain and removal of mucosal Na^+ reduce CO_2 production by toad bladder by similar amounts (Coplon and Maffly, 1972). If 75-80% of the Na^+ pool were of serosal origin, removal of mucosal Na^+ should have had a much smaller effect than did the ouabain. Third, removal of mucosal Na^+ has been reported to prevent the increased utilization of pyruvate characteristically elicited by addition of aldosterone (Sharp et al., 1966). If Na^+ actually entered the same intracellular pool far more rapidly from the serosal medium than from the mucosal bath, we would expect that removal of mucosal Na^+ would have only a modest effect on CO_2 production.

Among alternative interpretations of the data, the next simplest appears to be that addition of radioactive Na^+ separately to the mucosal and serosal baths results in the labelling of two relatively discrete pools. Since removal of mucosal Na^+ reduces net Na^+ transport and CO_2 production in toad bladder (Coplon and Maffly, 1972) and reduces net Na^+ transport and O_2 consumption in toad skin, a functionally similar epithelium (Danisi and Vieira, 1974), the intracellular pool labelled from the mucosal medium would be the pool of physiologic interest.

The data thus far obtained (Table 1) with the techniques described are entirely consistent with this view. Vasopressin is thought

Table 1.

Experimental Procedure	Total Intracellular Na^+ Content (mmoles/kg dry wt)			Intracellular Na^+ Content Labelled from Mucosal Medium (mmoles/kg dry wt)		
	C	E	Δ	C	E	Δ
Addition of Vasopressin (15)	176	202	26 ± 9	34	65	31 ± 6
Removal of Mucosal Na^+ (2)	173	140	-33 ± 7	37	2	-35 ± 2
Addition of Ouabain (10^{-2}M) (18)	164	253	89 ± 38	30	83	52 ± 4

to stimulate transepithelial active Na^+ transport primarily by increasing Na^+ entry into the cells from the mucosal medium (Civan et al., 1966; Civan and Frazier, 1968; Civan, 1970; Macknight et al., 1970; Macknight et al., 1971b; Yonath and Civan, 1971; Handler et al., 1972). In fact, vasopressin increases the size of both the mucosally labelled pool and the total intracellular Na^+ by the same amount (Macknight et al., 1971b). Similarly, removal of mucosal Na^+ appears to reduce the size of the mucosally labelled pool and the total intracellular Na^+ by the same amount (Macknight et al., 1975a). Finally, addition of ouabain, an agent thought to inhibit active Na^+ extrusion from the cell into the serosal medium

(Herrera, 1966), appears to increase the size of both the serosally-labelled and mucosally-labelled pools, but largely the latter (Macknight et al., 1975b). In fact, the fractional increase in size of the mucosal pool induced by ouabain is so much greater than that of the serosal pool, this datum alone excludes the simplest possibility that a single intracellular pool of Na^+ exchanges with the bathing media by simply exponential entry and exit processes at the two cell surfaces. Other, more complicated kinetic models are, of course, possible.

The Na^+ transport pool may be even slightly smaller than the size of the pool labelled from the mucosal medium (Macknight et al., 1975a). At a concentration of $10^{-4}M$, the diuretic amiloride would be expected to mimic removal of mucosal Na^+ in its effect on the Na^+ pool (Bentley, 1968). However, substitution of choline for mucosal Na^+ REDUCES THE MUCOSAL Na^+ pool by 36 ± 3 mmoles/kg dry weight from 38 to 2 mmoles/kg dry weight, whereas amiloride reduces the mucosal pool by 22 ± 6 mmoles/kg dry weight, a significantly smaller amount (Macknight et al., 1975a). This discrepancy is eliminated by subsequently washing the amiloride-treated tissue with isotope-free media over brief periods lasting no longer than 60 sec. This observation suggests that only some 2/3 of the mucosal pool is truly intracellular, and that the true transport pool is no larger than some 25 mmoles/kg dry weight of epithelial cell scrapings. Thus, the Na^+ transport pool seems to constitute roughly 15% of the intracellular Na^+ of the epithelial cells, which in turn constitutes only 13% of the total intracellular Na^+ contents of the entire toad bladder. In short, the transport pool constitutes only about 2% of the total tissue intracellular Na^+ contents. Similar conclusions have been suggested by studies of the sodium transport pool in another anuran epithelium, frog skin (Cereijido and Rotunno, 1967; Aceves and Erlij, 1971; Moreno et al., 1973; Cereijido, 1974).

The basis for the heterogeneity of the intracellular Na^+ contents is not yet entirely clear. Heterogeneity of the cell population constituting the surface epithelium may well play a role. Some evidence indeed supports this concept. Of the surface cells, only the granular cells and not the mitochondria-rich or goblet cells, respond to vasopressin by swelling when the mucosal medium is hypoosmotic (DiBona et al., 1969b) and by shrinking when the mucosal medium is hyperosmotic (Civan and DiBona, 1974). Whether only the mitochondria-rich cells (Scott et al., 1974) or both these cells and the granular cells (Handler and Preston, 1976) respond to neurohypophyseal hormone with an increased intracellular cyclic AMP content is not yet resolved.

In addition to the role of heterogeneity of cell population, measurements with ion-selective microelectrodes in a great variety of cells, including single cell preparations, suggest that heterogeneity of the intracellular Na^+ may be a general characteristic of biological cells (Lev and Armstrong, 1975). This raises the question whether intracellular binding and compartmentalization of

Na^+ may be playing significant roles. By the terms "binding" and "immobilization", we refer to an extended period of residence by an ion at a binding site which is several orders of magnitude longer than the ion's correlation time in dilute aqueous solution. By the term "compartmentalization", in the present context, we refer to an asymmetric distribution of free Na^+ ion between the cytoplasm and subcellular organelles. Over the past several years, my laboratory has been concerned with quantifying the possible roles of subcellular immobilization and compartmentalization.

SUBCELLULAR IMMOBILIZATION

Dr. Mordechai Shporer and I have been studying intracellular ^{23}Na with the technique of nuclear magnetic resonance (NMR) spectroscopy in an effort to quantify the degree of immobilization of intracellular Na^+. In principal, the rates of nuclear magnetic relaxation of the ^{23}Na nucleus depend both upon the value of the correlation time t_c characterizing the ion, and upon the Larmor resonance frequency ω_o. For example, the rate of longitudinal relaxation $(1/T_1)$ is related to t_c and ω_o by:

$$(1/T_1) = \frac{(e^2qQ)^2}{(10)} \left(1 + \frac{n^2}{3}\right) t_c \quad \frac{1}{1+\omega_o^2 t_c^2} \qquad (1)$$

where (e^2qQ) is the magnitude of the nuclear quadrupole constant characterizing the ^{23}Na and n is asymmetry factor characterizing the electric field gradient imposed on the ^{23}Na nucleus (Shporer and Civan, 1977). Assuming a conservatively low estimate for $\left[e^2qQ/\overline{1+(n^2)}\right]$ of 4.8×10^6 rad·sec^{-1}, the functional dependence of $(1/T_1)^{\frac{1}{3}}$ upon t_c is presented in Fig. 1 for $\omega_o = 9.86 \times 10^7$ rad· sec^{-1}. At this frequency, $(1/T_1)$ for the intracellular ^{23}Na of frog striated muscle has been measured to be 41 sec^{-1} (Shporer and Civan, 1974). From Fig. 1, it is clear that $(1/T_1)$ is not a single-valued function of the correlation time; when $(1/T_1) = 41$ sec^{-1}, two values of t_c are possible, either 1.74×10^{-11} sec or 5.78×10^{-6}. The smaller of the two values is only 2-6 times greater than the value of t_c for ^{23}Na in aqueous solution which is $3-7 \times 10^{-12}$ (Shporer and Civan, 1977); in this event, Na^+ could not be considered immobilized. On the other hand, the second possible value of t_c would represent a prolongation of t_c by six orders of magnitude, consistent with marked immobilization of the ^{23}Na.

These two possibilities may be distinguished by measuring the dependence of the rates of relaxation of intracellular ^{23}Na upon ω_o. In practice, the Larmor frequency can be changed simply by altering the magnetic field strength (H_o) imposed on the sample, since ω_o is proportional to H_o. As will be appreciated from Fig. 2, reducing ω_o by a factor of two from 9.86×10^7 rad·sec^{-1} should have no appreciable effect on $(1/T_1)$ if the ^{23}Na is free in solution. In contrast, if t_c is six orders of magnitude larger, halving ω_o should increase $(1/T_1)$ by a factor of four. Experimentally

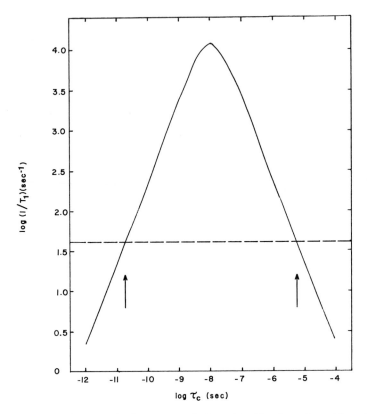

Fig. 1. Dependence of the rate of longitudinal relaxation $(1/T_1)$ of ^{23}Na on its correlation time (τ_c). The uninterrupted curve is a plot of eq. 1 taking $[(e^2qQ)^2(1+n^2)]$ to be 4.8×10^6 rad·sec^{-1}. The Larmor frequency used in the $\bar{3}$ calculation was 98.6×10^6 rad·sec^{-1}, one of the two frequencies employed in the study of intracellular ^{23}Na within frog striated muscle (Shporer and Civan, 1974). The interrupted horizontal line corresponds to the observed value of $(1/T_1) = 41$ sec^{-1} at this Larmor frequency. As indicated by the arrows, in principal two values of τ_c ($l.74 \times 10^{-11}$ and 5.78×10^{-6} rad·sec^{-1}) could be consistent with this measurement.

$(1/T_1)$ was found to be increased by only 4 sec^{-1} from a mean value of 41 sec^{-1} at the higher frequency, strongly suggesting that the great bulk of the intracellular Na$^+$ of frog striated muscle is not in bound form.

The NMR data may be subjected to a more extensive semiquantitative analysis in which the possibility of a rapid exchange between bound and free populations of ^{23}Na is considered. The results suggest that <1% of the intracellular ^{23}Na is truly bound (Shporer and Civan, 1974). Instead, interactions between intracellular Na$^+$ and macromolecular sites is likely to constitute a more non-specific

192

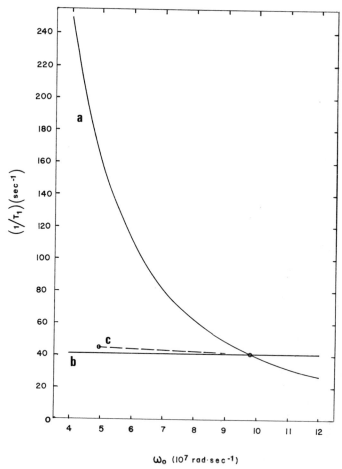

Fig. 2. Dependence of the rate of longitudinal relaxation $(1/T_1)$ of ^{23}Na on the Larmor frequency (ω_0). The uninterrupted curves are expressions of eq. 1, taking $[e^2qQ)^2(1+\eta^2)]$ to be 4.8×10^6 rad·sec^{-1}. From the datum of Fig.1, two possible values of the correlation time could characterize the intracellular ^{23}Na. In curve (a), τ_c has been taken to be 5.78×10^{-6} rad·sec^{-1}, corresponding to an immobilized species of ^{23}Na. In curve (b), τ_c has been taken to be 1.74×10^{-11} rad·sec^{-1}, corresponding to a free form of ^{23}Na. The two circles, arbitrarily connected with an interrupted line, present the values of $(1/T_1)$ observed at the two values of ω_0 used (Shporer and Civan, 1974). The absence of a significant dependence on ω_0 strongly suggests that the great bulk of the intracellular ^{23}Na is not immobilized.

coulombic interactions which do not immobilize the Na$^+$ nuclei.

SUBCELLULAR COMPARTMENTALIZATION

By exclusion, subcellular compartmentalization seems more likely than subcellular binding to be playing a significant role in producing heterogeneity of the intracellular Na^+. We have been studying the cell nucleus as one possible source for intracellular compartmentalization. This choice was dictated by a number of considerations.

First, measurements of electrical resistance have demonstrated that the nuclear envelopes of some cells constitute a significant permeability barrier to the free passage of ions (Loewenstein, W.R. and Kanno, 1963; Ito and Loewenstein, 1965). Therefore, the nuclear membrane could support gradients of ionic concentration between the nucleus and cytoplasm.

Second, some investigators have presented evidence suggesting that the endoplasmic reticulum may provide a direct path for exchange of ions between the nucleoplasm and extracellular fluid (Watson, 1955; Epstein, 1957; Palay, 1960; Siebert and Langendorf, 1970). If indeed present, such a pathway could lead to the establishment of gradients in ionic activity across the nuclear membrane.

Third, direct measurements of the ionic composition of the nucleoplasm and cytoplasm from several tissues have suggested that large concentration differences exist although agreement is far from complete among the various studies [results summarizes by Palmer and Civan (1976b)].

For these reasons, we have tried to determine the extent of compartmentalization of Na^+ (and also of K^+ and Cl^-) within the nucleus. The electrochemical activities of these ions have been measured directly with microelectrodes of the liquid resin type (Brown et al., 1970; Walker, 1971) in both the nucleus and cytoplasm. The salivary gland cells of Chironomus larvae are a particularly suitable epithelial preparation both because the large size of these cells permits simultaneous impalement of a single cell with several microelectrodes, and because the nuclear membrane is known to possess a relatively high electrical resistance.

Both micropipets and microelectrodes were used in the study carried out by Dr. Lawrence Palmer in my laboratory. Micropipets drawn from ordinary Corning 7740 glass capillaries were filled with 3 M KCl and used to measure electrical potential and to pass electrical current; the resistance of the pipets ranged from 5 to 20 Mohms. The ion-selective microelectrodes were made from similar micropipets by coating the inner surface with diethyldichlorosilane, baking the pipets, and introducing liquid ion exchange resins into the pipet tips. The selectivity characteristics of the microelectrodes were then determined by the characteristics of the liquid resins. When filled with a K^+-selective resin (Corning 477317) obtained commercially (Corning Glass, Corning, N.Y.), the K^+ - microelectrode was some 70 times more selective for K^+ than for Na^+. When filled with a Na^+-selective resin made by dissolving $K(\phi Cl)_4 B$ in triethylhexyl phosphate, the Na^+-microelectrode was some three times more selective for Na^+ than for K^+. Thus, we may calculate the electrochemical activity of intracellular Na^+ from measurements of: (1) the electrical potential with the open-tipped micropipets, (2) the electrochemical activity of K^+ with the K^+-selective microelectrodes, and (3) the potential signal from the Na^+-selective microelectrodes, reflecting the activities of both the Na^+ and K^+.

Lev (1964), and the later Armstrong and Lee (1971) and Lee and Armstrong (1972) used similar systems to measure the intracellular activities of Na^+ and K^+. Their microelectrodes were made from Corning 27-4 glass, which displays a moderate selectivity for K^+. Random differences in electrode sensitivity were exploited in order to obtain relatively greater or lesser selectivities for K^+ over Na^+. The system we use constitutes a technical advance. Because of the considerably greater selectivity of the K^+ - and Na^+-selective microelectrodes than was possible with those fashioned from K^+- sensitive glass capillary.

It is possible to make a more highly selective glass microelectrode from NAS 11-18 glass (Hinke, 1961). However, because of the high electrical resistivity of the glass (Kostyuk et al., 1969) this approach is practicable only when an appreciable length of Na^+-sensitive glass is incorporated into the electrodes. On the other hand, a short enough length must be used, so that the entire ion-selective surface is introduced into the cell during the impalement. Thomas (1970) has approached the problem by designing an electrode made from the same NAS 11-18 glass, but where the ion-sensitive tip is recessed behind the insulating glass tip. This imaginative design necessarily introduces a finite dead space, slowing the response time of the electrodes, and potentially perturbing the ionic composition of small cells. Quite apart from the technical problem of the dead space (which is of little importance in studying large cells) the resin electrodes have been, in our hands, simpler to fabricate and more consistent and reliable in use.

In each experiment, a 3-electrode system was used. One micropipet was placed in the nucleus, one micropipet in the cytoplasm, and one ion-selective microelectrode was placed alternately in the cytoplasm and nucleus. The position of the microelectrode was verified both visually and by measuring the electrical resistance interposed between the electrode and the micropipet in the cytoplasm. In an initial series of experiments, we found no significant differences in electrical potential or electrochemical potential of K^+ between the nucleoplasm and cytoplasm (Palmer and Civan, 1975). Subsequent studies using Cl^- (Palmer and Civan, 1976a) and Na^+-selective (Palmer and Civan, 1976b) microelectrodes have demonstrated that the electrochemical and chemical activities of Cl^- and Na^+, respectively, are also the same on both sides of the nuclear envelope.

From these data, we have concluded that Na^+, K^+ and Cl^- are not sequestered in free form within the nucleus of the <u>Chironomus</u> salivary gland cell. Since the resistance of the nuclear membrane of these cells is known to be appreciable, it is likely that this observation may be extended to other cells, as well.

In conclusion, it seems likely that the Na^+ transport pool constitutes only a fraction of the total intracellular Na^+ content of the transporting cells of epithelial tissue. Heterogeneity of cell population may be partly responsible. Fractional binding of intracellular Na^+ is likely to be very slight (<1%). Subcellular compartmentalization in organelles other than the nucleus may well be playing a significant role.

SUMMARY

The Na^+ transport pool may be defined as that amount of intracellular Na^+ which is readily available for active extrusion from the transporting cells. We have studied the transport pool of the urinary bladder of the toad, a useful model of transepithelial active Na^+ movement. From analysis of the surface epithelial cells, we have found that only about 13% of the total intracellular Na^+ of the tissue is present in the transporting cells. Of the intracellular Na^+ content of these cells, only some 20% is labelled from the ucosal medium, while 80% is labelled from the serosal medium.

Interpreting these data within the simplest possible framework of a 3-compartment system, we could conclude that considerable recycling of Na^+ occurs at the serosal aspect of the transporting cells; the rate of Na^+ entry from the serosal medium would be some four times greater than that from the mucosal bath. However, this interpretation is not consistent with other published data suggesting that removal of mucosal Na^+ entirely abolishes the m tabolic activity associated with Na^+ extrusion through the pump. If most of the Na^+ destined for extrusion enters from the serosal medium, removal of mucosal Na^+ would be expected to have only a modest effect on metabolic activity related to the pump.

Of the various possible alternative interpretations, the next simplest appears to be that the intracellular Na^+ labelled separately from the mucosal and serosal media may constitute relatively discrete pools. The pool of physiologic interest would be that labelled from the mucosal bath. Data obtained thus far with vasopressin, amiloride and ouabain are consistent with this hypothesis.

Part of the heterogeneity of the intracellular Na^+ may reflect heterogeneity of the cell population within the surface epithelium. Some evidence indeed supports this concept. However, a great variety of cells, including single cell preparations, display a heterogeneity of intracellular Na^+ raising the question whether subcellular binding and compartmentalization may be playing significant roles.

The degree of subcellular binding of Na^+ within cells in general has been estimated by studying frog striated muscle with pusled NMR spectroscopy. From the modest effect of magnetic field strength on the nuclear relaxation rates of ^{23}Na, it seems likely that <1% of intracellular Na^+ is bound.

The possibility of subcellular compartmentalization within the nucleus has been studied by using Na^+- and K^+- sensitive intracellular microelectrodes. In one of the very few preparations where this approach is technically feasible, the salivary gland cells of Chironomus, nuclear compartmentalization appears to be insignificant.

We conclude that the Na^+ transport pool of toad bladder constitutes only a fraction of the total intracellular Na^+ within the transporting cells. This may reflect heterogeneity of cell population and subcellular compartmentalization in organelles other than the nucleus.

ACKNOWLEDGEMENTS

Supported in part by Research Grants from the National Institutes of Health (1 R01 AM 16586-03) and the National Science Foundation (BMS 73-01161 A03) during the tenure of an Established Investigatorship (70-148) from the American Heart Association.

REFERENCES

Aceves, J., Erlij, D. (1971) *J. Physiol. 212:* 195.
Armstrong, W. McD., Lee, C.O. (1971) *Science 171:* 413.
Bentley, P.J. (1968) *J. Physiol. 195:* 317.
Brown, A.M., Walker, J.L., Sutton, R.B. (1970) *J. Gen. Physiol. 56:* 559.
Cereijido, M., Rotunno, C.A. (1967) *J. Physiol. 190:* 481.
Cereijido, M., Rabito, C.A., Rodriguez Boulan, E., Rotunno, C.A. (1974) *J. Physiol. 237:* 555.
Civan, M.M., Kedem, O., Leaf, A. (1966) *Am. J. Physiol. 211:* 569.
Civan, M.M., Frazier, H.S. (1968) *J. Gen. Physiol. 51:* 589.
Civan, M.M. (1970) *Am. J. Physiol. 219:* 234.
Civan, M.M., DiBona, D.R. (1974) *J. Membrane Biol. 19:* 195.
Coplon, N.S., Maffly, R.H. (1972) *Biochim. Biophys. Acta 282:* 250.
Danisi, G., Vieira, F.L. (1974) *J. Gen. Physiol. 64:* 372.
DiBona, D.R., Civan, M.M., Leaf, A. (1969a) *J. Cell Biol. 40:* 1.
DiBona, D.R., Civan, M.M., Leaf, A. (1969b) *J. Membrane Biol. 1:* 79.
Epstein, M.A. (1957) *J. Biophys. Biochem. Cytol. 3:* 851.
Gatzy, J.T., Berndt, W.O. (1968) *J. Gen. Physiol. 51:* 770.
Handler, J.S., Preston, A.S., Orloff, J. (1972) *Am. J. Physiol. 222:* 1071.
Handler, J.S., Preston, A.S. (1976) *J. Membrane Biol. 26:* 43.
Herrera, F.C. (1966) *Am. J. Physiol. 210:* 980.
Hinke, J.A.M. (1961) *J. Physiol. 156:* 314.
Ito, S., Loewenstein, W.R. (1965) *Science 150:* 909.
Kostyuk, P.G., Sorokina, Z.A., Kholodova, Yu.D. (1969) In *Glass Microelectrodes* (Lavallee, M., Schanne, O.F., Hebert, N.C., eds.) p. 322, Wiley & Sons, N.Y.
Lee, C.O., Armstrong, W. McD. (1972) *Science 175:* 1261.

Lev, A.A. (1964) *Nature* *201:* 1132.
Lev, A.A., Armstrong, W. McD. (1975) *Current Topics in Membranes and Transport 6:* 59.
Lipton, P., Edelman, I.S. (1971) *Am. J. Physiol. 221:* 733.
Loewenstein, W.R., Kanno, Y. (1963) *J. Gen Physiol. 46:* 1123.
Macknight, A.D.C., Leaf, A., Civan, M.M. (1970) *Biochim. Biophys. Acta 222:* 560.
Macknight, A.D.C., DiBona, D.R., Leaf, A., Civan, M.M. (1971a) *J. Membrane Biol. 6:* 108.
Macknight, A.D.C., Leaf, A., Civan, M.M. (1971b) *J. Membrane Biol. 6: 127.*
Macknight, A.D.C., Civan, M.M., Leaf, A. (1975a) *J. Membrane Biol. 20:* 365.
Macknight, A.D.C., Civan, M.M., Leaf, A. (1975b) *J. Membrane Biol. 20:* 387.
McIver, D.J.L., Macknight, A.D.C. (1974) *J. Physiol. 239:* 31.
Moreno, J.H., Reisin, I.L., Rodriguez Boulan, E., Rotunno, C.A., Cereijido, M. (1973) *J. Membrane Biol. 11:* 99.
Palay, S.L. (1960) *J. Biophys. Biochem. Cytol. 7:* 391.
Palmer, L.G., Civan, M.M. (1975) *Science 188:* 1321.
Palmer, L.G., Civan, M.M. (1976a) *Clin. Res., In press.*
Palmer, L.G., Civan, M.M. (1976b) *J. Membrane Biol. In press.*
Scott, W.N., Sapirstein, V.S., Yoder, M.J. (1974) *Science 184:* 797.
Sharp, G.W.G., Coggins, C.H., Lichtenstein, N.S., Leaf, A. (1966) *J. Clin. Invest. 45:* 1640.
Shporer, M., Civan, M.M. (1974) *Biochim. Biophys. Acta 354:* 291.
Shporer, M., Civan, M.M. (1977) *Current Topics in Membranes and Transport, In press.*
Siebert, G., Langendorf, H. (1970) *Naturwiss. 57:* 119.
Thomas, R.C. (1970) *J. Physiol. 210:* 82P.
Walker, J.L. (1971) *Analytical Chem. 43:* 89A.
Watson, M.L. (1955) *J. Biophys. Biochem. Cytol. 1:* 257.
Yonath, J., Civan, M.M. (1971) *J. Membrane Biol. 5:* 366.

PASSIVE WATER TRANSPORT ACROSS EPITHELIA

Ernest M. Wright

University of California School of Medicine

INTRODUCTION

Central to discussions of mechanisms of water transport across epithelial tissues is the permeability of the epithelium to water. Questions frequently raised include: i) what is the magnitude of the water permeability coefficient; ii) are there any differences between the various measurements of water permeability, e.g. diffusional and osmotic permeability coefficients; iii) what is the route of water permeation across the epithelium - is it through the cells where water permeates through two plasma membranes and the cell interior or does water bypass the cells and proceed through the junctions between cells; and iv) how does water permeate through the barriers - are there small aqueous channels or is the mechanism a simple solubility diffusion process. Although answers to these fundamental questions have their own intrinsic value, they are also highly relevant to the complex problems posed by "active" fluid transport across epithelia. For example, the actual magnitude of the osmotic water permeability coefficient is a central issue in deciding between alternate models for solute-linked water flow across the gall bladder, intestine, choroid plexus and renal proximal tubule (see, Wright, Wiedner and Rumrich, 1977). In this chapter I will survey the information available so far about the permeability of epithelial membranes to water, with particular emphasis on the studies carried out in my own laboratory on the gall bladder and urinary bladder.

ORGANIZATION OF EPITHELIA AND ROUTES OF PERMEATION

The salient features of simple epithelia such as the gall bladder, urinary bladder and choroid plexus are illustrated in Figure 1. The epithelium is composed of a single layer of cells joined together at

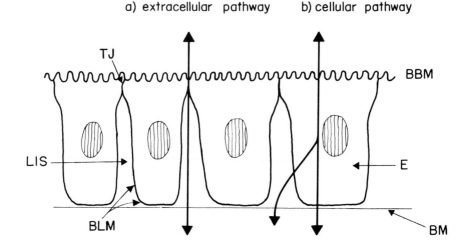

Fig. 1. Pathways for permeation across epithelia. The diagram shows some general structural features of epithelial tissues such as the intestine, gall bladder, renal tubule, urinary bladder and choroid plexus. The epithelium (E) is shown as a single layer of cells resting upon a basement membrane (BM), and joined together at the apical surface by tight junctions (TJ). These junctions are continuous around the circumference of each cell. Solutes and water cross the epithelium by either the extracellular pathway (a) or the cellular pathway (b). The extracellular path is through the tight junctions and lateral intercellular spaces (LIS), whereas the cellular pathway is across the brush border membrane (BBM), cytoplasm and basal-lateral membrane (BLM). The lateral intercellular spaces are common to both pathways.

the apical, or brush border, surface by junctional complexes that are continuous around the entire circumference of each cell. These junctions are referred to as tight junctions by virtue of the fact that they restrict the movement of large protein molecules across the epithelium (see McNutt and Weinstein, 1973, for a review of cell junctions). All epithelial cells rest upon a thin basement membrane composed of collagen fibrils, and a supporting layer of tissue composed of connective tissue, blood vessels and smooth muscle. The thickness of the supporting layer ranges from a few microns in the toad bladder to over 300 microns in some regions of the gastroin-testional tract.

These are two potential pathways for the permeation of water and solutes across an epithelium, i.e. the cellular and extracellular pathways (see Figure 1). In the former, water and solutes permeate across two plasma membranes and the cell interior, while in the latter they simply pass through the junctions and lateral intercellular spaces. At least in the case of small ions, progress has been made in elucidating the importance of the shunt path. For example, in "leaky" epithelia such as the gall bladder, intestine, renal proximal tubule and choroid plexus, as much as 95% of the passive ion fluxes occur through the junctions and lateral intercellular spaces, but in "tight" epithelia such as the frog skin and urinary bladder, passive ion permeation across the junctions is relatively unimportant (see Frömter and Diamond, 1972; Erlij, 1976; Lewis and Diamond, 1976).

So far all the information about the water permeability of epithelia comes from black box experiments where it is virtually impossible to assess the relative contributions of the two paths. Two black box approaches are used; the first is to measure the unidirectional fluxes of isotopes (THO or D_2O) across the epithelium to obtain the diffusional water permeability (P_d), and the second is to measure volume flows across the epithelium in response to osmotic gradients to obtain the osmotic water permeability (P_{os} or L_p). The importance of the extracellular route to these measurements depends upon the area of the junctions relative to the cell membranes, the permeabilities of the membranes and the junctions, and, for osmotic flows, the solute reflection coefficients at each barrier. Although there are no definitive answers to this question, my personal opinion is that water permeation via the junctions is insignificant. This stems from the fact that the area of the junctions represent less than 0.004% of the total area of the epithelium, and calculations which show that less than 10% of the osmotic water flow occurs through the junctions in the gall bladder (Wright, Smulders and Tormey, 1972). Furthermore, some experimental observations suggest that the junctional reflection coefficients for the solutes used to generate osmotic flows, e.g. sucrose, are significantly less than 1 (see Wright, Smulders and Tormey, 1972; Smulders and Wright, 1971; Wright and Pietras, 1974; van Os, de Jong and Slegers, 1974).

One way to resolve this problem is to measure the water permeability of the plasma membranes directly, and compare the results with the overall P's for the intact epithelium. At this juncture the only conclusion that can definitely be drawn is that there are two possible pathways for water permeation across epithelia, and that the lateral intercellular spaces present a final common pathway irrespective of whether water reaches the spaces via the cells or via the junctions. Under certain circumstances flow along the lateral spaces is the rate limiting step in water permeation across highly permeable epithelia, namely when the lateral spaces collapse (Wright, Smulders and Tormey, 1972).

WATER IS JUST ANOTHER NON-ELECTROLYTE

It should be recognized that many of the problems associates with water permeation across epithelia are essentially the same as those for non-electrolytes in general. A summary of the patterns of non-electrolyte permeation across one epithelium, the toad bladder, is presented in Figure 2. This shows the permeability coefficients of

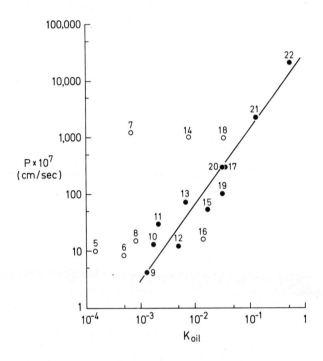

Fig. 2. Non-electrolyte permeation across the toad urinary bladder. Permeability coefficients (P) are plotted against their olive oil/water partition coefficients (K_{oil}). The line of identity was drawn to fit all points except the small molecules (MV < 60 cc/mole) and the branched compound iso-butyramide. The non-electrolytes are urea (5), ethylene glycol (6), H_2O (7), acetamide (8), 1,3-propanedic (9), 1,2-propanediol (10), 1,4-butanediol (11), nicotinamide (12), 1,6-hexanediol (13), methanol (14), n-butramide (15), isobutyramide (16), 1,7-heptanediol (17), ethanol (18), antipyrine (19), caffeine (20), n-propanol (21) and n-butanol (22). Taken from Bindslev & Wright (1976).

18 non-electrolytes (including water) plotted against their olive oil/water partition coefficients. The P's range over 5 orders of magnitude, and there is a general correlation between P's and partition coefficients; for the larger solutes P α $K_{oil}^{1.3}$. However, molecules with a molecular volume less than 60cc/mole are deviant in that they are more permeant than predicted from their K's. For example, water (Wright, Smulders and Tormey, 1972) and methanol (Pietras and Wright, 1975) are 2-3 orders of magnitude more permeabl than expected. In the case of small molecules the most dominant factor controlling permeation appears to be their size; P's are inversely proportional to the size of the non-electrolyte (P α MV^{-2}. Similar observations have been reported for other epithelial membran

and single cells (see Wright and Pietras, 1974).

The correlation between P's and K's in membranes is interpreted as showing that non-electrolytes permeate across the plasma membranes of epithelia by a solubility diffusion mechanism (Diamond and Wright, 1969; Bindslev and Wright, 1976; Wright and Bindslev, 1976). The common explanation for the deviant behaviour of small polar non-electrolytes, e.g. water, is that they bypass the membrane lipid and permeate through small aquous pores, but in the toad bladder, where small polar *and* small non-polar molecules both permeate more rapidly than expected, there is evidence that this is simply due to the presence of structural defects in the quasi cystalline membrane lipids (Bindslev and Wright, 1976; Wright and Bindslev, 1976). These considerations, and the fact that antidiuretic hormone (ADH) increases the water permeability of the toad bladder more than an order of magnitude, strongly suggests that the major pathway for water transport across this epithelium is through the cells rather than through the junctions.

WATER PERMEABILITY OF EPITHELIA

Granted the problem of the route and mechanism of water permeation across epithelia is part of the more general question of non-electrolyte permeation, what is the range of water permeability encountered among epithelia? The osmotic and diffusional water permeabilities of some epithelial membranes are listed in Table 1 along with the range of values in the literature for black lipid membranes and single cells. For convenience both measurements of water permeability are quoted in the same units (cm, sec^{-1}), and included in the final column is the ratio P_{os}/P_d.

The first point to note is that the values for both P_{os} and P_d for all five epithelia lie close to the range obtained for lipid bilayers and single cells. P_{os} ranges from 0.39×10^{-3} to $20 \times 10^{-3} cm, sec^{-1}$ and P_d ranges from 0.13×10^{-3} to $1 \times 10^{-3} cm, sec^{-1}$. This is consistent with the view that the major pathway for water permeation across epithelia is through the plasma membranes rather than through the junctions; one would expect the permeability of epithelia to be much greater than single cells if permeation across the junctions made any significant contribution to the overall flow. Finally, the observation that ADH increases water permeability of the toad urinary bladder by almost two orders of magnitude indicates that subtle changes in the state of the plasma membranes produce radical changes in permeability (Pietras and Wright, 1975; Bindslev and Wright, 1976).

The second point to note is that the ratio P_{os}/P_d is close to unity in lipid bilayers and some single cells (see also Table 5.5, in House, 1974), and this is widely interpreted as showing that both the diffusional and osmotic flux of water across the membranes occurs by a solubility diffusion process. However, in other single cells, e.g. red blood cells, and epithelia the ratio is significantly greater than 1. In fact, ratios as high as 299 have been reported

Table 1. WATER PERMEABILITY OF EPITHELIA

	OSMOTIC P_{os}(cm/sec $\times$ 10^3)	DIFFUSIONAL P_d	RATIO P_{os}/P_d	REFERENCES
Lipid bilayers	0.2 - 6	0.2 - 2	1	1
Single cells	0.2 - 40	0.2 - 6	1-4	1
Toad urinary bladder -ADH	0.39	0.13	~3	1-4
+ADH	20	>1	<20	
Frog choroid plexus	2.2	0.07	32	2,5
Frog urinary bladder	3.2	0.11	29	6,7
Rabbit gall bladder	>>2.9	0.20	>>15	2,8
Frog gall bladder	>>15	0.68	>>22	6,9

1) House, 1974; 2) Wright and Pietras, 1974; 3) Pietras and Wright, 1975; 4) Hays and Franki, 1970;
5) Wright and Wiedner, unpublished observations; 6) Wiedner and Wright, unpublished observations;
7) Pietras, Naujokaitis and Szego, 1976; 8) Wright, Smulders and Tormey, 1972; 9) Moreno, 1975.

for certain epithelia (see Table 9.5 in House, 1974). It has been argued that ratios of P_{os}/P_d greater than 1 indicate the presence of aqueous pores in membranes; if water traverses a membrane via aqueous channels the osmotic water permeability will be greater than the diffusional water mpermeability owing to frictional inter- actions between water molecules in the channel. However, to account for a P_{os}/P_d ratio of 30 (Table 1) the channel radius would need to be about 25 Å - a value which is incompatible with the patterns of solute permeability observed in these epithelia, e.g. in the toad bladder although ADH increases P_{os}/P_d from 3 to 23 (Hays and Franki, 1970), there is no significant increase in the permeability of the bladder to mannitol or galactose (solutes with molecular radii of 4 Å) [see also Table 9.9 in House, 1974].

QUESTIONS ABOUT PERMEABILITY MEASUREMENTS

How reliable are these estimates of P_{os} and P_d in epithelia? This question arises because high ratios of P_{os}/P_d are mostly seen in membranes with large absolute values of the permeability coeffi- cients, and where there are thick unstirred layers adjacent to the membrane. Unstirred layers under these circumstances are known to lead to serious errors in permeability measurements in both artificial and biological membranes. A common assumption made in the interpretation of permeability measurements is that the solutions on each side of the membrane are perfectly stirred, i.e. the concentration of the solution adja- cent to the membrane is identical to the bulk phase concentration. In practice it is difficult to achieve perfecting stirring, and in many cases where the rate of transport across the membrane is high the concentration of the solute at the membrane interface differs significantly from the bulk phase concentration (see Figure 3). The concentration of the solute on one side of the membrane falls below the bulk phase concentration, and on the other side increases above the bulk phase concentration. In the steady state, where the flux of the solute across the two unstirred layers and across the membrane are equal, the concentration gradient across the membrane is reduced by an amount $\phi \frac{\delta_1 \delta_2}{D_1 D_2}$, where D_1 and D_2 are the diffusion coefficients in the unstirred layers, δ_1 and δ_2 are the thickness of the two unstirred layers and ϕ is the steady state flux of the solute across the membrane. It follows that the measured permea- bility coefficient, i.e. calculated from the steady state flux and the bulk phase concentration gradient (Δc^b) is related to the true membrane permeability coefficient (P_m) by the relation

$$1/P = 1/P_m + \delta_1/D_1 + \delta_1/D_1 + \delta_2/D_2$$

when $C_2 = 0$. This expression equates the overall resistance to permeation with the sum of the resistances offered by the two unstirred layers and the membrane. Thus in the presence of thick unstirred layers and a highly permeable membrane the flux of solute, or water, may be limited by diffusion across the unstirred layers.

ERNEST M. WRIGHT

membrane

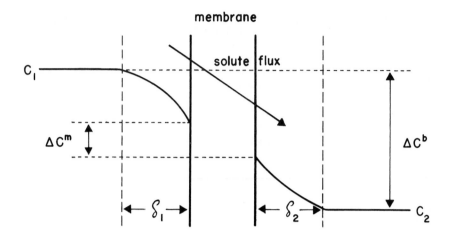

Fig. 3. The effect of unstirred layers on a solute flux across a membrane. Shown is the flux of the solute across a membrane and its associated unstirred layers, δ_1 and δ_2. The bulk phase concentration gradient is $C_1 - C_2 = \Delta c^b$, and the actual concentration gradient across the membrane is Δc^m.

In the toad bladder in the absence of ADH, where vigorous stirring reduces the thickness of the unstirred layers to 160μ (Bindslev and Wright, 1976), the resistance offered by the unstirred layers amounts to 20% of the total resistance, but in the presence of ADH the flux of water is completely limited by the unstirred layers. Consequently in the presence of hormone the value of P_d is grossly underestimated, and P_{os}/P_d is overestimated.

Similar problems exist for the other epithelia where the unstirred layers are much thicker than those in the urinary bladders under our experimental conditions: 900μ in the choroid plexus, and 345-800μ in the gall bladder (see Smulders and Wright, 1971; Wright and Pietras, 1974; Bindslev, Tormey and Wright, 1974; Wiedner and Wright, 1975). This is particularly acute in gall bladders where in addition the water permeability is high. In fact, van Os & Slegers (1971) find that unstirred layers account for 74-100% of the total resistance to THO diffusion across the rabbit gall bladder. [Under their experimental conditions they find that the unstirred layer thickness is 1420μ and that the true membrane permeability to water is about 1-2 x 10^{-3}cm,sec^{-1}. It should be noted that they estimated the thickness of the unstirred layers from butanol fluxes assuming that the flux of this solute was completely unstirred layer limited, and that their thick unstirred layers were due in part to the sintered glass they used to support the tissue.] Estimates of P_d for epithelial membranes, particularly for highly permeable epithelia, should therefore only be considered as minimum values until we are confident that unstirred layer effects have

ERNEST M. WRIGHT

membrane

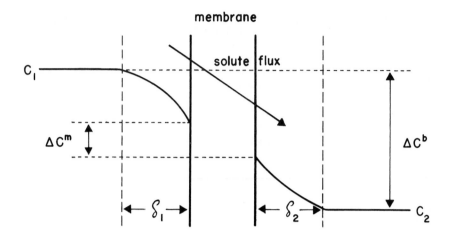

Fig. 3. The effect of unstirred layers on a solute flux across a membrane. Shown is the flux of the solute across a membrane and its associated unstirred layers, δ_1 and δ_2. The bulk phase concentration gradient is $C_1 - C_2 = \Delta c^b$, and the actual concentration gradient across the membrane is Δc^m.

In the toad bladder in the absence of ADH, where vigorous stirring reduces the thickness of the unstirred layers to 160μ (Bindslev and Wright, 1976), the resistance offered by the unstirred layers amounts to 20% of the total resistance, but in the presence of ADH the flux of water is completely limited by the unstirred layers. Consequently in the presence of hormone the value of P_d is grossly underestimated, and P_{os}/P_d is overestimated.

Similar problems exist for the other epithelia where the unstirred layers are much thicker than those in the urinary bladders under our experimental conditions: 900μ in the choroid plexus, and 345-800μ in the gall bladder (see Smulders and Wright, 1971; Wright and Pietras, 1974; Bindslev, Tormey and Wright, 1974; Wiedner and Wright, 1975). This is particularly acute in gall bladders where in addition the water permeability is high. In fact, van Os & Slegers (1971) find that unstirred layers account for 74-100% of the total resistance to THO diffusion across the rabbit gall bladder. [Under their experimental conditions they find that the unstirred layer thickness is 1420μ and that the true membrane permeability to water is about 1-2 x 10^{-3}cm,sec^{-1}. It should be noted that they estimated the thickness of the unstirred layers from butanol fluxes assuming that the flux of this solute was completely unstirred layer limited, and that their thick unstirred layers were due in part to the sintered glass they used to support the tissue.] Estimates of P_d for epithelial membranes, particularly for highly permeable epithelia, should therefore only be considered as minimum values until we are confident that unstirred layer effects have

been accounted for. Likewise, estimates of P_{os}/P_d for highly permeable membranes can only be taken as upper estimates, and caution
should be exercised in the interpretation of these high ratios.

A second unstirred layer effect poses difficulties in obtaining
accurate estimates of the osmotic permeability of epithelia. This
arises because the flow of water across a membrane tends to reduce
the solute concentration on one side of the membrane and increase
the solute concentration on the other (see Figure 4). The actual
concentration of the solute adjacent to the membrane (C_m) is related
to the velocity of water flow (V) and the unstirred layer thickness
(δ) by the relation

$$C_m = C_b \ e^{\pm \ V\delta/D}$$

where C_b is the bulk phase concentration and D is the solute diffusion coefficient in the unstirred layer (Dainty, 1963). Thus the

membrane

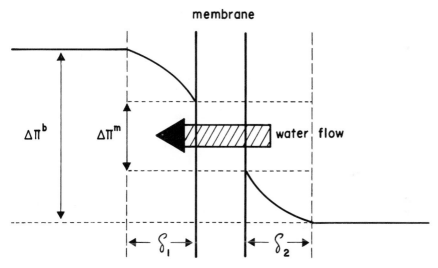

Fig. 4. The effect of water flow on solute concentration profiles
in the unstirred layers adjacent to a membrane. An osmotic flow,
in response to a bulk phase osmotic gradient $(\Delta\pi^b)$, is shown to
enhance the solute concentration at the membrane solution interface on one side of the membrane, and to reduce the concentration
at the other side of the membrane. These sweeping away effects
reduce the effective osmotic gradient across the membrane from
$\Delta\pi^b$ to $\Delta\pi^m$.

effective concentration gradient of the solute across the membrane
is not $(C_b^1 - C_b^2)$, but $(C_m^1 - C_m^2) = [(C_b^1 \ e^{\pm V\delta_1/D}) - (C_b^2 \ e^{\pm V\delta_2/D})]$.
Consequently, the effective osmotic pressure across the membrane
in the steady state is less than the osmotic gradient between the
bulk solutions on each side of the membrane. This leads to an underestimate of the real P_{os} if the bulk phase concentrations are used

as the driving force for water flow across the membrane. Further-more, one can predict that the instantaneous rate of osmotic water flow across a membrane is higher than the steady state flow, because it takes a finite time for the water flow to change the solute con-centrations in the unstirred layers. Osmotic flow transients have been studied in detail in the rabbit gall bladder (Wright, Smulders and Tormey, 1972; van Os, 1974) and renal cortical collecting tubule (Schafer, Patlak and Andreoli, 1974).

When considering sweeping away effects in epithelia three addi-tional factors have to be taken into account: i) the unstirred layer on one side of the epithelium may be much thicker than on the other side, e.g. in the gall bladder the thickness of the mucosal unstirred layer varies between 60 and 100, while the thickness of the serosal unstirred layer varies between 300 and 800μ (see Fig. 5 Diamond, 1966b; Smulders and Wright, 1971; Brindslev, Tormey and Wright, 1974; Wiedner and Wright, 1975). This is due to the thick supporting layer on the serosal side of the epithelium. A direct consequence of this asymmetry is that the osmotic transients depend upon the direction of flow (see Schafer, Patlak and Andreoli: (1974 Figure 4). ii) The velocity of the water flow may be different in the unstirred layer on each side of the epithelium due to the morph-ology of the tissue. In tissues like the gall bladder the lateral intercellular spaces represent a common pathway for water transport across the epithelium irrespective of the route of permeation (see Figure 1). Since the lateral spaces only occupy a fraction of the area of the epithelium (see Schafer, Patlak and Andreoli, 1974) the velocity of flow water in the spaces will be greater than the muco-sal unstirred layer. So the sweeping away effects will be greater in the lateral spaces than in the mucosal unstirred layer; and iii) solutes other than that used to generate the osmotic gradient are usually present on each side of the epithelium, e.g. in gall bladder experiments physiological salt solutions (~ 150mM NaCl) bath both sides of the tissue, and sucrose is added to one side to produce osmotic water flow (see Figure 5). This means that water flow across the membrane changes the concentration of all solutes in the unstirred layers, and in the case of the gall bladder experiments sweeping away effects generate a salt gradient across the epithelium which is in the opposite direction to the sucrose gradient. A fur-ther implication is that the rate of osmotic flow depends not only on the bulk phase osmotic gradient across the tissue, but also the absolute osmolarity of the bathing solutions. In fact, the steady state flow in response to a given osmotic gradient falls as the osmolarity of the bathing solutions increases, and this explains the observation by Diamond (1966a) that the gall bladder L_p decrease with increasing osmolarity.

A diagramatic representation of sweeping away effects observed in the rabbit gall bladder are shown in Figure 5. In this experi-ment the gall bladder was first incubated with identical salt con-centrations in the mucosal and serosal compartments, and then sucrose was added to the mucosal fluid to generate an osmotic volume flow from serosa to mucosa. In the steady state the flow of water

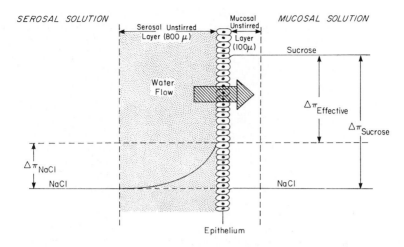

Fig. 5. A diagrammatic representation of the unstirred layers and osmotic water flows across rabbit gall bladder. This shows the gall bladder wall separating the mucosal and serosal solutions, and the sucrose and NaCl concentration profiles across the tissue. The bulk phase mucosal and serosal solutions contain identical concentrations of NaCl. The mucosal solution also contains 50 mM sucrose which generates an osmotic flow $(\sigma_{suc} L_p \Delta\pi_{suc})$ from serosa to mucosa. This flow of water has no significant effect on the sucrose and NaCl concentrations in the mucosal unstirred layer, but in the serosal unstirred layer the flow produces an enhancement of the NaCl concentration adjacent to the serosal face of the epithelium $[C_m = C_{se}(V\delta_s/D)$, where C_m and C_s are the NaCl concentrations adjacent to the epithelium and in the bulk phase serosal solution, respectively, V the velocity of water flow, δ_s the thickness of the serosal unstirred layer, and D the NaCl-free solution diffusion coefficient]. The local salt gradient across the epithelium creates a diffusion potential which represents most, if not all, of the so-called streaming potential; the magnitude of the p.d. being given by the constant field equation. The salt gradient produces a local osmotic gradient across the epithelium which is opposite to the sucrose gradient, i.e., the effective osmotic pressure across the epithelium is $\Delta\pi_{effective} = \Delta\pi_{sucrose} - \Delta\pi_{NaCl}$. Therefore, the flow of water across the gall bladder in the steady state is given by $(\sigma_{suc} L_p \Delta\pi_{suc}) - (\sigma_{NaCl} L_p \Delta\pi_{NaCl})$. From Wright, Smulders and Tormey, 1972.

across the epithelium is shown to: i) increase the NaCl concentration in the serosal unstirred layer; ii) reduce slightly the NaCl concentration in the mucosal unstirred layer. The effect here is smaller than on the serosal side owing to asymmetry of the unstirred layers and the fact that the velocity of water flow is lower in the mucosal unstirred layer; and iii) reduce the sucrose concentration in the mucosal unstirred layer. This effect is also relatively small owing to the low concentration of sucrose, and this is confirmed by measurement of sucrose unidirectional fluxes in the absence and presence of sucrose concentration gradients (see Figure 1, and Smulders, Tormey and Wright, 1972). The net result is that the effective osmotic gradient across the epithelium ($\Delta\pi_{effective}$) is not $\Delta\pi_{sucrose}$, but $\Delta\pi_{sucrose} = \Delta\pi_{NaCl}$.

How big are these effects in the gall bladder? An actual experiment is shown in Figure 6. The absence of active transport and

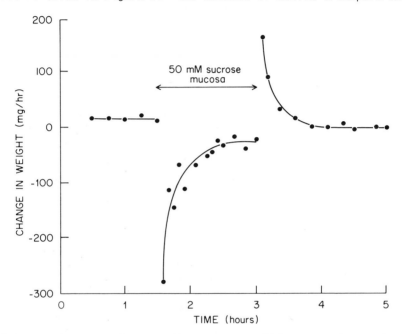

Fig. 6. *The rate of water flow across rabbit everted gall bladder as a function of time. Water flows were measured gravimetrically every 5 min and the rate of change in weight (in mg/hr) is plotted on the ordinate against time on the abscissa; positive changes in weight mean that water is flowing from mucosa to serosa and negative changes mean that water is flowing serosa to mucosa. The serosal solution (luminal) was regular Ringer's solution throughout and the external solution (mucosal) was also Ringer's solution except during the time indicated, where it was replaced with a solution identical in ionic composition but containing in addition 50 mM sucrose. From Wright, Smulders and Tormey (1972).*

osmotic gradients there is virtually no water flow across the gall bladder, but addition of sucrose to the mucosal fluid water generates a flow into the hypertonic solution. The initial rate of flow is high, and this rapidly falls to a steady state value about one tenth of the initial rate. The initial and steady state flows correspond to osmotic water permeabilities of 30×10^{-3} and $3 \times$

10^{-3}cm,sec^{-1} respectively. On returning a sucrose free solution to the mucosal surface of the gall bladder there is a rebound of water flow and this decays back to zero with a time course similar to that observed for the on response. This rebound behaviour originates with the polarization of salt in the serosal unstirred layer during water flow across the tissue, and when sucrose is rapidly removed from the mucosal fluid this leaves the NaCl gradient across the membrane intact, which in turn draws water back into the serosal compartment until the salt gradient dissipates.

This experiment indicates that estimates of the osmotic water permeability of the gall bladder using steady state flows are at least an order of magnitude too low. An accurate estimate of the true P_{os} is difficult to deduce from the experiment shown in Figure 6 owing to the poor time resolution and accuracy of the gravimetric method used to measure water transport. However, modelling of the magnitude and duration of the osmotic transients suggests that the true P_{os} is about 60×10^{-3}cm,sec^{-1} (see Schafer, Patlak and Andreoli, 1974).

Similar experiments on toad bladders and renal cortical collecting tubules in the presence of ADH show that the steady state osmotic flows also lead to underestimates of the true osmotic permeability of the epithelium, but here the effects are small (Schafer, Patlak and Andreoli, 1974; and R. Hays, personal communication). The smaller effects are probably due to thinner unstirred layers.

A final consideration is that flow of water across an epithelium may produce dramatic changes in permeability. In "leaky" epithelia large volume flows cause changes in the lateral intercellular spaces (see Figure 1, and the article by Tormey in this volume), and these are accompanied by changes in the permeability of ions, water and non-electrolytes (Smulders, Tormey and Wright, 1972; Wright, Smulders and Tormey, 1972; Brindslev, Tormey and Wright, 1974; Wright and Wiedner, unpublished). For example, in the rabbit gall bladder osmotic flow of water from the serosa to mucosa can reduce the value of L_p three fold (Wright, Smulders and Tormey, 1972). In "tight" epithelia osmotic gradients can also produce very large increases in ion, water and non-electrolyte permeability (see Bindslev, Tormey, Pietras and Wright, 1974), but in this case they are due to increases in the permeability of the cell junctions and/or plasma membranes. Consequently, experiments should be designed to minimize these effects when studying the permeability of epithelial tissues.

CONCLUSION

In epithelial membranes where the absolute water permeabilities are low, i.e. $P_{os} < 5 \times 10^{-4}$cm,sec^{-1} and $P_d < 1 \times 10^{-4}$cm,sec^{-1}, relatively accurate estimates of these parameters can be obtained if care is exercised to minimize the contributions of the unstirred layers. However, in more permeable epithelia the estimates of both P_{os} and P_d are unreliable, and should be considered minimum values. This is particularly true when the unstirred layers adjacent to the epithelium are large because of the morphology of the tissue or the poor design of the experiment. Unfortunately, the very permeable epithelia are the most interesting because these are specialised for water transport, and the actual magnitude of the water permeability is central in deciding between alternate theories of water transport (see Wright, Wiedner and Rumrich, 1977). Progress in obtaining answers to these fundamental questions about water transport largely depends on the development of new methods to overcome the problems generated by high flows and large unstirred layers.

ACKNOWLEDGEMENTS

These studies were supported by a grant from the United States Public Health Service (NS-09666).

REFERENCES

Bindslev, N., Tormey, J. McD., Pietras, R.J., & Wright, E.M. (1974 *Biophys. Acta 332,* 286.
Bindslev, N., Tormey, J. McD., & Wright, E.M. (1974) *J. Membrane Biol. 19:* 357.
Bindslev, N. & Wright, E.M. (1976) *J. Membrane Biol.* In the Press
Dainty, J. (1963) *Advanc. Botan. Res. 1:* 279.
Diamond, J.M. (1966a) *J. Physiol. 183:* 58.
Diamond, J.M. (1966b) *J. Physiol. 183:* 83.
Diamond, J.M. & Wright, E.M. (1969) *Ann. Rev. Physiol. 31:* 581.
Erlij, D. (1976) *Kidney Internat. 9:* 76.
Frömter, E. & Diamond, J.M. (1972) *Nature 235:* 9.
Hays, R.M. & Franki, N. (1970) *J. Membrane Biol. 2:* 263.
House, C.R. (1974) *Water Transport in Cells and Tissues.* Arnold, London.
Lewis, S.A. & Diamond, J.M. (1976) *J. Membrane Biol. 28:* 1.
McNutt, N.S. & Weinstein, R.S. (1973) *Prog. Biophys. Molec. Biol. 26:* 45.
Moreno, J.H. (1975) *J. Gen. Physiol. 66:* 117.
Pietras, R.J., Naujokaitis, P.J., & Szego, C.M. (1976) *Mol. Cell. Endo. 4:* 89.
Pietras, R.J., & Wright, E.M. (1975) *J. Membrane Biol. 22:* 107.
Schafer, J.A., Patlak, C.S., & Andreoli, T.E. (1974) *J. Gen. Physiol. 64:* 201.
Smulders, A.P., Tormey, J. McD., & Wright, E.M. (1972) *J. Membrane Biol. 7:* 164.
Smulders, A.P. & Wright, E.M. (1971) *J. Membrane Biol. 5:* 298.
van Os, C.H. (1974) Thesis, University of Nijmegen.
van Os, C.H., de Jong, M.D., & Slegers, J.F.G. (1974) *J. Membrane Biol. 15:* 363.
van Os, C.H. & Slegers, J.F.G. (1971) *Biochem. Biophys. Acta 291:* 197.
Wiedner, G. & Wright E.M. (1975) *Pflügers Arch. 358:* 27.
Wright, E.M. & Bindslev, N. (1976) *J. Membrane Biol.* In the Press

Wright, E.M. & Pietras, R.J. (1974) *J. Membrane Biol. 17:* 293.
Wright, E.M., Smulders, A.P., & Tormey, J. McD. (1972) *J. Membrane Biol. 7:* 198.
Wright, E.M., Wiedner, G., & Rumrich, G. (1977) *Exp. Eye Res.* In the Press.

IONIC ACTIVITIES AND SOLUTE TRANSFER IN EPITHELIAL CELLS
OF THE SMALL INTESTINE

W. McD. Armstrong

Indiana University School of Medicine

INTRODUCTION

The measurement of intracellular ionic activities with ion-selective microelectrodes is rapidly increasing in importance as a tool in electrophysiological research. In particular, there is a growing awareness of the information which this technique can provide about the mechanisms underlying the translocation of ions across cell membranes and more complex structures such as epithelial cell layers. By a logical extension, increasing knowledge of the activities of individual ions in the cell interior and of the way in which these activities respond to changes in the extracellular environment (Lee and Armstrong, 1972; Neild and Thomas, 1974; Kline and Morad, 1976) can be expected to yield valuable insights into the behavior, in biological systems, of water and certain non-ionized solutes whose movements depend upon, or are associated with, concomitant net movements of ions.

It is not intended herein to attempt a comprehensive description of intracellular ionic activity measurements and their implications for water and solute movement in cells and tissues. Several aspects of this question have been discussed at length in recent reviews (Lee and Armstrong, 1975; Edzes and Berendsen, 1975). Rather, attention will be focussed minaly on recent studies of intracellular K^+, Na^+ and Cl^- activities in one epithelial system, the mucosal cell layer of the small intestine. These studies, so far, are relatively few in number. Nevertheless, it is hoped that this arbitrarily restricted survey will serve as an example of the utility, in a more general sense, of intracellular ionic activity measurements, and will help to stimulate interest in such measurements as a tool in future studies of epithelial function. Because the techniques for measuring intracellular ionic activities are in a state of active development (Kessler et al.,1976a,b) some discussion of these techniques, parti-

215

cularly as they relate to the problem of determining ionic activitie in small cells like the absorptive cells of the small intestine, seems pertinent.

THE MEASUREMENT OF IONIC ACTIVITIES IN EPITHELIAL CELLS IN THE SMALL INTESTINE - SOME TECHNICAL CONSIDERATIONS

A wide variety of microelectrodes for the determination of intra-cellnlar ionic activities has been described in the literature. These range from relatively large electrodes suitable for impaling correspondingly large cells, e.g. crab muscle fibers and squid giant axons (Caldwell, 1954, 1958; Hinke, 1959, 1961) to ultrafine micro-electrodes which permit the determination of ionic activities in much smaller cells such as those of cardiac muscle (Walker and Ladle, 1973; Lee and Fozzard, 1974; Kline and Morad, 1976) and epi-thelial tissues (Janancek et al., 1968; Lee and Armstrong, 1972; Khuri et al., 1972). The properties and applications of many of these microelectrodes are summarized in a recent review (Lev and Armstrong, 1975).

The small size of the epithelial cells of the small intestine imposes severe limitations on the size and configuration of any ion-selective microelectrode which is used to measure their inter-nal ionic activities. The problems involved can, perhaps, best be illustrated by considering the nature of the electrical potential registered by such a microelectrode following its insertion into a cell. Briefly, if the microelectrode is exclusively sensitive to one ionic species $\underline{j}$ (or so highly selective for $\underline{j}$ that other ions, at the concentrations at which they normally occur in cyto-plasm, do not significantly affect the electrode potential), the total electrode potential following impalement of a cell is given by

$$E_t = E_m + E_0 + S \log a^i_j \qquad 1.$$

In Equation 1, E_m is the cell membrane potential, E_0 is a constant for a given electrode and corresponds to the potential which the electrode would register in a solution containing $\underline{j}$ at unit activity a^i_j is the intracellular activity of $\underline{j}$ and S is the slope (in mV) of the electrode potential as a function of log a_j. Theoretically, S has the value $\ln(RT/zF)$ where z is the valence of $\underline{j}$ and R,T and F have their usual meaning but, in practice, providing that S (which together with E_0, is normally determined by calibrating the elec-trode in a series of standard solutions for which a_j is known) is constant over the range of interest, divergence by a few mV from this theoretical value is acceptable.

It is clear from Equation 1 that the accurate determination of a^i_j in any cell requires first that the ion-selective microelectrode whîch is used to measure it must be of such a size and shape that cell impalement can be accomplished without undue damage to the membrane. This ensures that the value of E_m recorded as part of E_t (Equation 1) is close to its true value and, in effect, limits the tip diameter of microelectrodes which can be used successfully with cells such as those of the intestinal mucosa to 1μm or less. Second, an accurate, independent estimate of E_m is needed in order to solve equation 1 for a^1_j. There are two ways in which this second requirement can be met. In large cells such as the giant nerve and muscle fibers of some invertebrates, it is a comparatively easy matter to insert a second microelectrode of the conventional open-tip variety (Ling and Gerard, 1949) into the cell which contains the ion-selective microelectrode and record both potentials simultaneously. In smaller cells this is not always feasible and one may be forced, first to record a representative selection of cell potentials with the ion-selective microelectrode, and then to record independently a series of membrane potentials in the same preparation with an open-tip microelectrode. In this situation, both E_t and E_m appear as the average values of unpaired sets of observations in the solution of Equation 1.

In tissues such as the sartorius muscle of the frog, where the resting potential is well defined and does not vary greatly between individual fibers under constant experimental conditions, independent recording of E_m has proved quite satisfactory for the determination of intrafiber K^+ and Na^+ activities (Lev, 1964; Armstrong and Lee, 1971). In the small intestine, however, there is considerable variation in the magnitudes of the membrane potentials recorded for individual epithelial cells, both in the same animal species and between different animals. Further, there does not yet appear to be complete agreement between different workers on the criteria that define the validity of membrane potential measurements in this tissue, so that the degree to which the divergent values reported in different studies (e.g. White and Armstrong, 1971; Rose and Schultz, 1971; Barry and Eggenton, 1972) are real (i.e. reflect differences between different animal species, or heterogeneity in the cell population samples) and the extent to which they reflect different degrees of technical imperfection in the method of measurement (Pederson, 1976) remains a moot question. For these reasons, liquid ion-exchanger microelectrodes of the type described by Walker (1971; see Fig. 1A) appear, outside the present author's laboratory, to be the only type of microelectrode so far used to any great extent to monitor intracellular ionic activities in intestinal mucosa. In addition to their intrinsic ease of construction, these microelectrodes readily lend themselves to fabrication in a double-barrelled form (Khuri et al., 1972; White, 1976a) in which one barrel is the ion-selective microelectrode, the other being an open-tip microelectrode for recording membrane potentials. In this form both microelectrodes can be simultnaeously inserted into a single epithelial cell and one can be reasonably certain that the value of E_m recorded by both barrels is essentially the same. However, as will be discussed further, it should be emphasized that the use of double-barrelled microelectrodes in no way diminishes the need for adequate impalement techniques which avoid excessive damage to the cell membrane.

So far, the use of liquid ion-exchanger microelectrodes for activity determination in intestinal epithelia has been limited to the

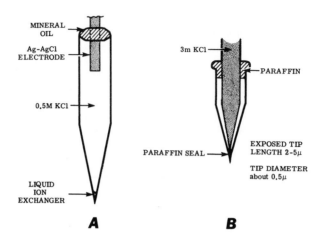

Fig. 1. A. Liquid ion-exchanger microelectrode for measuring intracellular K⁺ or Cl⁻ activities (Modified from Walker, 1971. Reproduced with permission). B. NAS 27-4 glass membrane microelectrode for simultaneous recording of K⁺ and Na⁺ activities (From Lee and Armstrong, 1974. Reproduced with permission).

determination of K^+ and Cl^- activities (Henriques de Jesus et al., 1974; Zeuthen and Monge, 1975, 1976; White, 1976) since no suitable Na^+-selective liquid ion-exchanger has been generally available. There are, happily, indications that this may not be the case much longer and that a number of Na^+-selective liquid ion-exchangers are now beginning to be used for intracellular activity measurements (Kessler et al., 1976, and personal communication from Dr. M. M. Civan). Consequently, little information appears to have been added as yet to the data reported by Lee and Armstrong (1972) and by Armstrong et al., (1973) for Na^+ activities in the epithelial cells of isolated bullfrog (*Rana catesbeiana*) small intestine. Since these Na^+ activities were, in a certain sense, obtained as a kind of "by-product" during the simultaneous determination of K^+ activities in the same tissues, some account of the methodology involved seems appropriate. Glass membrane microelectrodes, made from Corning NAS 27-4 K^+-selective glass (Eisenman et al., 1957) and filled with 3M KCl were used in these studies. Details of their construction are given by Lee and Armstrong (1974). Briefly, they had tip diameters <1µm and exposed tip lengths 2-5µm. Shielding was affected by enclosing all but the terminal 2-5µm of the sealed K^+-selective glass probe in a slightly wider tapered capillary drawn from a non-selective glass (Corning Code 7740). The annular space between the open tip of the outer glass shield and the inner K^+-selective microelectrode was sealed with paraffin. Fig. 1B is a schematic representation of one of these microelectrodes.

It should be apparent (Fig. 1B) that these electrodes are not readily adaptable to a double-barrelled configuration. Hence, in those studies in which they were used, it was necessary to measure E_m separately with conventional open-tip microelectrodes. Thus, the estimates of K^+ and Na^+ activities (a^i_K and a^i_{Na}) in epithelial cells of bullfrog small intestine obtained with these electrodes were subject to the uncertainties in the measurement of E_m which have already been discussed. In order to minimize these uncertainties, it was required that all the estimates of E_m which were included in final computation of a^i_K and a^i_{Na} conform to the rather rigorous criteria for acceptability previously established by White and Armstrong (1971) for this tissue under similar experimental conditions. Recent measurements by White (1976, 1976a) of a^i_K in isolated *Amphiuma* small intestine appear to confirm the validity of Lee and Armstrong's (1972) procedure. In White's studies, a^i_K was estimated by two methods. One of these involved penetration of the epithelial cells with single-barrelled liquid ion-exchanger K^+-selective microelectrodes and independent measurement of E_m with conventional glass microelectrodes, utilizing criteria for acceptability which were virtually identical to those of White and Armstrong (1971). In White's second method, single epithelial cells were impaled with double-barrelled microelectrodes. Following equilibration of the tissue with a sodium chloride Ringer solution containing 2.5 mM K^+, the mean values for a^i_K obtained by these two methods, 41.6 ± 1.5 (S.E.) and 38.5 ± 2.4 mM, where in very satisfactory agreement.

A second point of interest concerning the measurement of a^i_K with NAS 27-4 glass membrane microelectrodes is this. As supplied by the manufacturer this glass has a nominal selectivity for K^+ over Na^+ about 10/1. During the process of pulling and sealing, the selectivity of the glass at the tip of the capillary microelectrode is usually changed. These changes are random and variable so that the finished microelectrodes have K^+/Na^+ selectivities which range from values close to that of the original glass to values of 2/1 or less (Lev, 1964; Armstrong and Lee, 1971). This has two important consequences. First, it is obvious that, with these microelectrodes, the contribution of intracellular Na^+ to the overall intracellular electrode potential cannot be ignored. Second, because of this fact, it is possible by using two glass membrane microelectrodes which have widely different K^+/Na^+ selectivities and which have linear responses to both $\log a_K$ and $\log a_{Na}$, to measure simultaneously both a^i_K and a^i_{Na} in the same cell. This will be apparent from the following equations (Nicolsky, 1937; Eisenman et al., 1957) which express the behavior of such a pair of electrodes in a solution containing both K^+ and Na^+:

$$(E-E_0)/S = \log (a_K + k_{KNa}\, a_{Na}) \qquad\qquad 2.$$

$$(E'-E'_0)/S' = \log (a_K + k'_{KNa}\, a_{Na}) \qquad\qquad 3.$$

In these equations, E and E' are the potentials registered (with respect to, say, a grounded reference electrode) by each microelectrode following immersion in the test solution. E_0, E'_0, S and S', as well as k_{KNa} and k'_{KNa} are readily obtained by calibrating both

219

microelectrodes in pure KCl and pure NaCl standard solutions (Lee and Armstrong, 1974). It will at once be apparent that, providing E_m, as well as the total potential registered by each microelectrode inside a cell, is known, Equation 1 can be replaced by a pair of equations analogous to Equations 2 and 3 and that these equations can be solved for a^i_K and a^i_{Na} in the cell interior.

Two criticisms of this method for determining a^i_K and a^i_{Na} may be advanced. First, it is somewhat tedious since it involves the use of two K^+-selective microelectrodes, in addition to a conventional open-tip microelectrode, in each set of measurements. Second, since a^i_{Na} is normally much smaller than a^i_K, and since nearly all microelectrodes made from NAS 27-4 glass have some degree of selectivity for K^+ over Na^+, the contribution of Na^+ to the total intracellular electrode potential will usually be considerably less, and thus more sensitive to errors in measurement, than that of K^+. This limits the accuracy of estimates of a^i_{Na} by this method. However, since microelectrodes made from highly Na^+-selective glasses (e.g. NAS 11-18) do not appear, even in the ingenious "recessed-tip" configuration devised by Thomas (1970), to perform satisfactorily in small cells such as cardiac muscle fibers and epithelial cells of the small intestine (personal communications from Drs. C.O. Lee and J. F. White) it remains, in certain cases, the only feasible approach to the measurement of a^i_{Na}, at least until highly Na^+-selective liquid ion-exchangers become generally available.

K^+ ACTIVITIES IN EPITHELIAL CELLS OF THE SMALL INTESTINE

Three groups of studies may be considered under this heading. They are: those from the present author's laboratory (Lee and Armstrong, 1972; Armstrong et al., 1973), those performed by Zeuthen and Monge (1975,1976), and the recent studies of White (1976,1976a).

Lee and Armstrong (1972) estimated a^i_K in epithelial cells of isolated bullfrog small intestine, following equilibrium in an isotonic sodium sulfate medium, to be 85 mM. When the actively transported sugar analog 3-O-methyl glucose was added, at a concentration of 26 mM, to the solution bathing the mucosal surface of the cells, a^i_K fell to 64 mM. This decrease was considered to be satisfactorily explained by the increase in cell volume brought about by 3-O-methyl glucose accumulation and consequent entry of water into the cells under these conditions (Csaky and Esposito, 1969; Armstrong et al., 1970). Thus, no specific change in K^+ transport or accumulation as a result of mucosal sugar entry was considered necessary to account for these findings.

Concerning the physical state of intracellular K^+, it appeared that the results of Lee and Armstrong (1972) were consistent with the idea that all or virtually all the cytoplasmic K^+ was in a "free" or osmotically active state. Indeed, the ratio of the mean a^i_K value found in this investigation to an earlier estimate of C^i_K, the cytoplasmic K^+ concentration under the same conditions (Armstrong et al., 1970), was very close to unity. This is higher than the theoretical value of the mean cytoplasmic activity coefficient (calculated on the assumption that all the cell water is free to act as solvent water for K^+ and other ions) under these conditions. The latter parameter is about 0.8. This discrepancy was considered (Lee and Armstrong, 1972) as indicating that a significant fraction (about 16 per cent) of the cell water was not available as solvent water for intracellular ions. A more recent determination of C^1_K under the same conditons but with more rigorous control of tissue oxygenation (Armstrong et al., 1975) indicates that the ratio a^1_K/C^1_K is about 0.75 rather than unity and cells into question Lee and Armstrong's (1972) estimate of "non-solvent" cell water. Nevertheless, this revised value for a^1_K/C^1_K, which is in very good agreement with the values previously reported by other workers for a wide variety of cells (Lev and Armstrong, 1975), is still consistent with the existence, in an osmotically active state, of most of the cytoplasmic K^+ in the mucosal cells of bullfrog small intestine.

In contrast to these results, White's (1976, 1976a) measurements of a^1_K (~40mM) and C^1_K (146 mM) in epithelial cells of *Amphiuma* small intestine, following equilibration in a normal sodium chloride medium, give an a^1_K/C^1_K ratio of about 0.28. This ratio is much lower than the mean activity coefficient of the bathing medium (0.76) and was interpreted by White as indicating that a large fraction of the cytoplasmic K^+ in these cells is "bound" or compartmentalized in such a way that it is not detected by a K^+-selective microelectrode in the cytoplasm. The origin of the discrepancy between these results and those of Lee and Armstrong (1972) is not clear. It is unlikely that it could be due to an effect of intracellular Na^+ on White's estimate of a^1_K since liquid ion-exchanger electrodes of the kind he used have K^+/Na^+ selectivity ratios of 50/1 or more (Walker, 1971; White, 1976a), nor is it likely (White, 1976a) that it can be accounted for by the fact that White's media contained Cl^- as the major anion whereas those employed by Lee and Armstrong (1972) contained $-SO_4^=$. The a^1_K/C^1_K ratio found by White is, admittedly, much lower than the corresponding ratio in the majority of cells so far studied (Lev and Armstrong, 1975), but appears to be approached in at least one other epithelial system, the proximal tubular cells of rat kidney, for which Khuri (1976) reported an a^1_K/C^1_K ratio of about 0.4. Clearly, further studies are needed to determine whether there are, in fact, such wide fluctuations in the a^1_K/C^1_K ratios of epithelial cells in the small intestine of different animals as the studies of Lee and Armstrong (1972) and of White (1976, 1976a) seem to suggest.

Despite these differences, one important conclusion emerges rather unequivocally from the above studies. That is, a^1_K in epithelial cells of the small intestine far exceeds the activity required for electrochemical equilibrium. From the data of Lee and Armstrong (1972) one can estimate a K^+ equilibrium potential (E_K) of approximately 90 mV (inside negative). Under the same conditions the average mucosal membrane potential is about 40 mV (inside negative) and the serosal membrane potential (also inside negative) is some

35 mV greater. Similarly, in White's (1976a) studies, E_K was consistently greater than either the mucosal or the serosal membrane potential. Hence, there exists a clear necessity for inwardly directed pumping of K^+ across either the mucosal, the baso-lateral, or both membranes of the epithelial cells. Present evidence suggests that this function most probably resides in the baso-lateral membrane (Schultz et al., 1974). To what extent this K^+ transporting mechanism is linked to active Na^+ extrusion across this membrane is an intriguing question that awaits further study, particularly in view of the rather compelling evidence that the baso-lateral $Na+$ pump in the small intestine is, at least in part, electrogenic (Schultz et al., 1974; Armstrong, 1976).

Zeuthen and Monge (1975; 1976) recorded a^1_K in the epithelial cell layer of rabbit ileum in vivo with double-barrelled liquid ion-exchanger K^+-selective microelectrodes similar to those used by White (1976a). One barrel of the microelectrode was used to record membrane potentials. The other was used either to monitor a^1_K or for iontophoretic staining of the recording site. Animals were anethetized with sodium pentobarbitone and a loop of terminal ileum was removed from the abdominal cavity and supported on a warmed plate. An incision was made in the upper part of the serosa and a Perspex tube was pushed through this and pressed against the mucosa on the opposite side of the ileal lumen. This tube was filled with Ringer solution and the microelectrodes were advanced down its lumen in order to impale the epithelial cells.

The results obtained by these workers differ in one important respect from those already discussed. Although Zeuthen and Monge (1975) obtained an a^1_K value of about 40-50 mM in the mucosal end of the cell, they claim that a^1_K at the distal end is much higher (about 160 mM) and that there is thus a gradient of K^+ activity along the major axis of the intestinal epithelial cell. On the basis of their measurements of membrane potentials and two experiments in which a^1_{Cl} was measured with double-barrelled Cl^--selective liquid ion-exchanger microelectrodes, they further postulate an increasingly negative (with respect to the mucosal bathing solution) gradient of electrical potential along this axis together with a gradient for a^1_{Cl} which has the opposite orientation to that for a^1_K. Thus, a^1_{Cl} at the mucosal end of the cell was estimated to be about 80 mM. This decreased to about 10 mM in the neighborhood of the basement membrane.

Although the idea that there may be gradients of electrical potential and of ionic activities along the axis of net ionic transport in asymmetrical transporting cells such as those of the intestinal epithelium is intriguing, a careful scrutiny of the results of Zeuthen and Monge (1975; 1976) suggests that their conclusions should be received with caution, at least until further evidence for the reality of such intracellular ionic gradients is forthcoming. In the first place, the electrical potentials they recorded on first penetrating the mucosal membrane were exceedingly low (5-7 mV, inside negative). It is true, as Zeuthen and Monge (1975; 1976) point out, that similarly low mucosal membrane potentials have been reported for a number of intestinal preparations in vitro (Wright, 1957; Lyon and Sheerin, 1972; Barry and Eggenton, 1972), but, in no instance has satisfactory evidence been advanced to show that these low potentials were not due to excessive membrane damage by the penetrating microelectrodes. In particular, there is no published recording which shows that such a low potential was maintained at a stable value for more than a few seconds or made to undergo reversible increases or decreases in response to changes in the external conditions such as have been repeatedly demonstrated for mucosal membrane potentials of the order of 30-40 mV (White and Armstrong, 1971; Rose and Schultz, 1971; Armstrong et al., 1975). It seems entirely reasonable to suggest that mucosal membrane potentials in the range 5-10 mV are more analogous to "injury potentials" in the classical sense than to true membrane potentials.

The implications of this argument for intracellular ionic activity measurements are as follows. While it may be true, as White (1976a) suggests, that equilibrium between a K^+-selective microelectrode and cytoplasmic K^+ is rapidly established, the possibility exists, where the mucosal membrane has been excessively damaged, that a rapid loss of K^+ from the mucosal end of the cell will result in an artifactually low K^+ activity in this region within a very short time following impalement. Indeed, the reverse phenomenon, a rapid and transient increase of a^o_K in the extracellular space of frog ventricle, due to diffusive loss of K^+ from damaged cells, has been reported (Kline and Morad, 1976). Hence, part at least of the intracellular a^i_K gradient postulated by Zeuthen and Monge (1975; 1976) could arise from this cause.

Second, the absence of any time scale in the two sample recordings published by Zeuthen and Monge (1975; 1976) makes it difficult to evaluate accurately the quality of their electrical data. Apparently their microelectrode was advanced continuously through the cells at approximately $2\mu m$/sec with occasional pauses for 10-20 sec in those experiments in which iontophoretic injection of dyes was done. If one accepts their estimate of $40\mu m$ for the average length of the epithelial cells in rabbit terminal ileum, it would appear that the total duration of their impalements was just about sufficient, under favorable conditions, for the establishment of a stable membrane potential (White and Armstrong, 1971; Rose and Schultz, 1971; Lee and Armstrong, 1972; Armstrong et al., 1975; White, 1976a). Thus, it is entirely possible that the apparent dependence on depth of penetration of cell/lumen potential difference which they report in reality reflects the well known time dependence following impalement of the membrane potential in a variety of cells (See White and Armstrong, 1971 for a discussion of this phenomenon), and, unless E_m and E_K, or E_m and E_{Cl}, as measured by each barrel of the microelectrode, change synchronously, the change in E_m could affect the final calculated value of a^i_K or a^i_{Cl}.

Zeuthen and Monge (1975; 1976) and White (1976a) observed that, in the immediate vicinity of the mucosal membrane, a^o_K was often significantly higher than in the bulk of the mucosal bathing solution.

Zeuthen and Monge attributed this to the presence of a mucous layer on the lumenal surface of the intestinal epithelium. However, White's (1976a) observation that this local increase in a^o_K was greatly reduced at high lumenal perfusion rates suggests that it reflects, in part at least, the existence of an unstirred layer in the neighborhood of the villous surface of the intestine.

Na^+ ACTIVITY IN THE SMALL INTESTINE

So far, the studies of Lee and Armstrong (1972) and of Armstrong et al., (1973) with the isolated small intestine of the bullfrog appear to be the only ones which deal explicitly with a^i_{Na} in the epithelial cells of the small intestine and the implications of this parameter for the intestinal transport of Na+ and other solutes. Since the significance of these studies has been discussed in a recent review (Armstrong, 1976), a brief summary only of their more salient implications will be given here. The experimental conditions and technqiues used in these studies have already been discussed in sections II and III. In these experiments an average value of 14.4 mM was found for a^i_{Na}. Under identical conditions, C^i_{Na} is about 35-40 mM (Armstrong et al., 1970; 1975) so that a^i_{Na}/C^i_{Na} is about 0.4. This suggests that at least half the Na^+ content of the epithelial cells is in a form which is not detectable by a microelectrode inserted into the cytoplasm. As has already been pointed out, the average electrical potential differences across the mucosal and baso-lateral cell membranes under the conditions of Lee and Armstrong's (1972) measurements are about 40 and 45 mV respectively, and both these PDs. are oriented so that the cell interior is negative with respect to the external medium. Thus the direct measurement of a^i_{Na} demonstrates unequivocally that, for in vitro experiments in which the external Na^+ concentration approximates that of blood plasma, the Na^+ electrochemical potential gradient ($\Delta\bar{\mu}_{Na}$) is downhill from the external medium to the epithelial cell interior across both the apical and baso-lateral cell membranes. This confirms the conclusion, on thermodynamic grounds, that net Na^+ transport from the interior of the epithelial cell to the serosal medium is an uphill, energy requiring process (Schultz and Curran, 1968).

A knowledge of a^i_{Na} for the epithelial cells, together with the mucosal and serosal membrane potentials, enables both $\Delta\bar{\mu}_{Na}$ and $\Delta\mu_{Na}$ (the chemical potential gradient for Na^+) across either of these membranes to be accurately determined, since a^o_{Na} is easily measured. Armstrong et al (1973) utilized this fact to examine $\Delta\mu_{Na}$ and $\Delta\bar{\mu}_{Na}$ across the apical membrane of bullfrog small intestinal mucosa in relation to active sugar transport. It is well established that the active uptake of sugars, and a number of other nonelectrolytes, across the brush border membrane of intestinal epithelia is coupled to Na^+ transport, and much evidence has accumulated which suggests that a major, if not the only driving force for this accumulation is $\Delta\mu_{Na}$ or $\Delta\bar{\mu}_{Na}$ across the mucosal cell membrane (Schultz and Curran, 1970). The experiments of

Armstrong et al (1973) were addressed to an evaluation of the energetic adequacy of the transmucosal Na+ gradient for the accumulation of the actively transported sugar D-galactose. In these experiments the steady-state concentration of galactose in the epithelial cells was measured following uptake from a Ringer solution containing 2 mM of the sugar and was found to be 7.8 m.mol/l. cell water. The minimum work required to achieve this intracellular concentration ($\Delta\mu_{gal}$) was compared to $\Delta\mu_{Na}$; $\Delta\bar{\mu}_{Na}$ and $\Delta\bar{\mu}_K$ under the same conditions. The results obtained are summarized in Table 1. It is evident from this table that, if one assumes a 1/1 coupling

Table 1.

Minimum work ($\Delta\mu_{gal}$) required to attain a steady-state cell/medium

galactose concentration ratio in isolated mucosa of bullfrog small

intestine together with maximum work obtainable, under the same

conditions, from transmucosal gradients of chemical and electrochemical

potential for Na$^+$ ($\Delta\mu_{Na}$;$\Delta\bar{\mu}_{Na}$) and electrochemical potential for K$^+$ ($\Delta\bar{\mu}_K$).

$\Delta\mu_{gal}$ in joules/mole. Other parameters in joules/equiv. E_m is mucosal

membrane potential. (From Armstrong et al., 1973).

$\Delta\mu_{gal}$	(RT ln C_i/C_o)	3,400
$\Delta\mu_{Na}$	(RT ln a^i_{Na}/a^o_{Na})	2,700
$\Delta\bar{\mu}_{Na}$	($\Delta\mu_{Na}+E_mF$)	5,800
$\Delta\bar{\mu}_K$	(RT ln $a^i_K/a^o_K-E_mF$)	5,700

ratio between Na$^+$ entry and galactose uptake, as was found by Goldner et al (1969) for rabbit ileum, $\Delta\mu_{Na}$ is not alone sufficient to account for the steady-state intracellular galactose concentration. $\Delta\bar{\mu}_{Na}$ could provide sufficient energy for the observed accumulation of sugar. It has been proposed (Schultz and Curran, 1970) that the K$^+$ electrochemical gradient across the mucosal cell membrane may also be implicated in transmucosal sugar accumulation. In this case the combined electrochemical gradients for Na$^+$ and K$^+$ could easily provide the required energy (Table 1).

It should be emphasized that these results are not completely definitive since, as yet, the coupling ratio between Na$^+$ entry and galactose uptake in bullfrog small intestine has not been determined under the conditions of the experiments under review. They serve, however, to indicate how, when combined with other techniques, e.g. kinetic studies of the interaction between the transport of Na$^+$ and that of other solutes by the small intestine, the measurement of intracellular ionic activities can contribute to a fuller understanding of the transport of ions and non-electrolytes by this tissue.

Cl⁻ ACTIVITY IN INTESTINAL EPITHELIA

There is, at present, an increasing interest in intestinal Cl⁻ transport and in the role played by Cl⁻ ions in intestinal absorption and secretion (Schultz et al., 1974; Field et al., 1976). The state and distribution of Cl⁻ ions in the interior of the epithelial cells has, therefore, become a question of major concern, and a number of workers have begun to study Cl⁻ activities in these cells. Following the conclusion of Frizzell et al (1973), which was based on radiotracer experiments, that the electrochemical potential of freely exchangeable Cl⁻ in the epithelial cells of isolated rabbit ileum exceeds that in the external medium, Henriques de Jesus et al (1975) directly measured a^i_{Cl} in these cells with double-barrelled microelectrodes in which one barrel was used to record membrane potentials while the other barrel, containing Corning 477315 liquid ion-exchanger, as described by Walker (1971), was used to measure a^i_{Cl}. The results obtained by these workers are shown in Figure 2. from which it appears that, in media containing 20 mM Cl⁻, a^i_{Cl} was consistently higher than the value predicated from the corresponding Nernst equation whereas, in media which contained 145 mM Cl⁻, a^i_{Cl} tended to be lower than the predicted

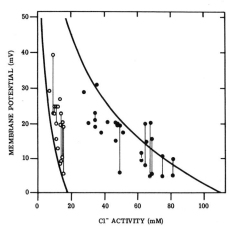

Fig. 2. Relationship between mucosal membrane potential and a^L_{Cl} in rabbit ileum immersed in physiological saline containing 20 (0) 145 (0) mM Cl⁻. Initial and final values of membrane potentials connected by vertical lines. Solid curves are theoretical values derived from the Nernst equation for media containing 20 or 145 mM Cl⁻ (From Henriques de Jesus et al., 1975. Reproduced with permission.

value. This was interpreted (Henriques de Jesus et al., 1975) as indicating that, at low external Cl⁻ concentrations, there was active absorption of Cl⁻ into the epithelial cells whereas, at high external Cl⁻ levels, there was active Cl- secretion.

As yet,little has been reported that would tend either to substantiate or refute these findings. Recently, White (1976) reported some preliminary measurements, with liquid ion-exchanger Cl⁻-selective microelectrodes, of a^i_{Cl} in epithelial cells of

Amphiuma intestine. These indicate that, in amphibian Ringer containing normal concentrations of Cl^- (about 100 mM), intracellular Cl^- activity is at or slightly below the level required for electrochemical equilibrium. However, both these experiments and those of Henriques de Jesus et al (1975) are subject to one serious criticism. It has been found (Saunders and Brown, 1975; Owen et al., 1975) that Cl^--selective microelectrodes in which Corning 477315 liquid ion-exchanger is used as the Cl^--sensing element yield erratic values for a_{Cl} in solutions containing mixtures of Cl^- and other anions. Further, when the composition of such mixtures is varied, at constant ionic strength, the calculated selectivity coefficients for these electrodes are not constant. These results have recently been confirmed and extended in the author's laboratory (Wojtkowski et al., 1976), and the list of anions which can interfere with the determination of a_{Cl} now includes such physiologically important anions as $-HCO_3^-$, $-H_2PO_4^-$, $-HPO_4^=$, propionate$^-$, and isethionate$^-$, as well as Br^- and I^-. In addition, the results obtained by Henriques de Jesus et al (1975) are, in many instances, complicated by low and unstable membrane potential measurements (Fig. 2). These criticisms, together with those already discussed in connection with their measurements of a^i_K, apply with equal force to the two experiments reported (in very general terms only) by Zeuthen and Monge (1975; 1976) on the measurement of a^i_{Cl} in rabbit ileum in vivo.

Recent studies in the author's laboratory have been concerned with the determination of a^i_{Cl} in epithelial cells of isolated bullfrog small intestine. Because of the above-mentioned shortcomings of liquid ion-exchanger Cl^--selective microelectrodes, and because an alternative type of microelectrode, the Ag/AgCl microelectrode of Neild and Thomas (1972) has been shown by these authors (Neild and Thomas, 1974) to give erroneous values for a^i_{Cl} in snail (*Helix aspera*) neurones, these studies were addressed initially to the design, fabrication, and evaluation of Cl^--selective microelectrodes which would yield reliable estimates of a^i_{Cl} in epithelial cells of the small intestine. The solid-state Cl^--selective microelectrode described by Wojtkowski et al (1976) and shown schematically in Fig. 3A appears to be eminently suitable for this purpose. Briefly, these electrodes are made by depositing a layer (~0.3μm thick) of spectroscopic grade silver on the outside of very fine tapered glass capillaries the tips of which had previously been sealed. All but the terminal 2-5μm of the silver coated tip is then isolated from the ambient medium by sealing the silver coated probe inside a slightly wider glass capillary as described by Lee and Armstrong (1974) for NAS 27-4 glass membrane microelectrodes (see also section II and Fig. 1B). These microelectrodes have slopes in the range 50-60 mV per decade change in a_{Cl}. The electrode response is rapid (10-20 sec), is unaffected by $-HCO_3^-$, $-H_2PO_4^-$ or $-HPO_4^=$, and remains essentially unchanged for at least 24 hr. Fig. 3B shows an oscilloscope recording of an impalement, with one of these microelectrodes, of an epithelial cell of isolated bullfrog small intes-

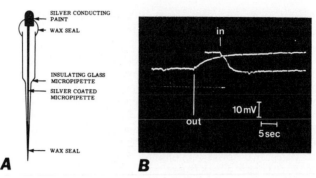

Fig. 3.   A.   The solid-state Cl⁻-selective microelectrode of Wojtkowski et al. (1976).   B.   Impalement, with this microelectrode, of an epithelial cell of bullfrog small intestine.

tine maintained in an oxygenated Ringer solution containing 5 mM K⁺ and 105 mM Cl⁻ (Quay and Armstrong, 1969).   In 44 successful impalements with 9 animals, a^i_{Cl} was found to be 71.0±4.7 (S.E.) mM.   This greatly exceeds the equilibrium value (31.5 mM) calculated from the Nernst Equation using the average mucosal membrane potential (24 mV) found in these experiments.   Since, under these conditions, the serosal membrane potential is usually not more than 1-3 mV greater than the mucosal potential, it is apparent that a^i_{Cl} exceeds the level required for electrochemical equilibrium across both the apical and baso-lateral cell membranes.

   Thus, in the frog intestine, as in rabbit ileum (Frizzell et al., 1973) the epithelial cells appear to have the ability to accumulate Cl⁻ against an electrochemical potential gradient.   In an earlier study from the author's laboratory (Quay and Armstrong, 1969) an energy requiring Cl⁻ pump across the brush border membrane from lumen to cell, together with a downhill movement of Cl⁻ across the baso-lateral cell membrane, was proposed as one possible mechanism for active transepithelial Cl⁻ transport in bullfrog small intestine.   The direct measurements of a^i_{Cl} summarized herein appear to provide strong support for this hypothesis.   There is not much evidence for active Cl⁻ absorption by the distal small intestine of several mammalian species, including man.   In rabbit ileum it appears that the primary active step in Cl⁻ absorption may be an energy requiring, uphill transport of Cl⁻ from the intestinal lumen to the interior of the epithelial absorptive cells (Frizzell et al., 1973).   Thus the results obtained in the present author's laboratory for bullfrog small intestine may reflect a component of intestinal Cl⁻ transport which is of rather general occurrence in the animal kingdom.

CONCLUSION

   Although the number of studies so far reported is small, there can be little doubt that the measurement, with ion-selective microelectrodes, of intracellular ionic activities will play an increasingly important role in future investigations of epithelial function. The capacity of this technique to provide unequivocal estimates of the electrochemical potential gradients for individual ions across both the apical and baso-lateral membranes of epithelial cells should contribute significantly to a more exact definition of the driving forces responsible for transepithelial transport of water, ions, and other solutes, and of the effects, on these forces, of agents such as drugs and hormones which modify absorption or secretion by epithelial layers. The development of new and improved types of microelectrodes coupled with further refinements in methods for positioning them in specific regions of the cell should enable the question of the occurrence of gradients of ionic activity inside the cell (Zeuthen and Monge, 1975; 1976) to be rigorously studied.

ACKNOWLEDGEMENTS

The support received from USPHS grants AM 12715 and HL 06308 in studies described herein and in the preparation of this review is gratefully acknowledged. The author wishes to thank his students, former students, and collaborators who contributed so much to the successful development, in his laboratory, of techniques for measuring intracellular activities. It is a particular pleasure to acknowledge the patient and painstaking work done by Drs. C. O. Lee and Wita Wojtkowski; also the enthusiastic and expert help and guidance of Mr. Charles Simmons of the Materials Laboratory, Delco Electronics, Kokomo, Indiana, in the field of solid-state technology. Finally, the author wishes to thank Dr. J. F. White who generously made available to him, before publication, his data on ionic activities in *Amphiuma* small intestine.

REFERENCES

Armstrong, W. McD. (1976) In *Intestinal Ion Transport*. (J.W.L. Robinson, ed) p. 19. MTP Press Ltd., Lancaster, England.
Armstrong, W. McD., Byrd, B.J., Cohen, E.S., Cohen, S.J., Hamang, P.M. & Myers, C.J. (1975) *Biochim. Biophys. Acta 401:* 137.
Armstrong, W. McD., Byrd, B.J. & Hamang, P.M. (1973) *Biochim. Biophys. Acta. 330:* 237.
Armstrong, W. McD. & Lee, C.O. (1971) *Science 171:* 413.
Armstrong, W. McD., Musselman, D.L. & Reitzug, H.C. (1970) *Am. J. Physiol. 219:* 1023.
Barry, R.J.C. & Eggenton, J. (1972) *J. Physiol., Lond. 227:* 201.
Caldwell, P.C. (1954) *J. Physiol., Lond. 126:* 169.
Caldwell, P.C. (1958) *J. Physiol., Lond. 142:* 22.
Csaky, T.Z. & Esposito, G. (1969) *Am. J. Physiol. 217:* 753.
Edzes, H.T. & Berendsen, H.J.C. (1975) *Ann. Rev. Biophys. and Bioeng. 4:* 265.
Eisenman, G., Rudin, D.O. & Casby, J.F. (1957) *Science 126:* 831.
Field, M., Brasitus, T.A., Sheerin, H.E. & Kimberg, D.V. (1976)

In *Intestinal Ion Transport* (J.W.L. Robinson, ed.) p. 233. MTP Press Ltd., Lancaster, England.

Frizzell, R.A., Nellans, H.N., Rose, R.C., Markscheid-Kaspi, L. Schultz, S.G. (1973) *Am. J. Physiol. 224:* 328.

Goldner, A.M., Schultz, S.G. & Curran, P.F. (1969) *J. Gen. Physiol. 53:* 362.

Henriques de Jesus, C., Ellory, J.C., & Smith, M.W. (1975) *J. Physiol., Lond. 244:* 31P.

Hinke, J.A.M. (1959) *Nature, Lond. 184:* 1257.

Hinke, J.A.M. (1961) *J. Physiol., Lond. 156:* 314.

Janancek, K., Morel, F. & Bourguet, J. (1968) *J. Physiol., Paris 60:* 51.

Kessler, M., Clark, L.C., Jr., Lubbers, D.W., Silver, I.A. & Simon, W. (1976a) In *Ion and Enzyme Electrodes in Biology and Medicine.* p. 158. (M. Kessler, L.C. Clark, Jr., D.W. Lubbers, I.A. Silver, and W. Simon, eds.) University Park Press, Baltimore, Md.

Kessler, M., Hajeck, K. & Simon, W. (1976b) In *Ion and Enzyme Electrodes in Biology and Medicine* (M. Kessler, L.C. Clark, Jr., D.W. Lubbers, I.A. Silver & W. Simon, eds.) p. 136. University Park Press, Baltimore, Md.

Khuri, R.N. (1976) In *Ion and Enzyme Electrodes in Biology and Medicine* (M. Kessler, L.C. Clark, Jr., D.W. Lubbers , I.A. Silver & W. Simon, eds.) p. 364. University Park Press, Baltimore, Md.

Khuri, R.N., Hahhar, J.J. & Aguilian, S.K. (1972) *J. Appl. Physiol. 32:* 419.

Kline, R. & Morad, M. (1976) *Biophys. J. 16:* 367.

Lee, C.O. & Armstrong, W. McD. (1972) *Science 175:* 1261.

Lee, C.O. & Armstrong, W. McD. (1974) *J. Membrane Biol. 15:* 331.

Lee, C.O. & Fozzard, H.A. (1974) *J. Gen.Physiol. 65:* 695.

Lev, A.A. (1964) *Nature, Lond. 201:* 1132.

Lev, A.A. & Armstrong, W. McD. (1975) In *Current Topics in Membranes and Transport* (A. Kleinzeller & F. Bronner, eds.), 6: 59. Academic Press, New York, N.Y.

Ling, G.N. & Gerard, R.W. (1949) *J. Cell. and Comp. Physiol. 34:* 383.

Lyon, I. & Sheerin, H.E. (1971) *Biochim. Biophys. Acta. 249:* 1.

Neild, T.O. & Thomas, R.C. (1972) *J. Physiol., Lond. 231:* 7P.

Neild, T.O. & Thomas, R.C. (1974) *J. Physiol., Lond. 242:* 453.

Nicolsky, B.P. (1937) *Acta Physicochem. USSR, - 7:* 597.

Owen, J.D., Brown, H.M. & Saunders, J.H. (1975) *Biophys. Soc. Abstr. p. 45a.*

Pedersen, O.H. (1976) *Physiol. Revs. 56:* 535.

Quay, J.F. & Armstrong, W. McD. (1969) *Am. J. Physiol. 217:* 694.

Rose, R.C. & Schultz, S.G. (1971) *J. Gen. Physiol. 57:* 639.

Saunders, J.H. & Brown, H.M. (1975) *Biophys. Soc. Abstr. p. 323a.*

Schultz, S.F. & Curran, P.F. (1968) In *Handbook of Physiology Section 6, Vol. III* (C.F. Code and W. Heidel, eds.), p. 1245 American Physiological Society, Washington, D.C.

Schultz, S.F. & Curran, P.F. (1972) *Physiol. Revs. 80:* 637.

Schultz, S.G., Frizzell, R.A. & Nellans, H.N. (1974) *Ann. Rev.*

Physiol. 36: 51.

Thomas, R.C. (1970) *J. Physiol., Lond. 210:* 82P.

Walker, J.L., Jr. (1971) *Anal. Chem. 43:* 89A.

Walker, J.L., Jr. & Ladle, R.O. (1973) *Am. J. Physiol. 225:* 263.

White, J.F. (1976) *Federation Proc. 35:* 464.

White, J.F. (1976a) *Am. J. Physiol.* (in press)

White, J.F. & Armstrong, W. McD. (1971) *Am. J. Physiol. 221:* 194.

Wojtkowski, W., Frey, K.F., Myers, C.J. & Armstrong, W. McD. (1976) *Physiologist 19:* 415.

Wright, E.M. (1966) *J. Physiol., Lond. 185:* 486.

Zeuthen, T. & Monge, C. (1975) *Phil. Trans. Roy. Soc. Lond. B. 71:* 277.

Zeuthen, T. & Monge, C. (1976) In *Ion and Enzyme Electrodes in Biology and Medicine* (M. Kessler, L.C. Clark, Jr., D.W. Lübbers, I.A. Silver & W. Simon, eds.), P. 345. University Park Press, Baltimore, Mc

ANATOMICAL METHODS FOR STUDYING TRANSPORT
ACROSS EPITHELIA

John McD. Tormey

UCLA *School of Medicine*

INTRODUCTION

The anatomy and physiology of epithelia are inextricably related.
Nevertheless, only within the last decade have physiologists stopped treating transport across epithelia as across black-box membranes, and focused on the relationship between subcellular structure
and function. In keeping with this trend, many of the presentations
at this Symposium revolve around such questions as the localization
of transport mechanisms, the routes by which substances cross epithelia, the nature of tight junctions, and the functional significance
of intracellular space geometry.

Anatomical methods, therefore, must now be numbered among the
many modern biophysical tools required to analyze transport across
cells. This paper elaborates on this theme by discussing three
anatomical methods which have had or soon will have wide application
to localizing and understanding transport functions. There are (I)
traditional electron microscopy, (II) autoradiographic localization
of transport enzymes, and (III) electron microprobe X-ray analysis
of diffusable ions in tissues. Rather than attempt a comprehensive
review, the discussion will be limited primarily to work from my
laboratory. The purpose if to impart some feeling for the strengths
and limitations of the methods and to indicate directions for the
future.

I. TRADITIONAL ELECTRON MICROSCOPY

Traditional electron microscopy involves fixation of cells by
aqueous solutions of chemical agents, followed by embedding in
plastic and by examination of thin sections in a conventional transmission electron microscope. This method has made substantial con-

tributions simply by describing the ultrastructure of transporting cells and thereby providing physiologists with the ground-rules to be followed in speculating on transport mechamisms.

Beyond this, traditional electron microscopy can supply more direct information about the functional significance of the structures observed. The strengths and limitations of this approach are well illustrated by considering its application to the question, "What routes do various substances follow as they cross epithelia?"

A partial basis for answering this question was provided by experiments on gallbladders carried out ten years ago (Diamond and Tormey, 1966b; Kaye, Wheeler, Whitlock and Lane, 1966; Tormey and Diamond, 1967). The lateral intercellular spaces of the epithelium dilated hugely in the presence of active fluid transport (Figure 1), but the spaces were collapsed in various other physiological states

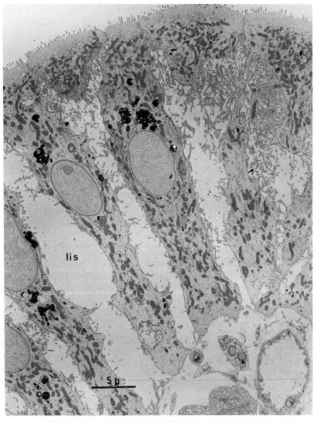

Fig. 1. Electron micrograph of rabbit gallbladder epithelium fixed during maximal fluid transport. Lateral intercellular spaces (lis) are markedly dilated in this condition, but are collapsed when fluid transport is minimal. Modified after Tormey and Diamond (1967)

associated with zero fluid transport. This was widely accepted as evidence that these spaces are the principal route of water and salt transport, and furthermore that they are the principal site for the osmotic coupling between active salt and passive water fluxes. Subsequently, similar observations were made in other tissues and with passive water flows [e.g. in isolated renal collecting ducts (Grantham, Ganote, Burg and Orloff, 1969)].

In spite of this success, many difficulties stand in the way of interpreting changes in cell geometry. Firstly, not all tissues involved in fluid transport show structural changes. For instance, in my work on the ciliary epithelium of the eye (Tormey, 1963), and in my unpublished work with Jared Grantham on isolated perfused renal proximal tubules, no correlation between fluid transport and tissue geometry was observed. Secondly, although water flow might cause the intercellular spaces to dilate, there is no good basis for determining how much water is going through them, i.e., whether it is 100%, or only 5% with the remainder crossing the epithelium by some other route. Thirdly, great care must be exercised to prevent tissue preparation artifacts. For instance, fixation with glutaraldehyde can set up transepithelial water flows that can drastically alter tissue geometry (Grantham, Cuppage and Fanestil, 1971; Smulders, Tormey and Wright, 1972). Some authors (Fredriksen and Rostgaard, 1974) have gone so far as to suggest that dilated intercellular spaces are the inevitable result of standard preparative techniques; however, my own unpublished experiments with both freeze-sectioned and intact tissue have confirmed the original observations with osmium-fixed gallbladders (Diamond and Tormey, 1966b; Kaye, Wheeler, Whitlock and Lane, 1966; Tormey and Lane, 1966; Tormey and Diamond, 1967). Lastly, a number of other factors can cause the spaces to dilate. For instance, externally applied hydrostatic pressures have been observed to change cell geometry (Tormey and Diamond, 1967; Grantham, Cuppage and Fanestil, 1971); therefore, pressures built up in the tissue outside of the epithelium could conceivably force intercellular spaces to dilate. Also, substances which stimulate water flow in toad urinary bladders can affect the tone of the underlying smooth muscle, and thereby cause intercellular spaces to dilate (DiBona and Civan, 1970).

When attempting to interpret the functional significance of changes in tissue geometry, we are therefore in a difficult position. The phenomena studied and the means by which they are studied are both so complex that interpretations have become insecure.

In light of this, the observations I made with A. Smulders and E. Wright (Smulders, Tormey and Wright, 1972; Wright, Smulders and Tormey, 1972) several years ago are particularly gratifying. We studied the effects of lateral intercellular space geometry on the passive permeation of gallbladders by a variety of substances, electrolytes, nonelectrolytes and water. Our approach places earlier interpretations on a more secure ground by circumventing many of the above difficulties.

Osmotically generated water flows were used to cause the lateral intercellular spaces of the bladder epithelium to open or close over a wide range. The question was, "Does permeation take place through these channels?" If so, changing their dimensions could have a significant effect on permeability.

The effects of water flow on structure are illustrated in Figs. 2 and 3. When 300 mM sucrose is added to the serosal side of the gallbladder, the resulting mucosa to serosa water flux causes large spaces to open up between the cells, averaging approximately 1.5 μm

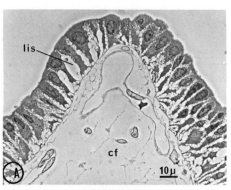
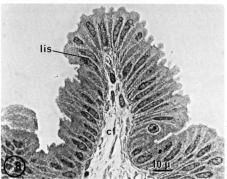
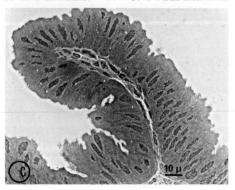

Fig. 2. High power light micrographs of rabbit gallbladders. Three osmotic water flow conditions are represented: (A) mucosa to serosa flow with the serosal bathing solution hypertonic with 300 mM sucrose; (B) zero water flow with both bathing solutions hypertonic by 150 mM; (C) serosa to mucosa flow with the mucosal solution hypertonic by 300 mM. The epithelium has widely dilated lateral intercellular spaces (lis) in (A); narrow, barely discernible spaces in (B); no detectable spaces in (C). Osmium-fixed, 0.5 μm plastic sections. Modified after Smulders, Tormey and Wright (1972).

Fig. 3. High magnification electron micrographs showing details of the lateral intercellular spaces. (A) depicts the same zero osmotic flow conditions shown in Fig. 2B. The channel is approximately 0.7 μm wide, but is effectively narrowed by cytoplasmic folds. (B) depicts an area of maximal channel closure produced by the same serosa to mucosa osmotic flow conditions as in Fig. 2C. The cytoplasmic folds are so highly interdigitated that the channel appears to be patent in only a few places. Modified after Smulders, Tormey and Wright (1972).

in width (Fig. 2A). In the absence of an osmotic gradient, the channels collapse to the point where they are barely resolved by light microscopy (Fig. 2B). Electron microscopy (Fig. 3A) revealed that these "collapsed" spaces were still approximately 0.3 μm wide, on the average. The most significant effect occurred when 300 μm sucrose is added to the mucosal bathing solution, thereby causing water to flow from serosa to mucosa. No intercellular spaces are resolved by light microscopy (Fig. 2C), and even with electron microscopy (Fig. 3B) they are so exceedingly narrow that they could be seen only with great difficulty. In fact the intercellular spaces were almost completely obliterated in many places, and on the average they were only about the width of a unit plasma membrane, i.e., about 0.005 - 0.010 μm wide.

Although the mechanism of these hugh changes was probably the development of positive or negative hydrostatic pressures within the spaces as a result of the direction of water flow, the morphological results might have other explanations. The important point is that the subsequent argument and conclusions were entirely independent of uncertainty in regard to the mechanism of anatomical change.

Changes in physiological parameters accompanied the changes in structure. Addition of sucrose to the serosal solution had no discernible effect on electrical conductance across the bladder. However, adding sucrose to the mucosal solution reduced the conductivity until at a 300 mM gradient, it was reduced to only about 1/3 of its original value (Fig. 4). Similar parallel effects of osmotic gradient were observed on hydraulic conductivity and on the permeability of such nonelectrolytes as sucrose and 1, 4-butanediol.

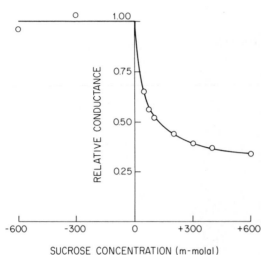

Fig. 4. Conductance of the gallbladder as a function of the osmotic gradient between mucosal and serosal solutions. The conductance on the ordinate, is the conductance relative to the average value obtained in the absence of gradients. The sucrose concentration, on the abscissa, is positive when sucrose is added to the mucosal solution and negative when sucrose is added to the serosal solution. Gradients which produce serosa to mucosa water flow cause a fall in conductance. Modified after Smulders, Tormey and Wright (1972).

Did altered cell geometry cause these physiological effects? An alternate interpretation is that they were due to water flow per se, i.e., due to solvent drag or unstirred layer effects. Evidence for the rejecting of such alternative explanations is the facts that (a) while water flow from serosa to mucosa had an effect, water flow in the opposite direction had no effect, and that (b) for any given flow conditions, the permeability was the same whether measured by mucosa to serosa or serosa to mucosa tracer fluxes.

Changes in the lateral space dimensions, on the other hand, could fully explain the results. The aggregate conductance of the channels was calculated from their length and aggregate cross sectional area. The partially closed channels found in the absence of any water flow accounted for no more than 3% of the total electrical resistance of the bladder. When water flow widened the channels further, the effect on overall conductance was therefore negligible. On the other hand, closing down the channels when the mucosal side was hypertonic accounted for most of the observed change in conductance. Similar calculations likewise predicted the effects of changing channel dimensions on the other physiological parameters.

The argument became more tenuous when attempting to decide by what route permeating substances reached the lateral spaces. As indicated in Figure 5, a substance could reach the space by directly crossing the so-called tight junction between the cells and thence

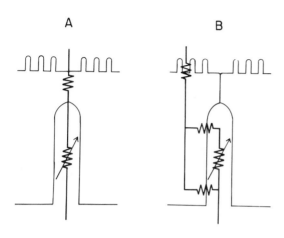

A   B

Fig. 5. Equivalent circuit representation of transepithelial permeation pathways. In (A) permeation is through the tight junction and along the lateral intercellular space. In (B) it is across the apical cell membrane, through the cytoplasm, and then through the lateral cell membrane in a distributed fashion. The quantitative effects of space closure (increasing resistance) depend on which route is followed.

traversing the entire length of the channel, or it could successively cross the apical cell membrane and cytoplasm, and then permeate the channel in a distributed fashion across the entire length of the lateral cell membrane. Consideration of the electrical analogues of these routes showed that the same degree of space closure would have a quantitatively different effect depending on which was followed. The best fit was obtained by presuming that electrical current and sucrose fluxes went via the tight junctions for the entire length of the channel, whereas fluxes of the more rapidly permeating, lipophilic substances, such as 1, 4-butanediol, went first through the cell. Additional considerations led to the conclusion that water also reached the lateral channels primarily by way of the cells (Wright, Smulders and Tormey, 1972).

In reaching the conclusion that the lateral intercellular space is the major, common pathway shared by transepithelial fluxes of water and solutes, many interpretive difficulties common to most previous electron micriscopic studies were avoided.

Nevertheless, the importance of conventional electron microscopic studies for future investigations is questionable. Even under the favorable circumstances just described, interpretation remains complex and less direct than desirable.

One area for future progress is more precise quantitation of geometric changes by morphometric analysis. Such data should prove helpful in validating physiological models. For instance, H. Blom and H.F. Helander (Personal communication) used morphometric analysis to conclude that the channel dimensions of gallbladders are indeed compatible with the standing gradient osmotic flow model (Diamond and Tormey, 1966a; Diamond and Bossert, 1967). This model, which has been widely accepted as an explanation for the coupling between solute and water transport, had been dismissed (Hill, 1975) on the grounds that the channels are of the wrong dimensions to support appropriate standing gradients.

II. AUTORADIOGRAPHIC LOCALIZATION OF ENZYMES

Whatever the future role of conventional electron microscopy, the main hope for progress appears to lie with other, more sophisticated approaches to localization of transport functions. One of the more successful of these has been using autoradiography of ouabain binding sites to localize Na-K ATPase. The validity of the method depends on the specificity of ouabain binding to the transport enzyme.

Autoradiography of binding sites was first employed by Stirling (1972) to demonstrate Na-K ATPase on the lateral membranes of the intestinal mucosal epithelium. This confirmed earlier predictions of where Na pumps should be localized in "forwards-transporting", i.e. absorptive, epithelia. As suggested in Fig. 6A, pumping Na into the intercellular spaces also provides a geometric basis for the coupling of active salt and passive water transport.

A   B   C

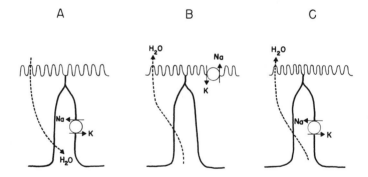

Fig. 6. Light microscopic autoradiogram of tritiated ouabain bound to choroid plexus epithelium. The region shown is a portion of a fold, in which the epithelium separates a large vessel packed with red cells (rbc) from cerebrospinal fluid (csf). The silver grains indicate ouabain localized exclusively to the apical membrane region of the epithelial cells. Modified after Quinton, Wright and Tormey, (1973).

Subsequently, Quinton, Wright and Tormey (1973) used this approach to localize Na-K ATPase in the choroid plexus. Our study included a separate biochemical demonstration on the specificity of ouabain binding to this tissue. The choroid plexus differs from intestine, gallbladder, renal proximal tubule, and other epithelia in that it secretes fluid from blood to lumen rather than absorbing it in the direction lumen to blood. Thus, choroid plexus is an example of a "backwards-transporting" epithelium. The Na-K ATPase was localized exclusively at the apical (luminal) membranes of the epithelium (Fig. 7). This orientation is appropriate to the direction of solute and fluid transport (cf. Fig. 6B). However,

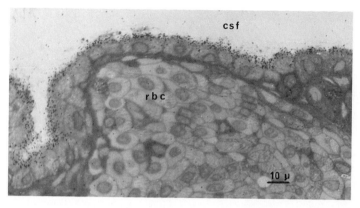

Fig. 7. Light microscopic autoradiograms of tritiated ouabain bound to human sweat glands. (A) is a portion of the reabsorptive duct. Transport proceeds from lumen (L) to blood (B). Almost all grains are localized close to the basal and lateral surfaces of the cells of both layers of epithelium. (B) is a portion of the secretory tubule. Transport proceeds in the opposite direction, from blood to lumen. However, almost all grains are localized to the basal and lateral surfaces of the dark and light cells. Few grains are associated with the lumen in either case. Modified after Quinton and Tormey (1977).

since the apical folds are very short, there is apparent difficulty in explaining the coupling of fluid transport in terms of the standing-gradient osmotic flow model. In other words, the anatomical compartment into which solute is actively transported is very small compared with the presumed spatial requirements (Diamond and Bossert, 1967) for osmotic equilibration. Perhaps, as suggested by E. Wright elsewhere in this Symposium, the unstirred layer adjacent to the apical surface of this epithelium is the "compartment" in which osmotic equilibration occurs.

Additonal considerations are raised by a study on isolated perfused sweat glands, which Quinton and Tormey (1977) have recently completed. The epithelium of the coiled secretory portion of the gland is another example of a backwards transporting epithelium. Sweat glands form a primary secretate into the lumen of the gland, moving salt and water from the basal-lateral to the apical side of the epithelium. On the other hand, the epithelium of the reabsorptive duct is a forwards transporting epithelium, moving a hypertonic

solution of selected solutes from the lumen back into the blood.
Comparison of ouabain binding sites in the two types of epithelium
(Fig. 8A, B) shows that the localization and orientation of Na-K
ATPase is essentially the same in each, irrespective of the direct-
ion of transport. Although localization of the enzyme on the latera

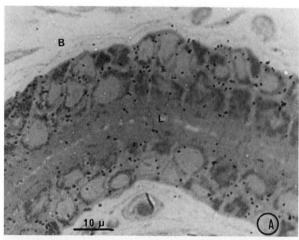

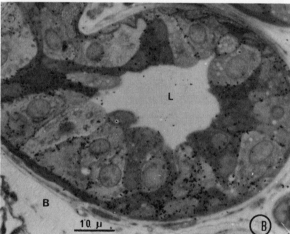

Fig. 8. Diagrams to
illustrate the loca-
tion of Na-K pumps
in three different
classes of epithelia.
(A) is typical of
"forward transporting"
epithelia, such as
intestine, gallblad-
der, and reabsorptive
duct of sweat gland.
Na is pumped out of
the cell into the
lateral intercellular
space. It can read-
ily be seen that this
could create a local
osmotic gradient
within the space
that could drive
fluid transport from
the apical to the
basal side. (B) is
the case of the cho-
roid plexus, a "back-
wards transporting"
epithelium. Na
transport out of the
cell is again in the
direction of fluid
transport. Unlike
case (A), the dimen-
sions of the apical
microvilli are pro-
bably too small to
define an anatomical compartment within which osmotic equilibration
occurs. (C) illustrates the dilemma of another "backwards transporting" epithelium, namely, the secretory tubule of the sweat gland.
In this case the pump is transporting Na out of the cell in the
opposite direction from secretion, and it appears that some other
basis must be found to explain fluid transport.

membranes can be explained by current hypotheses for backwards
transporting epithelia (Diamond and Bossert, 1968), the orientation

cannot. There is good reason to believe that the enzyme is here oriented in the direction which pumps Na out of the cell (Quinton and Tormey, 1977). This leaves the problem (illustrated in Fig 6C) of explaining how pumping Na out of the cells into the lateral intercellular space can explain a net transport of both salt and water in the opposite direction. It may turn out that Na-K ATPase does not usually play a direct role in the secretion of electrolyte solutions, and that there is some other primary solute pump which remains to be identified and localized.

Thus an anatomical approach is forcing the reevaluation of current dogma regarding the nature and localization of the solute pumps involved in transepithelial fluid movement.

III. ELECTRON MICROPROBE X-RAY ANALYSIS

The solution of transport problems requires a precise knowledge of the distribution of electrolytes at the cellular and subcellular levels. The so-called electron microprobe X-ray analyzer is tailor-made for this purpose. On the one hand, it can form an image by scanning a fine beam of electrons over a specimen. On the other hand, it can perform a qualitative and quantitative elemental analysis of any given point, by stopping the beam at that point and measuring the X-rays produced by the electron bombardment. Sensitivities in the 10^{-18} gram range have been predicted.

Because successful application of microprobe analysis to sections of biological tissue has been handicapped by difficulties with specimen preparation, this method is just beginning to be fully exploited (see Hall, Echlin and Kaufman, 1974). The problem has been to obtain and analyze sections in such a way that ionic translocation does not occur. Because pitfalls abound, it is prudent to avoid the study of complex, unknown systems, until methods can be validated on simple systems whose ionic composition can be reliably determined by independent methods. Erythrocytes, because of their homogeneous anatomical structure and ease of direct chemical analysis, are ideal for this purpose.

Accordingly, we have been using erythrocytes to validate a specimen preparation technique for microprobe analysis (Tormey and Tormey, 1976). Unlike others who have reported microprobe analysis of whole-mounts of red cells, for example, Kirk, Crenshaw and Tosteson, (1974), our method employs thin sections. Suspensions of erythrocytes are incubated in a Ringer's solution which contains dextran. This layers of this suspension are formed on Millipore filters, which are then rapidly frozen in liquid propane. Fresh frozen sections of these filters with adherent red cells are made in a Christensen-Sorvall Frozen Thin Sectioner at -85°C. The 1/2 μm thick frozen sections are manipulated onto collodion films spanning the ends of carbon tubes. The sections are then sandwiched with a second collodion film, following the method of Dorge, Gehring, Nagel and Thurau (1974). The sandwiched sections are then freeze-dried at -75° for two days. Following a careful warm-up to room temperature, exposure to atmospheric moisture is scrupulously avoided.

The arrangement of the specimens in the microprobe is diagrammed in Figure 9. The section, whose thickness is grossly exaggerated,

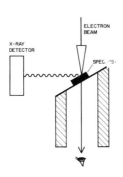

Fig. 9. Diagram to illustrate the arrangement of specimens in the microprobe. See text for description.

rests on a thin film over the end of a carbon tube. Most of the impinging electron beam passes through the section, permitting its visualization by scanning transmission electron microscopy. The X-rays are picked up by a nearby energy-dispersive detector, which subtends the relatively large solid angle of 0.25 steradian. After visualizing an area containing cells, the electron beam is repetitively scanned over a selected region in a short line pattern (Figure 10). A beam with a diameter of 0.1 μm is used to obtain enough current (0.5 nA for reasonable X-ray counting statistics. Intracellular ion concentrations are regarded as unknown and are determined by comparing X-rays produced by scans within the cell, with X-rays obtained from scans over nearby extracellular fluid (ECF). The ECF thus constitutes an internal standard of known

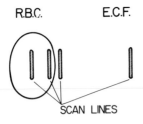

Fig. 10. Diagram to illustrate manner of analysis of sections. After visualizing a region containing red blood cells and extracellular fluid, the electron beam is adjusted to scan small regions in a short line pattern and the X-rays generated from each region are analyzed. Quantitative analysis is usually carried out by comparing a spectrum from the middle of a cell with one from ECF several microns away. Spatial resolution is checked by placing two scans close together on either side of the cell membrane.

244

composition. In sections this thin, corrections for atomic number, absorption and fluorescence effects are unnecessary. Spatial resolution is determined by comparing scans immediately adjacent the cell membranes with those from farther away.

Typical X-ray spectra obtained from a rabbit erythrocyte preparation are shown in Fig. 11. The upper trace is from a cell,

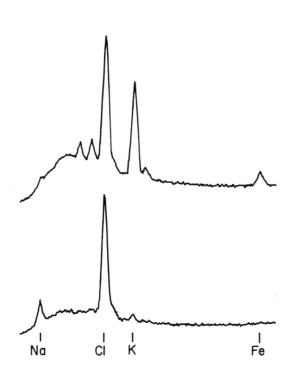

Fig. 11. Spectra obtained from a rabbit erythrocyte (upper trace) and from adjacent region of extracellular fluid (lower trace). X-ray energy on the abscissa is plotted against intensity on the ordinate. The characteristic locations of peaks from Na, Cℓ, K and Fe are indicated. Since both spectra were obtained under close to identical conditions, including total analysis time, the relative heights of the peaks for each element are indicative of relative concentration. See text for further details. Redrawn from X-Y output of multichannel analyzer.

whereas the lower is from ECF about 1 μm away. Each spectrum consists of a number of peaks superimposed upon a continuous background. The position of each peak on the horizontal energy scale is a characteristic of each element, while the height above background is proportional to the amount present. Quantitation is achieved by comparing peak heights between cells and ECF. The K ratio observed is 20:1, and, given an extracellular K of 6 mM, the intracellular K is 120 mM. On the other hand, intracellular Na is barely detectable and very low. Also there is a large intracellular Fe signal from hemoglobin, as well as substantial (unlabelled) intracellular phosphorous and sulfur signals. Cl, however, as always, proves a problem; here the ratio is 1:1, instead of a predicted 0.75:1. Results which Cl have been consistently inconsistent compared with those with Na and K, and this appears to result from differential volatilization under the electron beam. By contrast, results with K in such high K cells have often been extremely good. For instance, in one series, the probe measured 115 mM SE ± 2.3, as compared with a chemical analysis of 123 mM.

Spectra from a low K sheep erythrocyte are shown in Fig. 12.

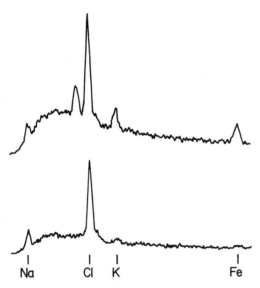

Fig. 12. Similar to Fig. 11, except obtained from low-K sheep erythrocyte preparation. The spectra resemble those from the rabbit, except that they reflect the relatively low intracellular K concentration and high intracellular Na concentration characteristic of these sheep cells.

The K ratio is reduced to nearly 4:1, and the Na ratio is increased to 0.8:1. In a series of experiments, the probe measured an intracellular K of 19.1 ± 1.5 compared with a chemical value of 20.2 mM. The probe simultaneously measured intracellular Na of 112 mM ± 5, compared with a chemical estimate of 100 mM. Most of the variance in these results was due to X-ray counting statistics. Mixtures of high K (human) and low K (sheep) erythrocytes have also been successfully analyzed.

Two other parameters are important. A) Spatial resolutions of about 0.3 μm have been measured, i.e., the plasma membrane can be bridged with a pair of line scans separated by this distance and the step between intracellular and extracellular electrolyte concentrations seen. In fact the ratio is usually nearly right on. This is especially good considering a probing electron beam of 0.1 μm diameter. B) Considering the ease with which 5 mM K is detected and the volume of the section being excited by the electron beam, absolute sensitivity of K in the order of 10^{-18} g is demonstrated. Sensitivity for Na is about an order of magnitude lower.

Though impressive, these data should be regarded as preliminary. The results have not always been as good as presented here. Not only has Cl been very erratic over a two-fold range, but K has frequently been underestimated by about 30%. These inconsistencies probably arise from ion losses occurring while the specimen is being irradiated by the electron beam rather than during specimen preparation. In order to permit more reliable quantitative work, we are now searching for ways to overcome this. The approach most likely to succeed appears to be keeping the specimen at a cryogenic temperature while under the electron beam.

Although we are on the verge of having an exquisitely sensitive quantitative technique for localizing electrolytes with a spatial resolution approaching the best available by electron microscopic autoradiography, the millenium has not quite yet arrived.

SUMMARY

In summary, this paper has considered three anatomical methods which are important biophysical tools for analyzing fluid transport. As is true of all methods, each has its own limitations and is best used in conjunction with other, correlative techniques. Microprobe analysis, for instance, has enormous promise for the future, but will become particularly powerful when used in conjunction with the ion selective microelectrodes discussed by W. McD. Armstrong elsewhere in this Symposium. Such technical progress should prove to be the wellspring of new kinds of observations and new, more refined hypotheses regarding fluid transport.

ACKNOWLEDGMENTS:

Supported in part by the American Heart Association Greater Los Angeles Affiliate, Grant #510; USPHS, Grant #AM-12621; and a Research Scholar Award from the National Cystic Fibrosis Foundation to P.M. Quinton.

REFERENCES

Diamond, J.M., & Bossert, W.H. (1967) *J. Gen. Physiol. 50:* 2061.
Diamond, J.M., & Bossert, W.H. (1968) *J. Cell Biol. 37:* 694.
Diamond, J.M., & Tormey, J. McD. (1966a) *Fed. Proc. 25:* 1458.
Diamond, J.M., & Tormey, J. McD. (1966b) *Nature 210:* 817.
DiBona, D.R., & Civan, M.M. (1970) *J. Cell Biol. 46:* 235.
Dorge, A., Gehring, K., Nagel, W., & Thurau, K. (1974) In *Microprobe Analysis as Applied to Cells and Tissues* (Hall, T., Echlin, P. & Kaufman, R. Editors) pp. 337. New York: Academic Press.
Frederiksen, O., & Rostgaard, J. (1974) *J. Cell Biol. 61:* 830.
Grantham, J., Cuppage, F.E., & Fanestil, D. (1971) *J. Cell Biol. 48:* 695.
Grantham, J., Ganote, C.E., Burg, M.B., & Orloff, J. (1969) *J. Cell Biol. 41:* 562.
Hall, T., Echlin, P., & Kaufman, R., eds. (1974) *Microprobe Analysis as Applied to Cells and Tissues* 435 pp., New York: Academic Press.

Hill, A.E. (1975) *Proc. R. Soc. Lond., B., 190:* 99.

Kaye, G.I., Wheeler, H.O., Whitlock, R.T., & Lane, N. (1966) *J. Cell Biol. 30:* 237.

Kirk, R.G., Crenshaw, M.A., & Tosteson, D.C. (1974) *J. Cell Physiol. 84:* 29.

Quinton, P.M., & Tormey, J. McD. (1977) *J. Membrane Biol.* In press.

Quinton, P.M., Wright, E.M., & Tormey, J. McD. (1973) *J. Cell Biol. 58:* 724.

Smulders, A.P., Tormey, J. McD., & Wright, E.M. (1972) *J. Membrane Biol. 7:* 164.

Stirling, C.E. (1972) *J. Cell Biol. 53:* 704.

Tormey, J. McD. (1963) *J. Cell Biol. 17:* 641.

Tormey, J. McD., & Diamond, J.M. (1967) *J. Gen. Physiol. 50:* 2031.

Tormey, J. McD., & Tormey, A.A. (1976) *Fed. Proc. 35:* 781.

Wright, E.M., Smulders, A.P. & Tormey, J. McD. (1972) *J. Membrane Biol. 7:* 198.

A MECHANISM FOR COUPLING ISOTONIC SALT AND WATER TRANSPORT IN THE MAMMALIAN PROXIMAL STRAIGHT TUBULE

Thomas E. Andreoli,* James A. Schafer,* and Clifford S. Patlak[#]

*University of Alabama
#National Institutes of Health

INTRODUCTION

This chapter considers salt and water absorption from superficial proximal straight tubules isolated from rabbit renal cortex and perfused according to the technique of Burg et al. (1966). One of the questions we address is the mechanism of fluid transport in tubules exposed to isosmotic solutions at the same hydrostatic pressure. In 1962, Curran and MacIntosh showed that osmotic flow could occur between isosmotic, isobaric solutions separated by two membranes enclosing a central, well-stirred compartment; they argued that a similar mechanism might account for isotonic fluid absorption in epithelia, if active solute transport produced hypertonicity in the epithelial equivalent of the central compartment. Diamond and Bossert (1967) then reasoned that active salt transport from apical (luminal) solutions into unstirred intercellular spaces might raise the space osmolality, thus providing a driving force for osmotic flow from cells to spaces. By assigning active transport sites to the apical ends of intercellular spaces, and assuming a relatively high value for the diffusion resistance of intercellular spaces, Diamond and Bossert computed that intercellular spaces of actively transporting epithelia might contain a standing osmotic gradient, progressing from apical hypertonicity to isotonicity at the open ends.

In the mammalian proximal tubule, immunohistologic observations (Kyte, 1976) indicate that $(Na^+ + K^+)$-dependent ATPase activity is uniformly distributed along intercellular spaces. Furthermore, the transepithelial electrical resistance of isolated rabbit proximal straight tubules is approximately 5 ohm-cm^2 (Lutz, Cardinal and Burg, 1973), i.e., remarkably low. Since passive ion permeation in proximal straight tubules (Schafer, Troutman and Andreoli, 1974),

as in other epithelia (Ussing and Windhager, 1964; Frömter and
Diamond, 1972; Frizzell and Schultz, 1972), may involve an extracellular route, the diffusion resistance of the paracellular pathway of proximal straight tubules is probably also remarkably small.
Finally, the hydraulic conductivity of leaky epithelia is probably
much greater than had been previously envisioned (Diamond and
Bossert, 1967), since solute polarization within intercellular spaces
may result in substantial underestimates of transepithelial hydraulic conductivity from steady-state osmotic flows (Wright, Smulders
and Tormey, 1972; Schafer, Patlak and Andreoli, 1974). Taken together, these observations indicate that isotonic fluid transport
in proximal straight tubules needs to be assessed in light of the
possibility that the diffusion resistance of intracellular spaces
might not be sufficiently great to permit the formation of a standing osmotic gradient within an intercellular space, particularly
if active transport sites are uniformly distributed along the length
of an intercellular space.

The second major issue to be evaluated in this chapter concerns
Na^+ absorption in isolated proximal straight tubules. Since Na^+
absorption in proximal convolutions is accompanied by acidification
of tubular fluid (Gottschalk, 1963), fluid entering the in vivo
pars recta has a higher Cl^- and a lower HCO_3^- content than plasma
(Gottschalk, 1963; Malnic, Enokibara, Aires and Vieira, 1969;
Warren, Luke, Kashgarian and Levitin, 1970; Clapp, Watson and
Berliner, 1963; Bennett, Brenner and Berliner, 1963). In the in
vitro proximal convoluted tubule (Kokko, 1973), the in vitro proximal straight tubule (Schafer, Troutman and Andreoli, 1974; Schafer,
Patlak and Andreoli, 1975), and possibly the in vivo proximal convoluted tubule (Barratt, Rector, Kokko and Seldin, 1974), lumen to
bath Cl^- concentration gradients are attended by spontaneous lumen-
positive transepithelial voltages. Therefore it is particularly
relevant to inquire about the relative contributions of active and
passive transport processes to salt absorption in the isolated superficial proximal straight tubule, a region inaccessible to micropuncture, by exposing the tubule to luminal and peritubular solutions
comparable to those which might obtain in vivo.

Two classes of solutions (Schafer, Troutman and Andreoli, 1974;
Schafer, Patlak and Andreoli, 1975) will be referred to in this
chapter: Krebs-Ringer (KR) solutions containing either 113.6 mM
Cl^-, 25 mM HCO_3^-, pH 7.4 (HCO_3-KR); or 136.4 mM Cl^-, 3.8 mM HCO_3^-,
pH 6.6 (Cl-KR). The HCO_3-KR perfusate, which mimics glomerular
ultrafiltrate, contains 8.3 mM glucose and 5 mM alanine; the Cl-KR
perfusate resembles the luminal fluid which might be seen by the
in vivo pars recta (Gottschalk, 1963; Malnic, Enokibara, Aires and
Vieira, 1969; Warren, Luke, Kashgarian and Levitin, 1970; Clapp,
Watson and Berliner, 1963; Bennett, Brenner and Berliner, 1963),
and contains 13.3 mM urea as an isosmotic replacement for alanine
plus glucose. The bathing solutions uniformly contain 6% albumin,
8.3 mM glucose, and 5 mM L-alanine. In certain instances, acetate
in either the perfusing or bathing solutions is replaced with the

impermeant, non-actively transported isethionate anion; these solutions are denoted by an asterisk.

PASSIVE PERMEABILITY PROPERTIES

Figure 1 (Schafer, Troutman and Andreoli, 1974) represents the zero-current NaCl dilution potentials measured in these tubules at

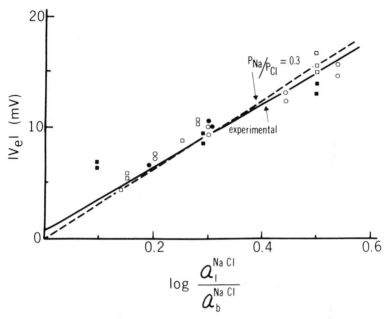

Fig. 1. *Zero-current NaCl dilution potentials. Circles: $J_v = 0$ at 21^0 C; squares, $J_v = 0$ with 10^{-4} M ouabain in the bath at 37^0 C. Open symbols: NaCl in the perfusate replaced isosmotically with mannitol; closed cymbols: NaCl in the bath replaced isosmotically with mannitol. The absolute magnitudes of V_e, denoted as $/V_e/$, are plotted in terms of equation 1. The solid line was drawn from a regression analysis of the data. The dotted line was drawn for a P_{Na}/P_{Cl} ratio of 0.3. Adapted from Schafer et al., 1974.*

zero net fluid absorption and a zero spontaneous transepithelial voltage. The experimental data are plotted according to a form of the Planck-Henderson equation:

$$1) \quad V_e = 2.3 \frac{RT}{F} \frac{P_{Na}/P_{Cl} - 1}{P_{Na}/P_{Cl} + 1} \log \frac{a^1_{NaCl}}{a^b_{NaCl}} \quad ,$$

where R, T and F have their usual meaning, and a^1_{NaCl} and a^b_{NaCl} refer to the NaCl activities in the lumen and bath, respectively. There is a very close correlation between the regression line for the experimental data and the theoretical line drawn from equation 1 for a P_{Na}/P_{Cl} ratio of 0.3; and for a given NaCl concentration ratio between lumen and bath, V_e is equal in magnitude but opposite in sign, depending on whether the NaCl concentration is reduced in the bath or in the lumen.

Two general conclusions (Schafer, Troutman and Andreoli, 1974) emerge from these data. Luminal and peritubular membranes of proximal tubules differ in their ionic permeability properties (Hoshi and Sakai, 1967; Boulpaep, 1967). And since the zero-current NaCl dilution voltages are symmetrical, it seems probable, in accord with comparable suggestions for other epithelia (Frizzell and Schultz, 1972; Frömter and Diamond, 1972; Frömter, Müeller and Wick, 1970; Barry, Diamond and Wright, 1971), that the NaCl dilution potentials are expressed across single symmetrical interfaces, presumably junctional complexes. Second, Fig. 1 illustrates that the P_{Na}/P_{Cl} ratio is constant over a five-fold variation in the ionic concentration of external solutions. But ionic partition into a membrane containing fixed charges should be regulated in part by Donnan effects, and hence should vary with the ionic concentrations in external solutions (Baker, Hodgkin and Meves, 1964). Thus it is difficult to reconcile the observations in Fig. 1 with ion permeation through junctional complexes containing a high density of fixed charges. Rather, the junctional complexes of these tubules (6), like those of gall bladder(Barry and Diamond, 1971) and rabbit ileum (Frizzill and Schultz, 1972), seem to contain weakly charged, or neutral, polar sites.

Table 1 summarizes the ionic permeability coefficients determined for these tubules in our laboratory (Schafer, Troutman and Andreoli, 1974; Schaefer, Patlak and Andreoli, 1975; Schafer and Andreoli, 1976) and by Warnock and Burg (1977). Essentially the same P_{Na} and P_{Cl} values have recently been obtained by Kawamura et al. (1975). It is clear that the P_{Na}/P_{Cl} ratio measured isotopically is very nearly equal to the one measured electrically (Fig. 1). The values for P_{HCO_3} computed by us (Schafer, Patlak and Andreoli, 1975), and by Warnock and Burg (1977), were estimated from bi-ionic voltages produced by luminal Cl-KR and bathing HCO_3-KR solutions, using the Goldman-Hodgkin-Katz equation and isotopic P_{Na} and P_{Cl} values. For the P_{Na} and P_{Cl} data listed in Table 1 and the Cl^-/HCO_3^- ratios in HCO_3-KR and Cl-KR solutions, one obtains a steep dependence of P_{HCO_3} on voltage, but for any bi-ionic voltage greater than 0.2 mV lumen positive, the P_{HCO_3}/P_{Cl} ratio is less than unity. Thus we conclude, either from our results (Schafer, Patlak and Andreoli, 1975) of from those of Warnock and Burg (1977), that junctional complexes of these tubules have an HCO_3^-/Cl^- selectivity ratio less than unity. The data listed in Table 1 provide a reasonable range of values for P_{HCO_3}.

A discrepancy arises by comparing the minimal transepithelial resistance of these tubules computed from P_i data with direct

Table 1. Isotopically and electrically determined ionic permeability coefficients.

ion	P_i	Reference
	cm sec^{-1} x 10^4	
Na^+	0.23	1
Cl^-	0.73	1
Acetate	0.14-0.18	2
Isethionate	0.02	3
HCO_3	0.04	3
HCO_3	0.20	4

The P_i values for Na^+, Cl^-, acetate, and isethionate were computed from tracer fluxes at zero volume flow. The P_{HCO_3} data were obtained from zero-current bi-ionic voltages as described in the text. (1) Schafer, et al., 1974; (2)Schafer & Andreoli, 1976; (3)Schafer et al., 1975; (4) Warnock & Burg, 1977.

electrical resistance measurements. The partial conductance of the i-th ion (g_i, ohm^{-1} cm^{-2}) may be expressed as:

$$2) \quad g_i = \frac{F^2}{RT} \cdot z_i^2 \cdot P_i \cdot [C_i] ,$$

where $[C_i]$ is the i-th ion concentration, and z_i is valence. Using P_{Na} and P_{Cl} from Table 1 and the Na^+ and Cl^- concentrations in HCO_3^--KR solutions, $(g_{Na} + g_{Cl})$ is 0.042 ohm^{-1} cm^{-2}, or a resistance of 23.8 ohm cm^2, i.e., a resistance appreciably greater than the value of 5 ohm-cm^2 measured by Lutz, Cardinal and Burg, 1973.

No complete explanation for this discrepancy exists. The resistance measurements of Lutz et al. (1973) coincide almost exactly with those of Boulpaep and Seely (1971) in proximal tubules of dog kidney. Futhermore, we (Schafer, Troutman and Andreoli, 1974) and Kawamura et al. (1975) have obtained nearly the same isotopic P_{Na} P_{Cl} values, and the isotopic (Table 1) and electrical (Fig. 1) P_{Na}/P_{Cl} ratios, coincide. However, Lutz et al. (1973) measured resistances using rabbit serum as a bath and a serum ultrafiltrate as a perfusate. Thus unidentified ions, present in the latter solutions but not in synthetic HCO_3-KR solutions, may account for the fact that measured electrical resistances are lower than those computed from $(P_{Na} + P_{Cl})$ and equation 2.

FLUID ABSORPTION AND TRANSEPITHELIAL VOLTAGE: SYMMETRICAL EXTERNAL SOLUTIONS

Fig. 2 (Schafer, Troutman and Andreoli, 1974; Schafer, Patlak and Androeli, 1975) summarizes some transport characteristics of superficial proximal straight tubules perfused and bathed with symmetrical HCO_3-KR solutions. At 37° C, J_v, the spontaneous rate of fluid absorption, is 0.43-0.46 nl min^{-1} mm^{-1}; and V_e, the spon-

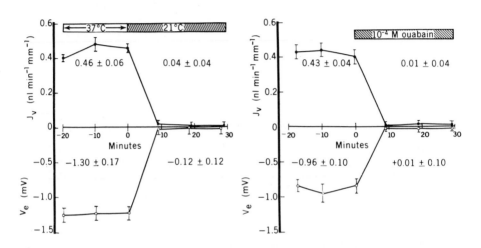

Fig. 2. Fluid absorption (J_v) and spontaneous transepithelial voltages (V_e) in tubules perfused and bathed in symmetrical HCO_3-KR solutions. Adapted from Schafer, Troutman & Androeli, 1974; Schafer, Patlak & Androeli, 1975.

taneous transepithelial voltage, is apprxoimately -1.0 mV lumen-negative. Fig. 3 also illustrates that both J_v and V_e become indistinguishable from zero, either when the system is cooled to 21° C, or at 37° C, when 10^{-4} M ouabain is added to the bathing solutions. Thus by conventional criteria, both J_v and V_e for these conditions depend on active, i.e., conservative, transport processes.

ISOTOPIC Na$^+$ FLUXES

Table 2 (Schafer, Troutman and Andreoli, 1974) shows that the observed net rate of Na$^+$ absorption is in excellent agreement with the net Na$^+$ flux predicted from J_v for isotonic Na$^+$ absorption. Likewise, the difference between the lumen to bath Na$^+$ fluxes, measured either in the absence or in the presence of 10^{-4} M ouabain (i.e., either in the presence or absence of fluid absorption), is also in agreement with the value for J_{Na}^{net} predicted from J_v for an

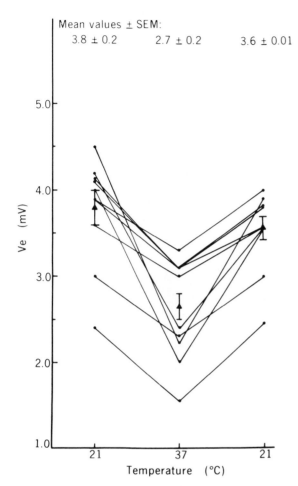

Mean values ± SEM:
3.8 ± 0.2   2.7 ± 0.2   3.6 ± 0.01

Fig. 3. The effect of temperature on V_e. The lumen and bath contain, respectively, Cl-KR* and HCO_3-KR* solutions. Each line connects data points on an individual tubule. The upper part of the figure presents the mean values ± SEM of the voltage measurements. Adapted from Schafer, Patlak & Andreoli, 1976.

isotonic transport process (Schafer, Troutman and Andreoli, 1974). In other words, the fluid absorbate is, within experimental error, an isotonic Na^+ solution.

THE COUNTER-IONS FOR Na^+ ABSORPTION

Table 3 summarizes the contributions of Cl^-, HCO_3^- and acetate as counter-ions for Na^+ absorption from symmetrical HCO_3-KR solutions (Schafer and Andreoli, 1976): passive net Cl^- flux accounts for approximately one-fourth of net Na^+ flux; half of net Na^+ transport is coupled to carbonic anhydrase-mediated HCO_3^- salvage; and acetate absorption, occurring via an undefined (but energy-dependent and carbonic anhydrase-insensitive) luminal efflux process, accounts for the remainder of net Na^+ absorption. As a consequence of HCO_3^- and acetate absorption, the tubular fluid Cl^- concentration rises from 116.6 ± 0.9 (SEM) mM in the prefusate to 127.6 ± 2.3 mM

Table 2.   The contribution of net sodium flux to net fluid absorption

$J_{Na}^{l \to b}$	$J_{Na}^{b \to l}$	J_{Na}^{net} (Observed)	J_{Na}^{net} (Predicted)	J_v
eq sec^{-1} cm^{-2} × 10^9				cm^3 sec^{-1} cm^{-2} × 10^5
4.9 ± 0.3	3.3 ± 0.4	1.6 ± 0.1 ($p < 0.05$)	1.7	1.16

The perfusate and bath are HCO$_3$-KR solutions at 37° C.   The values for unidirectional ^{22}Na$^+$ fluxes are expressed as mean values ± SEM.   The mean J_v value for the tubules where ^{22}Na$^+$ fluxes are measured is 0.49 nl min $^{-1}$ mm^{-1}; since the mean luminal diameter of these tubules is 22.3 × 10^{-4} cm (6), J_v (nl min^{-1} mm^{-1}) may be converted to J_v (cm^3 sec^{-1} cm^{-2}) by the factor 2.38 × 10^{-5} cm min mm nl^{-1} sec^{-1}.   The predicted net Na$^+$ flux is computed, for an isotonic fluid transport process, as the product of J_v and the Na$^+$ concentration in the perfusate.   The system has been gassed with 95% O$_2$-5% CO$_2$.   Adapted from Schafer, Troutman and Andreoli, 1974.

Table 3. The counter-ions for Na^+ absorption from symmetrical HCO_3-KR solutions

Ion	Method of Flux Measurement	Transport Process	J_i^{net}/J_{Na}^{net}
Cl^-	Chemical balance	Passive	27.5
HCO_3^-	Carbonic anhydrase inhibition	Carbonic anhydrase-mediated	47-60
Acetate	Tracer flux	Active, undefined luminal efflux	27.5

Adapted from Schafer and Andreoli (1976).

in the collected fluid, when fluid absorption is in the range of 0.4 nl min^{-1} mm^{-1} (Schafer and Andreoli, 1976). In short, the isolated superficial proximal straight tubule exposed to symmetrical HCO_3-KR solutions generates transepithelial Cl^- concentration gradients (Table 3) similar to those observed during in vivo micropuncture (Gottschalk, 1963; Malnic, Enokibara, Aires and Vieira, 1969; Warren, Luke, Kashgarian and Levitin, 1970; Clapp, Watson and Berliner, 1963; Bennett, Brenner and Berliner, 1968).

FLUID ABSORPTION AND TRANSEPITHELIAL VOLTAGES: ASYMMETRICAL SOLUTIONS

We consider next the salt and water transport in superficial proximal straight tubules exposed to solutions similar to those one might reasonably expect this nephron segment to see in vivo: a Cl-KR perfusate and an HCO_3-KR bath. In many instances, these same solutions, but with acetate replaced by the impermeant anion isethionate (denoted by an asterisk), have also been used.

J_v AND V_e WITH ASYMMETRICAL SOLUTIONS

The relevant observations on spontaneous rates of fluid absorption and transepithelial voltages with a Cl-KR or a Cl-KR* perfusate and an HCO_3-KR or HCO_3-KR* bath, are shown in Table IV and Fig. 3. Table IV illustrates clearly that omitting acetate from the perfusing and bathing media has no significant effect on J_v. A comparison of Table IV and Fig. 3 with Fig. 2 indicates three important phenomena. At 37^0 C, V_e is 2.7 mV lumen-positive when the perfusate and bath contain Cl-KR* and HCO3-KR*, respectively (Fig. 3), instead of approximately 1.0 mV lumen-negative, as occurs when the perfusate and bath contain symmetrical HCO_3-KR solutions (Fig. 2). In the latter instance, cooling to 21^0 C (or adding 10^{-4} M ouabain to the bath at 37^0 C) inhibits fluid absorption completely, and V_e becomes indistinguishable from zero (Fig. 2) But with a Cl-KR*

Table 4. Fluid absorption during tubular exposure to asymmetrical solutions with or without acetate.

Perfusate	Bath	T	J_v	ΔJ_v
		$^\circ$C	nl min^{-1} mm^{-1}	
Cl-KR	HCO$_3$-KR	37	0.41 + 0.04	
				0.01 ± 0.01 (p $\geq$ 0.5)
Cl-KR*	HCO$_3$-KR*	17	0.39 ± 0.04	
				0.03 ± 0.01 (p < 0.02)
Cl-KR*	HCO$_3$-KR*	21	0.16 ± 0.01	

The data are paired J_v measurements carried out on individual tubules for each of the three indicated conditions, and varied at random in the experiments. The data are expressed as mean values ± SEM. Adapted from Schafer, Patlak and Andreoli (1976).

perfusate and HCO$_3$-KR* bath, cooling to 21° C produces a 60% reduction in J_v (Table IV) and V_e rises approximately 1.0 mV (Fig. 3).

Virtually identical effects obtain by adding 10^{-4} M ouabain to the bath at 37° C, when the perfusate and bath contain Cl-KR and HCO$_3$-KR and HCO$_3$-KR solutions, respectively (Schafer, Patlak and Andreoli, 1975). Thus we have argued (Schafer, Patlak and Andreoli, 1975, 1976) that, for a Cl-KR* (or Cl-KR) perfusate and an HCO$_3$-KR* (or HCO$_3$-KR) bath, J_v and V_e values observed at 21° C [or, at 37°C in the presence of 10^{-4} M ouabain (Schafer, Patlak and Andreoli, 1975)] depend on dissipative driving forces.

Cl$^-$ BALANCE DURING FLUID ABSORPTION LINKED TO PASSIVE AND ACTIVE ION FLUXES

Table V (Schafer, Patlak and Andreoli, 1975, 1976) indicates the Cl$^-$ contribution to isotonic Na$^+$ absorption from asymmetrical solutions. During fluid transport linked to dissipative ion flows (i.e., at 21° C), Cl$^-$ is virtually the sole anion in an isotonic fluid absorbate. At 37° C, where active plus passive Na$^+$ transport occur, Cl$^-$ is the counter-ion for approximately 85% of Na$^+$ absorption, and it is probable that the remaining 15% of Na$^+$ absorption is referable to carbonic anhydrase-mediated HCO$_3^-$ salvage [(Schafer and Andreoli, 1976; Schafer, Patlak and Andreoli, 1976), Table 3].

ANALYSIS OF SALT AND WATER ABSORPTION FROM ASYMMETRICAL SOLUTION

We consider now quantitative aspects of salt and water absorption from tubules exposed to a Cl-KB* perfusate and an HCO$_3$-KR* bath. Na$^+$, Cl$^-$ and HCO$_3^-$ are the principal constituents of these

Table 5. *Net Cl⁻ absorption during fluid absorption coupled to active plus passive, or passive ion flows*

T°	Perfusate	Bath	J_v	J_{Na}	J_{Cl}/J_{Na}
C°			$cm^3\ sec^{-1}\ cm^{-2} \times 10^6$	$eg\ sec^{-1}\ cm^{-2} \times 10^9$	%
37	Cl-KR*	HCO₃-KR*	9.76	1.52	85.3
			+ 1.2	+ 0.08	+ 3.4
21	"	"	3.8	0.59	≈100
			+ 0.23	+ 0.04	

Mean values + SEM. Adapted from Schafer, Patlak and Andreoli, 1975, 1976.

solutions, so we evaluate the isotonic (Tables 2, 3, 4) transport of three ions: $Na^+ = 1$; $Cl^- = 2$, and $HCO_3^- = 3$. The data for analyzed are summarized in Table 4: J_v, J_{Na} and J_{Cl} are from Table 5; J_{HCO_3} is computed as $(J_{Na}-J_{Cl})$; and the V_e values are from Fig. 3. Transport processes occurring at 21° C are taken to be dissipative, while those at 37° C are considered to depend on dissipative plus conservative processes.

Table 6. Transport data: Cl-KR* perfusate and HCO₃-KR* bath

	37° C	21° C
J_v (cm³ sec⁻¹ cm⁻² x 10⁶)	9.76	3.8
J_{Na} (eq sec⁻¹ cm⁻² x 10⁹)	1.52	0.59
J_{Cl} (eq sec⁻¹ cm⁻² x 10⁹)	1.29	0.59
J_{HCO_3} (eq sec⁻¹ cm⁻² x 10⁹)	0.23	≈0
V_e (mV, lumen to bath)	2.7	3.7

The transport data at 37° C are considered to depend on active plus passive transport processes. The data at 21° C are considered to depend only on passive transport processes. Adapted from Table 5, Fig. 3, and Schafer, Patlak and Andreoli, 1975, 1976.

MODEL

The model (Schafer, Patlak and Andreoli, 1975, 1976) is illustrated schematically in Figure 4. An unstirred channel of unspecified geometry, and lacking fixed charges, separates the luminal and bathing solutions, and is bounded laterally by a cellular compartment. One cm² of luminal membrane area contains n channels. Each channel is separated from the lumen by the junctional complex at x = 0, and from the bath by the peritubular basement membrane at x = L. The concentrations (eq cm⁻³) of the i-th ion in lumen, cell, channel and bath are designated as, respectively, C_i^l, C_i^c, C_i^x, and C_i^b. Fluxes are positive for increasing x.

Fluid may enter a channel either through an extracellular (J_v^e) or a cellular route (J_v^c), and all fluid, designated J_v^L, leaves the channel at x = L. Thus we define a parameter B as:

$$3) \qquad B = \frac{J_v^e}{J_v^L} \, .$$

Ions enter the channel through an extracellular (J_i^e) or a cellular route (J_i^c); all of the ion flux, designated J_i^L, leaves the channel at x - L.

THE DIFFUSION RESISTANCE FOR THE PARACELLULAR PATHWAY

260

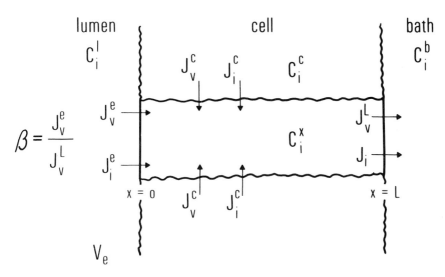

Fig. 4. Schematic model for analysis of the experimental data. Adapted from Schafer, Patlak & Andreoli, 1975, 1976.

The electrical resistance of these tubules perfused in serum ultrafiltrate and bathed in rabbit serum is approximately 5 ohm-cm^2 (Lutz, Cardinal and Burg, 1973), and a 7.5×10^{-4} cm thick layer [i.e., the thickness of the tubular epithelium (Schafer, Troutman and Andreoli, 1974)] of HCO_3^--KR buffer has a resistance of 0.045 ohm-cm^2. Thus for extracellular passive ion permeation, the diffusion resistance of lateral spaces is no more than 111 times greater than that of an equivalent thickness of HCO_3^--KR buffer, if junctional complexes make no contribution to diffusion resistance. Accordingly we have defined (Schafer, Patlak and Andreoli, 1975) a dimensionless parameter α as:

4) $1 \leq \alpha \leq 111$,

where the limits of α assign all ionic diffusion resistance either to junctional complexes, for $\alpha = 1$, or to the intercellular spaces, for $\alpha = 111$.

The ionic permeability coefficient P_i may be related to α according to the expression (Schafer, Patlak and Androeli, 1975):

5) $\dfrac{1}{P_i} = \dfrac{1}{P_i^j} + \dfrac{\alpha L}{D_i}$,

where P_i^j (cm sec^{-1}) is the permeability coefficient of the i-th ion across junctional complexes, D_i (cm^2 sec^{-1}) is the free diffusion coefficient, and L is the tubular thickness. The values of P_i^j in terms of α are shown in Table 7: for Na$^+$ and HCO$_3^-$, virtually

Table 7. The relationship between P_i^j *and* α

α	P_{Na}^j	P_{Cl}^j	$P_{HCO_3}^j$	$P_{HCO_3}^j$
			cm sec^{-1} x 10^4	
1	0.23	0.73	0.04	0.20
25	0.23	0.78	0.04	0.20
50	0.24	0.84	0.04	0.21
75	0.25	0.91	0.04	0.22
100	0.26	1.0	0.04	0.24

The values of P_{Na}^j, P_{Cl}^j *and* $P_{HCO_3}^j$ = 0.04 x 10^{-4} *cm sec*$^{-1}$, *for* α = 1, *are those observed in this laboratory (Ref. 1 & 2 in Table 1).* $P_{HCO_3}^j$ *of 0.20 x 10^{-4} cm sec^{-1} at* α = 1 *is that obtained by Warnock and* [3]*Burg (1977, Table 1).* P_i^j *as a function of* α *was computed according to Equation 5, using L - 7.5 x 10^{-4} cm sec^{-1}. Adapted from Schafer, Patlak and Andreoli, 1974, 1975.*

all of the resistance to passive ion permeation is referable to junctional complexes; for the more permeable Cl$^-$ species, inter-cellular spaces make a relatively small contribution to diffusion resistance when α exceeds 25. Table 7 lists both values for P_{HCO_3} cited in Table 1, because of the uncertainties, discussed in connection with Table 1, in measuring P_{HCO_3} accurately.

A major concern arises with respect to defining an upper limit for α as in Equation 4. As indicated above, the minimal transepithelial electrical resistance value computed from equation 2 and the P_{Na} and P_{Cl} values in Table 1 is approximately 23.8 ohm-cm^2. The latter value would yield an upper limit of approximately 528, rather than 111, for the parameter α. Thus an alternate method, independent of electrical measurements, will be required for a more explicit determination of the upper limit to α.

ASSUMPTIONS

Following previous analyses (Schafer, Patlak and Andreoli, 1975, 1976), we set the following assumptions:
(a) The channels are uniform. (b,c,d)
(b) Passive ion permeation in these tubules involves an extra-cellular route (Schafer, Troutman and Andreoli, 1974), so that junctional complexes are appreciably more permeable to ions than cellular membranes. Thus the membranes separating the cellular and channel compartments are assigned unity reflection coefficients for Na$^+$, Cl$^-$ and HCO$_3^-$.
(c) Since peritubular basement membranes of these tubules are moderately permeable to albumin (Welling and Grantham, 1972), we assume a zero reflection coefficient for Na$^+$, Cl$^-$ and HCO$_3^-$ at x = L.

And since the hydraulic conductivity of peritubular capillary basement membranes in these tubules exceeds 4×10^{-2} cm sec^{-1} atm^{-1} (Welling and Grantham, 1972), a negligible pressure gradient, 0.1-0.2 cm H_2O, accounts for volume flow between intercellular spaces and bath^2 (Schafer, Patlak and Andreoli, 1975, 1976). Thus hydrostatic pressure terms are not considered in the present analysis.

(d) The cells are in osmotic equilibrium with the external solutions. Since a unity reflection coefficient for ions has been assigned to the membranes separating cells and channels, the cellular osmolality is taken to be twice the cation content of external solutions.

THE FLOW-DIFFUSION EQUATIONS

The flux of the i-th ion through a plane at x normal to the direction of solute transport is, for one cm^2 of luminal surface area (Schafer, Patlak and Andreoli, 1975, 1976):

6) $\quad J_i^X = -\frac{D_i}{\alpha''}\left[\frac{d\ C_i^X}{dx} + \frac{z_i\ F}{RT}\ C_i^X\ \frac{dV^X}{dx}\right] + J_v^X\ C_i^X$,

where J_v^X is the net volume flux crossing x, and V^X is the voltage at x. The electroneutrality conditions are:

7) $\quad J_1^X = J_2^X + J_3^X$

and

8) $\quad C_1^X = C_2^X + C_3^X$.

The gradient of volume flow is (Schafer, Patlak and Andreoli, 1975, 1976):

9) $\quad \dfrac{d\ J_v^X}{dx} = \dfrac{P_f^C}{\alpha''}\ \overline{V}_w \left[\ \sum\limits_{i=1}^{3}\ (C_i^X - C_i^C)\ \right]$,

where $\overline{V}_w$ is the partial molar volume of water, and P_f^C (sec^{-1}) is:

10) $\quad P_f^C = \dfrac{P_f^C}{L}$,

where P_f^C is the osmotic water permeability coefficient (cm sec^{-1}) of the membrane between cells and channel. Since volume flow enters the channels through either a cellular or a paracellular route, P_f^C is:

11) $\quad P_f^C = P_f^O - P_f^j$,

where P_f^O is the observed transepithelial water conductance (cm sec^{-1}). Welling (L. W. Welling, Veterans Administration Hospital, Kansas City, Kansas; personal communication) has measured P_f^O in superficial proximal straight tubules by hydraulic conductivity measurements, and has obtained $P_f^O = 0.2535 \pm 0.027$ cm sec^{-1} (SEM).

By using this value, we have:

$$12) \quad P_f^c = \frac{0.2535 - P_f^j}{L} \quad .$$

J_i^x may be expressed (Schafer, Patlak and Andreoli, 1976) as:

$$13) \quad J_i^x = J_i^p + \tau_i (N_i - M_i),$$

where J_i^p is the passive flux of the i-th ion crossing x, τ_i is the net active transport (Eq/sec·cm of channel length) of the i-th ion in one cm^2 of luminal surface area, and M_i and N_i are, respectively, the beginning and end of the active transport sites along the channel length. Accordingly we have:

$$14) \quad \tau_i = 0 \text{ for } \{ {}_{N_i}^{0} \begin{array}{c} \leq X \leq M_i \\ \leq X \leq \end{array} \} \quad .$$

Thus the gradient of ion flux is:

$$15) \quad \frac{d \ J_i^x}{dx} = \tau_i \quad ,$$

and J_i^a (Eq sec^{-1} cm^{-2}), the rate of active ion absorption, is:

$$16) \quad J_i^a = \tau_i (N_i - M_i).$$

The boundary conditions for the system are across the junctional complexes. J_i^p is:

$$17) \quad J_i^P = - P_i^j (\Delta C_i^j + \frac{z_i F}{RT} \bar{C}^j \Delta V^j) + \beta j_v (1 - \sigma_i^j) \bar{C}_i^j \quad ,$$

where ΔC_i^j and ΔV^j are the concentration difference for the i-th ion and the voltage, respectively, across junctional complexes, σ_i^j is the reflection coefficient of the i-th ion across junctional complexes; and $\bar{C}_i^j$ is the mean i-th ion concentration across the junctional complexes. The volume flow across junctional complexes is:

$$18) \quad \beta j_v = P_f^j \bar{V}_w \sum_{i=1}^{3} \sigma_i^j \Delta C_i^j \quad .$$

THE CASE OF PASSIVE ION FLOWS

We now consider the transport data listed in Table 6 for 21° C (Schafer, Patlak and Andreoli, 1975); for these circumstances, τ_i in equation 15 is zero. Equations 6-16 may be integrated numerically, to yield: C_i^0, the solute concentration at x = 0; ΔV^j, the voltage across junctional complexes; and $P_f^'$. The required inputs

264

are C_i^b, D_i, J_i, L, V_e, β, α, and the cellular osmolality. Next, using D_i^o, ΔV^j and the values of P_i^j in terms of α (Table 7), one may compute σ_i^j and P_f^j, both in terms of α, from the simultaneous solution of Equations 17 and 18.

Figure 5 illustrates the values of C_{Na}^o and C_{Cl}^o computed for $1 \leq \alpha \leq 100$ when the parameter β was taken to be unity: Na^+ accumulation in intercellular spaces does not occur, and C_{Na}^o and C_{Cl}^o are virtually identical to C_{Na}^b and C_{Cl}^b. Although not shown in the figure, V^o is the same as V_e within 0.05 mV (Table 6). The fact that the computed osmolality of the channels is virtually identical to that of the bath requires, for a σ_i^c of unity, that fluid transport linked to passive ion flows enters intercellular spaces by an extracellular route. In accord with this view, the integrated flow-diffusion equations yield non-real values of P_f^c when the parameter

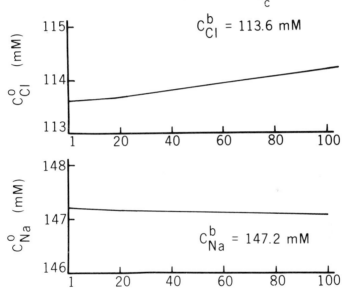

Fig. 5. The relationship between C_{Cl}^o, C_{Na}^o, and α, computed as described in the text for the 21° C transport data listed in Table 6. Adapted from Schafer, Patlak and Andreoli, 1975.

β is taken to be less than unity. For a β value of unity, P_f^c is evidently indeterminate by this approach.

Table 8 and Figure 6 illustrate the values of σ_i^j and P_f^j, respectively, computed using Equations 17 and 18 and the values of C_i^o (Fig. 5) and ΔV^j computed from solution of the flow-diffusion equations. Table 8 shows that, for $1 \leq \alpha \leq 75$, σ_{Cl}^j is consistently less than $\sigma_{HCO_3}^j$. Fig. 6 illustrates that only values of $\alpha \leq 65$ predict P_f^j values ≤ 0.2535 cm sec^{-1}; for $\alpha > 65$, P_f^j becomes infinite and cannot be used for computation. As indicated in

Table 8.   Relationship between σ_1^j and α

α	σ_{Na}^j	σ_{Cl}^j	$\sigma_{HCO_3}^j$
1	0.9	0.78	0.97
25	0.95	0.83	0.97
50	0.96	0.88	0.97
75	1.0	0.95	0.97

The values of σ_i^j were computed for the transport data at 21^0 C listed in Table 6 from simultaneous solution of Equations 17 and 18 using the C_i^0 values in Fig. 5 and the P_i^j values in Table 7; 0.04×10^{-4} cm sec^{-1} was used as the value for $P_{HCO_3}^i$. Adapted from Schafer, Patlak and Andreoli, 1975.

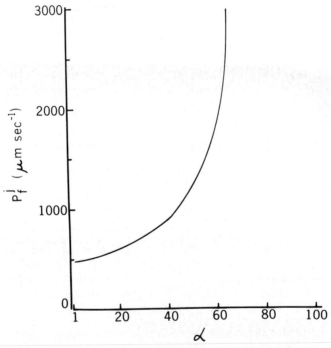

Fig. 6.   The relationship between P_f^j and α, computed from simultaneous solution of Equations 17 and 18 for the C_i^0 data shown in Fig. 5, and the transport data at 21^0 C listed in Table 6.   Adapted from Schafer, Patlak and Andreoli, 1975, 1976.

connection with equation 12, P_f^o for these tubules is 0.2535×10^{-4} cm sec^{-1}. Thus only the restricted range of:

19) $1 \leq \alpha \leq 65$

is consistent simultaneously with: the observed rates of fluid absorption coupled to passive salt flows (Table 6); measured trans-epithelial electrical resistances (Lutz, Cardinal and Burg, 1973); measured ionic permeability coefficients (Tables 1, 7); a measured P_f^o of 0.2535 cm sec^{-1}; and any finite value of P_f^1.

THE CASE OF ACTIVE PLUS PASSIVE ION FLOWS

Approximation of Active Transport Rates

It is evident from Equations 13-16 that, for transport at 37° C, computation of the solute concentration profiles in intercellular spaces requires an assessment of J_i^a. It is possible (Schafer, Patlak and Andreoli, 1976) to obtain a reasonable estimate of J_i^p, the rate of net passive ion absorption, and hence to J_i^a as $(J_i - J_i^p)$, by expressing equation 17 in terms of transepithelial driving forces:

$$20) \quad J_i^p \simeq - P_i \left\{ (C_i^b - C_i^1) + \frac{z_i F}{RT} \left(\frac{C_i^1 + C_i^b}{2} \right) V_e \right\} + J_v (1 - \sigma_i^j) \left(\frac{C_i^1 + C_i^b}{2} \right).$$

i.e., where P_i (Talbe 7 and σ_i^j (Table 8) are for $\alpha = 1$, and β is tacitly assumed to be unity.

This approach may be justified by a number of considerations. There are at least three instances where Equation 20 describes adequately the measured rates of net passive ion transport in these tubules. These include: net passive Cl^- absorption from tubules perfused and bathed with symmetrical HCO_3-KR solutions at 37° C (Schafer and Andreoli, 1976); and net passive Na^+ and Cl^- absorption during fluid absorption coupled to passive ion transport [(Schafer, Patlak and Andreoli, 1975); Fig. 5; Tables 7 and 8]. In these situations, diffusion accounts for 75-100% of ion flux (Schafer, Patlak and Andreoli, 1975; Schafer and Andreoli, 1976), and Table 7 indicates that, for $1 \leq \alpha \leq 65$, junctional complexes constitute the major resistance for transepithelial ionic diffusion. So it may be that, as in Fig. 5, intercellular spaces are very nearly in equilibrium with bathing solutions, and equation 20 may provide a reasonable method for assessing J_i^p.

Table 9 lists the values of J_i^p and J_i^a computed according to equation 20 for the transport data at 37° C in Table 6. Table 9 indicates that 34% of net Na^+ absorption was passive, in accord with the observation that 41% of net fluid absorption was insensitive to cooling (Tables 4,6). Net Cl^- transport in these tubules (Schafer and Andreoli, 1976) is passive; and Table 9 shows that, within experimental error, Equation 20 accounted for net Cl^- absorption. Finally, Table 9 illustrates that the net effect of assigning the higher value of 0.2×10^{-4} cm sec^{-1} to P_{HCO_3} is to

267

increase the diffusional backleak of HCO_3^- into the lumen, and hence $J_{HCO_3}^a$; for these conditons, the calculated active HCO_3^- transport is approximately 80% of the net rate of active Na^+ transport.

EVALUATION IN TERMS OF FLOW-DIFFUSION EQUATIONS

Equations 6-16 may now be integrated numerically for the 37^0 C case, subject to the boundary equations 17 and 18. The required inputs are: C_i^l, C_i^b, and D_i; P_i^J in terms of the variable parameter α (Table 7); J_i, J^i and V_e^i (Table 6); P_f^c in terms of α (Equation 12 and Fig. 6); and τ_i, in terms of N_i and M_i (Equations 13-16 and Table 9). The varying parameters are: α and $(N_i - M_i)$. Solution of these equations yields: C_i^o, σ_i^J, and ΔV^J. Since Equation 20 equates C_i^b and V_e with, respectively, C_i^o and ΔV^J in Equation 17, solution of the flow-diffusion equations provides a way of evaluating the validity of this assumption. And yet another means for testing Equation 20 obtains by comparing the σ_i^J values used in Equation 20 with those obtained from solution of the flow-diffusion equations

Fig. 7 illustrates the values of $(C_i^o - C_i^b)$ computed from numerica integration of the flow-diffusion equations. The values of J_i^a were obtained from Table 9, using the value of $P_{HCO_3} = 0.04 \times 10^{-4}$ cm sec^{-1}; when $P_{HCO_3} = 0.2 \times 10^{-4}$ cm sec^{-1} was used, C_i^o for Na^+, Cl^- and HCO_3^- varied by less than 0.01 mM from the C_i^o values shown in Fig. 7; thus virtually identical values for C_i^o were obtained using

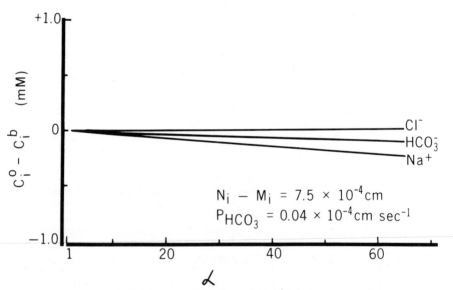

Fig. 7. The relationship between $(C_i^o - C_i^b)$ *and* α, *computed as described in the text for the transport data at* 37^0 *C listed in Table 6. Adapted from Schafer, Patlak and Andreoli, 1976.*

the higher and lower values for P_{HCO_3}. It is evident from Fig. 7 that C_{Na}^a, C_{Cl}^a, and $C_{HCO_3}^o$ were very nearly the same as C_{Na}^b, C_{Cl}^b, and $C_{HCO_3}^b$, respectively. Thus, the intercellular spaces were in virtual equilibrium with the bathing solutions, and varying P_{HCO_3} between 0.04×10^{-4} cm sec^{-1} and 0.2×10^{-4} cm sec^{-1} had essentially no effect on C_i^o. Evidently, these results are entirley consistent with the use of equation 20 as a valid approximation for estimating J_i^p. It also follows directly that, since the intercellular spaces were isosmotic with the external solutions, net fluid absorption involved an extracellular rather than a transcellular route.

Table 10 lists the values of σ_i^j computed by integrating the flow-diffusion equations for the conditions listed in Fig. 7, using both of the P_{HCO_3} in Table 7. The computed values of σ_i^j were nearly the same using the two different values of P_{HCO_3}, and very similar to the σ_i^j values computed for the case of fluid absorption coupled to passive ion flows (Table 8). Thus one can assount for isotonic fluid absorption from asymmetrical solutions (i.e., a Cl-KR* perfusate and an HCO_3-KR* bath) on the basis of unequal reflection coefficients for Cl^- and HCO_3^-, whether a lower (0.04×10^{-4} cm sec^{-1}) or higher (0.2×10^{-4} cm sec^{-1}) P_{HCO_3} is chosen. Finally, when the interval M_i to N_i in Equation 16 is taken to be from zero to 0.2×10^{-4} cm for both Na^+ and HCO_3^-, the flow-diffusion equations yield non-real values for $\sigma_{HCO_3}^j$. Thus in terms of the present model, localization of the active transport sites for for Na^+ and HCO_3^- at the apical ends of intercellular spaces is an unlikely possibility.

CONCLUDING REMARKS

The model described in this chapter (Fig. 5) assesses the mechanisms of ion and fluid absorption from tubules exposed to asymmetrical acetate-free solutions at 37^0 C, with particular regard to the use of Equation 20 for a quantitative analysis of passive ion transport, and the driving forces for net fluid absorption. The results in Table 7, taken together with the upper limit of 65 for α defined in Fig. 6, indicate that junctional complexes constitute the bulk of the transepithelial resistance to ionic diffusion. Thus Equation 20 describes passive ion transport across junctional complexes in terms of bulk phase driving forces. Or stated in another way, by equating α with unity, Equation 20 equates net dissipative ion flux across junctional complexes with net passive transepithelial ion flux.

The argument may have a more general applicability to passive ion transport in mammalian proximal renal tubules. Sauer (1973) has deduced, from thermodynamic considerations, that an expression formally identical to Equation 20 may be used to assess passive ion transport in renal tubules. And Frömter, Rumrich and Ullrich (1973); Ullrich (1973) and Neumann and Rector (1976) have used the equivalent of Equation 20 to describe the passive components of ionic fluxes measured during the in vivo micropuncture of rat proximal convolutions.

269

Table 9. The contributions of active and passive transport processes in net ion absorption

ion	J_i	J_i^P	J_i^a
		Eq sec^{-1} cm^2 x 10^9	
Na$^+$	1.52	0.51	1.01
Cl$^-$	1.28	1.20	-
HCO$_3^-$			
(using P_{HCO_3} = 0.04 x 10^{-4} cm sec^{-1})			
	0.24	-0.103	0.34
(using P_{HCO_3} = 0.20 x 10^{-4} cm sec^{-1})			
	0.24	-0.582	0.82

The values of J_i are from Table 6 for the 37° C case. J_i^P has been computed according to equation 20 using P_i and σ_{ji}^a (for α = 1) taken from Tables 7 and 8, respectively; the two indicate values of P_{HCO_3} were used. A negative sign denotes bath to lumen flux. Adapted[3] from Schafer, Patlak & Andreoli, 1976.

A comparison of Tables 4, 6 and 9 indicates that, in empirical terms, there is reasonable agreement between the experimental data and the values of J_i^P predicted from Equation 20. Likewise, the results obtained from Equation 20 also converge with predictions from solution of the flow-diffusion equations. When the J_i, J_i^P, and J_i^a values in Table 9 are used with the numerically integrated flow-diffusion equations: the computed concentrations of Na$^+$, Cl$^-$ and HCO$_3^-$ in intercellular spaces (Fig. 7) are virtually identical to those in the bathing solutions; and the predicted values for σ_{ji}^a (Table X) are very similar to those used in Equation 20 (Table 7), whether one chooses P_{HCO_3} = 0.04 x 10^{-4} cm sec^{-1} or P_{HCO_3} = 0.20 x 10^{-4} cm sec^{-1}.

Taken together, these observations are consistent with the possibility that Equation 20 and the results listed in Table 9 provide a reasonable estimate of the magnitudes of active and passive ion transport in these tubules, at least for the case of a Cl-KR* perfusate and an HCO$_3^-$KR* bath. Likewise, we may summarize the observations in Figs. 5 and 7 and Table 10 by saying that fluid absorption coupled either to passive, or to passive plus active, salt flows in produced by the osmotic pressure gradient arising across junctional compleses owing to the fact that the lumen and bath contain asymmetrical solutions, with respect to Cl$^-$ and HCO$_3^-$ concentrations, and to the fact that junctional complexes are more

Table 10. *Values of* σ_i^j *corresponding to* c_i^o

α	σ_{Na}^j	σ_{Cl}^j	σ_{HCO_3}
Using $P_{HCO_3} = 0.04 \times 10^{-4}$ cm sec^{-1}			
1	.91	.65	.96
20	.92	.70	.97
40	.97	.75	.97
60	.98	.86	.98
Using $P_{HCO_3} = 0.20 \times 10^{-4}$ cm sec^{-1}			
1	.91	.65	.95
20	.92	.70	.98
40	.94	.75	.99
60	.97	.85	1.02

The values of σ_i^j *have been computed for the conditions listed in Fig. 7, using each of the indicated values of* P_{HCO_3}, *and the corresponding transport rates for* HCO_3^- *from Table 9. Adapted from Schafer, Patlak and Andreoli, 1976.*

permeable to Cl$^-$ than to HCO_3^-. Put another way, it is probable that, for the latter condition, net fluid absorption traverses an extracellular route and occurs when intercellular spaces are isomotic to external solutions.

The $\sigma_{HCO_3}^j$ values listed in Table 10 were computed by assuming that active transport sites for Na$^+$ and HCO_3^- were uniformly distributed along lateral intercellular spaces. But when the active transport sites were localized to the first 0.2 μM of channel length, non-real values for $\sigma_{HCO_3}^j$ obtain (cf. above). These results are at least consistent with recent observations (Kyte, 1976) indicating that (Na$^+$ + K$^+$)-dependent ATPase activity is uniformly distributed along the intercellular spaces of canine cortical proximal convoluted tubules.

We emphasize the tentative nature of the present model. It is not possible using present experimental techniques to determine whether a fluid absorbate is isotonic or, as suggested (Sackin and Boulpaep, 1975) for the proximal convoluted tubule of *Necturus,* very slightly hypertonic. It is interesting to note in this context that the electrical resistance of the *Necturus* proximal convoluted tubule is 80 ohm-cm^2 (Boulpaep and Seely, 1971): if this value were used to define an upper limit to α, a significant resistance to transepithelial Na$^+$ diffusion would be due to intercellular spaces and intercellular space hypertonicity might obtain. But as indicated in connection with Fig. 6, only an upper limit of $\alpha = 65$ is consistent simultaneously with electrical resistance measurements, isotopic permeability coefficients, and either predicted or measured hydraulic conductivity coefficients in these tubules. Thus the present model, depicting virtual equilibrium between bath and intercellular spaces, provides a reasonable way of accounting for the experimental data.

Finally, it is evident that the present model is not adequate to account for the rates of fluid absorption observed when these tubules are exposed to symmetrical solutions. For example, when the perfusate and bath are summetrical HCO_3^--KR solutions, J_v at 37^0 C in these tubules is approximately 0.4 nl min^{-1} mm^{-1} but the collected fluid $[Cl^-]$ only rises to 127.6 mM (Schafer and Andreoli, 1976). Put another way, during spontaneous fluid absorption from symmetrical HCO_3^--KR solutions, the driving force for fluid absorption generated by a rising intraluminal $[Cl^-]$ is smaller than that produced by using a Cl-KR* perfusate. Thus it is evident that additional driving forces must exist for the case of fluid absorption from tubules perfused and bathed with symmetrical solutions.

REFERENCES

Baker, P.F., Hodgkin, A.L., & Meves, H. (1964) *J. Physiol. (Lond.) 170:* 541.

Barratt, L.J., Rector, F.C., Jr., Kokko, J.P., & Seldin, D.W. (1974) *J. Clin. Invest. 53:* 454.

Barry, P.H., & Diamond, J.M. (1971) *J. Membr. Biol. 4:* 295.

Barry, P.H., Diamond, J.M. & Wright, E.M. (1971) *J. Membr. Biol. 4:* 358.

Bennett, C.M., Brenner,B.M., & Berliner, R.W. (1968) *J. Clin. Invest. 47:* 203.

Boulpaep, E.L. (1967) In *Transport und Funktion Intracellulärer Elektrolyte* (F. Krück, editor) Urban & Schwarzenberg, München. pp. 98.

Boulpaep, E.L. & Seely, J.F. (1971) *Am. J. Physiol. 221:* 1084.

Burg, M., Grantham, J., Abramow, J. & Orloff, J. (1966) *Am. J. Physiol. 210:* 1293.

Clapp, J.R., Watson, J.F., & Berliner, R.W. (1963) *205:* 273.

Curran, P.F., & MacIntosh, J.R. (1962) *Nature 193:* 347.

Diamond, J.M., & Bossert, W.H. (1967) *J. Gen. Physiol. 50:* 2061.

Frizzell, R.A., & Schultz, S.G. (1972) *J. Gen. Physiol. 59:* 318.

Frömter, E., & Diamond, J. (1972) *235:* 9.

Frömter, E., Müeller, C.W., & Wick, T. (1970) In *Electrophysiology of Epithelial Cells* (G. Giebish, editor) F.K. Schattauer-Verlag, Stuttgart.p.119.

Frömter, E., Rumrich, G., & Ullrich, K.J. (1973) *Pflügers Arch. 343:* 189.

Gottschalk, C.W. (1962-63) *Harvey Lecture Ser. B. 58:* 99.

Hoshi, T., & Sakai, F. (1967) *Jap. J. Physiol. 17:* 627.

Kawamura, S., Imai, M., Seldin, D.W., & Kokko, J.P. (1975) *J. Clin. Invest. 55:* 1269.

Kokko, J.P. (1973) *J. Clin. Invest. 59:* 1362.

Kyte, J. (1976) *J. Cell Biol. 68:* 304.
Lutz, M.D., Cardinal, J., & Burg, M.B. (1973) *Am. J. Physiol. 225:* 729.
Malnic, G., Enokibara, H., Aires, M.M., & Vieira, F.L. (1969) *Pflugers Arch. 309:* 21.
Neumann, K.H., & Rector, F.C. (1976) *J. Clin. Invest. 58:* 1110.
Sackin, H., & Boulpaep, E.L. (1975) *J. Gen. Physiol. 66:* 671.
Sauer, F. (1973) In *Handbook of Physiology. Section B. Renal Physiology.* (J. Orloff and R. W. Berliner, editors). Washington, D.C.: American Physiological Society, p. 399.
Schafer, J.A., & Andreoli, T.E. (1976) *J. Clin. Invest. 58:* 500.
Schafer, J.A., Patlak, C.S., & Andreoli, T.E. (1974) *J. Gen. Physiol. 64:* 201.
Schafer, J.A., Patlak, C.S., & Andreoli, T.E. (1975) *J. Gen. Physiol. 66:* 445.
Schafer, J.A., Patlak, C.S., & Andreoli, T.E. (1976) *Amer. J. Physiol: Renal, Fluid and Electrolyte Physiol.* Submitted for publication.
Schafer, J.A., Troutman, S.L., & Andreoli, T.E. (1974) *J. Gen. Physiol. 64:* 582.
Ullrich, K.J. (1973) In *Handbook of Physiology Section 8, Renal Physiology.* (J. Orloff and R.W. Berliner, editors) Washington, D.C.: American Physiological Society. p. 377.
Ussing, H.H., & Windhager, E.E. (1964) *Acta Physiol. Scand. 61:* 484.
Warnock, D.G., & Burg, M.B. (1977) *Am. J. Physiol: Renal, Fluid and Electrolyte Physiol.* In press.
Warren, Y., Luke, R.G., Kashgarian, M., & Levitin, H. (1970) *Clin. Sci. 38:* 375.
Welling, L., & Grantham, J. (1972) *J. Clin. Invest. 51:* 1063.
Wright, E.M., Smulders, A.P., & Tormey, J. McD. (1972) *J. Memb. Biol. 7:* 198.

INTRODUCTORY REMARKS OF THE GENERAL SESSION

A. Kleinzeller

University of Pennsylvania

It gives me great pleasure to open this last session of our symposium. If at all required, the past three sessions have supplied evidence that the basic problems as well as mechanisms of dealing with the water relations in membrane transport are very similar in all living matter. This certainly does not come as a surprise. After all, it is more than 25 hundred years ago that Thales of Miletus showed his intuitive philosophical insight by pointing out παντα νδωρ εστι i.e., All Things - - that is, living things -- are Water. We are now trying to provide the phenomenology and eventually the molecular mechanisms, for his statement.

Since on this planet life without water does not appear to exist, living matter strives to maintain the cellular water at a level consistent with the performance of its basic functions. Please note that by this statement I have declared my bias in favor of the view that the activity of cellular water is rather close to that predictable for aqueous solutions of bulk cellular solutes. To my mind, convincing evidence for opposite views (see e.q., Ling, 1969) is not available. The basic problem plant and animal cells face is how to deal with major gradients of water activity across their membranes under varying external conditions. This is the broad subject of this general session. The mechanisms available to cells when dealing with major gradients of water activity across cell membranes to be rather restricted.

Some cells are capable of decreasing the permeability of their membranes to water to rather low levels. Thus, the extremely low permeability of the trout egg to water prevents the cells from drowning in its natural habitat, as shown by Prescott and Zeuthen (1952; See also Gray, 1932; Krogh and Ussing, 1937). Restrictions to water (and solute) movement into cells may also be brought about by a rigidity of the interphase between the cells and their immediate environment (e.g., cell walls and, possibly, rigid cell membranes). As we have heard from Dr. Gutknecht, this mechanism entails the establishment of major hydrostatic pressure gradients across the cell membrane. A low water permeability of the cell membrane also allows both plants and animals to minimize water losses and thus survive under conditions of extremely low partial pressures of water vapor, e.g. in deserts.

The second basic mechanisms available to cells to maintain their cell water are so called solute and possibly also water pumps. As we heard the day before yesterday, such pumps allow brine shrimps to survive even in brines of 3 M NaCl.

An analysis of these phenomena involves not only the understanding of the relationships between the physiocochemical properties of electrolyte solutions, their flows across membranes and related phenomena, but also physiological mechanisms by which the basic mechanisms involved can be regulated. The possible regulation of the membrane water permeability in the renal tubule by the action of the antidiuretic hormone may serve here as an example (Morgan and Berliner, 1968; Rocha and Kokko, 1974). By discussing these problems we then hope to put our fingers on points of joint interest to physiologists of the plant, invertebrate and mammalian denominations with the aim of provoking further search for the understanding of the molecular mechanisms involved.

This brings me to the second point I wish to make. By listening to the achievements in the study of water relations in membrane transport in plant, invertebrate and mammalian cells, we may appreciate the wisdom of A. V. Hill (1950) when he challenged primarily biochemists to learn to chose for their investigations that biological object which most clearly shows the phenomenon of interest. Papers reported in this symposium in the past three sessions, as well as reports to be given today, indeed show that the membrane physiologist has a wide choice of biological material to select from.

REFERENCES

Gray, J. (1932). *J. Exp. Biol.* 9, 277.
Hill, A. V. (1950). *Biochim. Biophys. Acta 4*, 4.
Krogh, A., & Ussing, H. H. (1937). *J. Exp. Biol. 14*, 35.
Ling, G. N. (1969). *Int. Rev. Cytology 26*, 1.
Morgan, T., & Berlinger, R. W. (1968). *Am. J. Physiol. 215*, 108.
Prescott, D. M., & Zeuthen, E. (1952). *Acta Physiol. Scand. 28*, 77.
Rocha, A. S., & Kokko, J. H. (1974). *Kidney Internat. 6*, 379.

ENERGY COUPLING IN ION AND WATER FLUXES ACROSS PLANT MEMBRANES

J. B. Hanson

University of Illinois

INTRODUCTION

In view of the other papers given at this symposium, it will not be necessary to belabor the fact that the water content of plant cells is osmotically generated. Metabolically produced organic solutes plus inorganic ions accumulated from the soil solution lower the osmotic potential (or in the older terminology, increase the osmotic pressure). Water enters rapidly until the pressure potential (formerly, turgor pressure) offsets the osmotic potential. The tensile strength of the restraining wall provides osmotically inflated cells which give the turgidity needed to display leaves, force roots through the soil, open stomates, etc.

During growth, the wall resistance is lowered in hormonally-mediated reactions, permitting cell extension driven by the internal pressure. When water becomes limiting, there are partially compensating increases in solute concentration, permitting a positive osmotic gradients and turgor pressure to be maintained. In short, the plant functions with an osmotically driven hydraulic system, rather than muscles. The system is well regulated.

Pressure-producing osmotic systems require semi-permeable membranes in addition to restraining walls. Water flows rapidly through the cell membranes, but solutes do not. In consequence, the ions which are chiefly responsible for the osmotic gradient must be transported and maintained at higher concentrations in the cell by some special membrane mechanism. A host of observations shows that the mechanism is linked to respiratory energy. Hence, the hydraulic system is powered by energy-linked ion transport systems in the membrane.

A somewhat different situation exists with the organelles in the cytoplasm. Mitochondria and chloroplasts have no restraining wall and must maintain their volume by equalization of their osmotic

concentrations with the surrounding cytosol. If solutes are pumped in, or leak in, an osmotic equivalent must be pumped out. Here,
too, there is energy linkage, as indicated by the in vivo swelling
of chloroplasts in the dark and shrinkage in the light. The volume
changes arise primarily from K^+ and Cl^- fluxes (Nobel, 1975). For
plant mitochondria there are no corresponding in vivo studies, but
in vitro investigations show that the passive swelling which occurs
in penetrating salts of strong acids (KCl, KNO_3) can be reversed
by respiration or ATP hydrolysis (Stoner and Hanson, 1966).

The purpose of this short paper is to review the energy-linked
membrane transport systems involved in these ion and water fluxes.
Water fluxes can be disposed of by simply stating that they are
osmotic (transpiration excluded) and follow ion fluxes. Much of
the comment will be reserved for the properties of the plant cell
membranes, but the task will be made easier by prior consideration
of Mitchell's chemiosmotic hypothesis (Mitchell, 1966) as applied
to mitochondrial transport.

ENERGY-LINKED MEMBRANE TRANSPORT

Figure 1 is a simple schematic representation of the hypothesis.

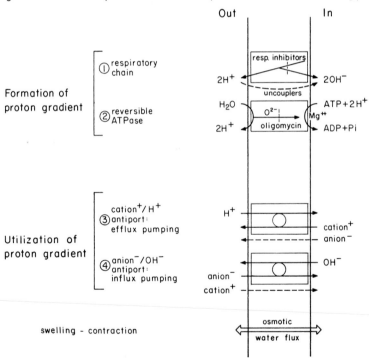

Fig. 1. Chemiosmotic hypothesis for formation of proton gradients
across the inner mitochondrial membrane, and utilization of the resulting proton motive force in ion transport. From Hanson and Koeppe
(1975).

A proton gradient is formed by a "loop" in the respiratory chain or by ATP hydrolysis. The resulting proton motice force, or electrochemical gradient of protons, can be used to transport salts in or out of the mitochondrion. For efflux pumping the chemical potential is collapsed in the exchange of cations (eg: K^+, Na^+) for H^+, with anions effluxing down the electrical gradient through undefined pathways. Conversely, for influx transport there is an anion $^-/OH^-$ exchange accompanied by cation influx. This beautifully simple hypothesis has been widely adopted to explain experimental results on ion transport in both animal (Chappell and Haaroff, 1966; Brierley, 1976) and plant (Hanson and Koeppe, 1975) mitochondria.

An illustrative example is given in Figure 2 (Hensley and Hanson, 1975). The curves are for oxygen concentration and percent transmission for corn mitochondria suspended in buffered sucrose plus 1 mM $MgSO_4$. Oligomycin is present to block the ATPase and rotenone to block endogenous substrate oxidation. NADH is used as substrate

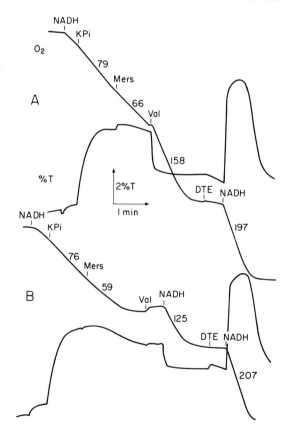

Fig. 2. Respiration-dependent shrinkage (K phosphate expulsion) induced by valinomycin in K phosphate-loaded, mersalyl-blocked corn mitochondria suspended in 0.2 M sucrose, 1 mM MgSO₄, 1 mg/ml BSA, 15 μM rotenone, 1 μg/ ml oligomycin and 10 mM TES buffer, pH 7.6. In part A, additions were 1.5 μmole NADH, 10 mM KH₂PO₄, 25 μM mersalyl, 0.35 μg valinomycin, 0.5 mM DTE. Respiration ceases after second addition of NADH due to O₂ exhaustion. Figures along oxygen trace are μmoles O₂/ min/mg protein. In part B, NADH additions were 0.5 μmole, 0.25 μmole and 0.5 μmole; other additions as in part A. From Hensley and Hanson (1975).

because it is not transported, and by addition of small amounts respiration can be pulsed.

Addition of K phosphate leads to rapid salt accumulation and osmotic swelling as indicated by the upward deflection of the transmission trace. There is a small increase in respiration rate associated with the transport (Hanson et al., 1972). In less than one minute the swelling ceases and the curve comes to a plateau. Since the respiration continues unabated it appears that this cannot be due to stoppage of the phosphate transport system (ie: reactions 1 and 4, Figure 1).

Is this then a pump-and-leak system? This was checked with mersalyl; a water soluble mercurial which very effectively blocks the phosphate transporter without penetrating into the mitochondrion [eg: phosphorylation at the expense of accumulated phosphate can proceed (Hanson et al., 1972)]. Mersalyl reduces the respiration to the original rate, but the back-leakage of salt is much too slow to offset the original pumping rate.

Is the fact that the mitochondria are still respiring responsible for the retention of the K phosphate? As shown in part B of Figure 2, the rate of leakage is not increased when the NADH is exhausted. Indeed, it seems to slow down slightly.

Addition of valinomycin, a lipid-soluble, K^+-binding ionophore which is widely used to equilibrate the electrochemical potential of K^+ across membranes, causes a rapid shrinkage but only if the mitochondria are respiring (cf: Figures 2A and 2B). Apparently, there is an avenue for energy-linked efflux pumping of the salt, (Figure 1, reaction 3) and this may normally operate to remove K phosphate in excess of the plateau level. Blocking the phosphate transporter leaves concentrations of salt at, or lower than, the plateau and the efflux mechanism is not very effective. It can be made effective, however, with valinomycin, which acts by mobilizing K^+ for transport. Salt exits rapidly until a new and lower steady-state is reached.

If the mersalyl-block is reversed with dithioerythritol (or other sulfhydryl protecting reagents) passive efflux via the phosphate transporter is initiated (Figure 2). In the absence of an electrochemical gradient of protons, the Pi $^-$/OH$^-$ antiporter (Figure 1, reaction 4) can work in reverse, driven by the outwardly directed electrochemical gradient of phosphate. This teaches us that the level of internal phosphate held in the normal respiring system must have as one parameter the proton motive force across the membrane.

Another parameter appears when more NADH is added and respiration resumes (Figure 2). Valinomycin is still present, and it leads to a new and higher steady state swelling. Allowing rapid equilibration of K^+ across the membrane permits more phosphate to be held in the mitochondrion before efflux equals influx. Both influx and efflux must be greater than without valinomycin because the rate of respiration is much higher -- greater proton pumping is required to keep reactions 3 and 4 operating (Figure 1). In effect, valinomycin uncouples the mitochondria in cyclic salt transport.

When respiration ceases the efflux of salt is very rapid. Without a proton motive force in opposition, Pi$^-$ and K$^+$ exit down their gradients, exchanging for OH$^-$ and H$^+$, respectively. The net result is one molecule of K phosphate lost, and one molecule of water entering (and rapidly equibrating). Figure 3 adds to these conclusions. In these experiments, Na phosphate was used, and part A

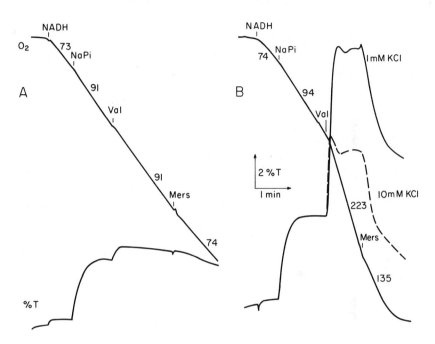

Fig. 3. *Induction of active shrinkage by blocking the phosphate transporter in valinomycin-treated corn mitochondria. Experimental conditions and additives as in Figure 2, but using low-K mitochondria and substituting NaH$_2$PO$_4$. Part A, no K added. Part B, 1 and 10 mM KCl present. From Hensley and Hanson (1975).*

shows that valinomycin is now ineffective; it does not bind Na$^+$. If KCl is added to the medium, however, as in Part B, valinomycin produces new and high steady state of osmotic swelling. With 1 mM KCl (optinal) the steady state is higher than with 10 mM. As the K$^+$ concentration rises so does the relative efficiency of the efflux pump; this is also true for increasing the concentration of valinomycin (Hensley and Hanson, 1975). Now if mersalyl is added to block the phosphate transport only the efflux pump remains and there is rapid energy-linked shrinkage, proportionately greater with 10 mM K$^+$. Stopping the influx pump also decreases the respiratory energy demand.

There is one last point to be made about the efflux pump. Not only valinomycin, but uncouplers as well can accelerate efflux pumping (Hensley and Hanson, 1975), and FCCP acts synergistically with valinomycin. As with valinomycin, respiration is required.

This will be enough on mitochondrial transport for the general point I want to make on membrane transport mechanisms and osmotic regulation. Plant membranes (like others) have effective systems not only for moving ions in, but also to transport them out. If we accept the chemiosmotic hypothesis, a single energy source, the proton motive force, will drive both influx and efflux. A balance is struck between these fluxes which establishes the osmotic content, which inturn governs the volume of mitochondria and chloroplasts and the pressure in plant cells. The old pump-leak hypothesis for volume and pressure control is not supported by experiment. There is a little non-specific leakage, and we can get more by abusing the membranes, but the evidence is that major passive efflux is only obtained through transport mechanisms (eg: the Pi^-/OH^- antiporter) under conditions where the sustaining proton motive force is lowered. With a normal energy supply the osmotic balance is achieved by the balance between energy-linked influs and efflux pumping of salts.

Figure 4 schematically depicts all the conclusions we reached with respect to corn mitohcondria. We believe the H^+/cation$^+$

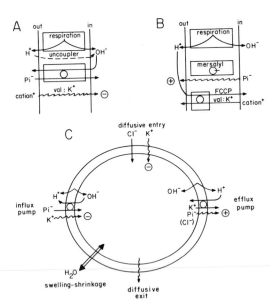

Fig. 4. Chemiosmotic model of potassium phosphate fluxes and steady-state swelling in corn mitochondria (to explain results in Figures 2 and 3). A: influx of phosphate by exchange. B: efflux of cation by exchange. C: steady state fluxes. From Hensley and Hanson (1975).

exchanger is instrumental in osmotic balance, coming into full operation under conditions of high K^+ content and swelling. It seems to be partially regulated by a protective lipid layer which offers resistance to H^+ and K^+ penetration, and which can be experimentally lowered by use of uncouplers and ionophores. Sodium can be substituted for K^+ by using suitable ionophores (Hensley and Hanson, 1975), and hence cation specificity is probably a function of the lipid barrier. Part C of Figure 4 schematically shows the influx and efflux pumps in operation at steady-state swelling.

SALT REGULATION BY PLANT CELLS

Let us move to consideration of plant cells. At first, it might appear that the restraining wall would obviate the need for efflux pumps. However, there is much evidence that plant cells regulate their ion content both qualitively and quantitively. There are upper limits to the amount and kind of ions accumulated. Roots grown in high salt media have a higher salt content than those grown in low salt -- a necessary condition for maintaining favorable osmotic gradients -- but under both conditions healthy plants reach a steady-state in salt content such that an increment of salt uptake is tied to an increment of growth. (There are some interesting deviations here, such as with halophytes which excrete Na^+ salts from their leaves, and the rapid uptake with vacuolar storage of NO_3^-, but as a generalization the above statement is valid.)

A recent paper by Glass (1975) reviews and reopens the problem of salt regulation. He shows that barley roots of low K^+ content exhibit rapid K^+ influx, while those of high K^+ content have much lower influx. The correlation between K^+ influx and K^+ content was best expressed as K^+ influx $/ \overline{(K^+ \text{ content})^2}$, which suggested allosteric inhibition of influx transport by high cytoplasmic K^+ concentration. Glass also speculated that high cytoplasmic K^+ content might repress carrier synthesis. [For completness, I can add that barley root cell potentials might be expected to vary (Pitman et al., 1971), and this might affect K^+ influx.]

Hence, one means of regulating salt content is probably by a negative feedback regulation of the influx transport mechanism. Cram (1975) has suggested that the turgor of the cell might play a role here. He also points out that with non-growing storage tissue at steady-state salt content there still remains an active influx of salt balanced by an equally large efflux as measured by radioactive tracers, and that this is not a pump-and-leak system. Biophysical analysis commonly shows a requirement for active Na^+ efflux at flux equilibrium (Clarkson, 1974), and plant roots which have accumulated Na^+ will actively extrude it on addition of K^+ to the medium (Jeschke, 1972). The high K^+/Na^+ ratio of most plant cells is not attained by a screening mechanism on entry so much as by elective efflux pumping of Na^+ or selective Na^+/K^+ exchange. However, except possibly for salt-tolerant species, a Na^+/K^+ ATPase resembling that of animals is not known for plants (Hall and Baker, 1975). Hodges (1973) has shown that the plasmalemma fraction from oat roots possesses a Mg-requiring, K^+-stimulating ATPase which correlates well with the activity of the roots in K^+ absorption.

It is widely believed on the basis of inhibitor and uncoupler experiments that ATP is the energy source for membrane transport, and chemisosmotic models have been presented to account for it (Smith, 1970; Hodges, 1973).   However, supporting data of the type available for mitochondria and chloroplasts has been meagre. One interesting recent development has been in the treatment of barley roots with fusicoccin, a fungal toxin. Pitman et al. (1975) find that fusicoccin activates K^+ uptake, proton extrusion and hyperpolarization in barley root cells. The authors believe that the results are best explained by a proton pumping mechanism which is accelerated by fucisoccin, hyperpolarizing the cell and causing K^+ influx (ie: a chemiosmotic model).

I can add to this general picture with some recent results from the thesis research of Dr. Willy Lin. The full account is published elsewhere (Lin and Hanson, 1976).

Figure 5 is a generalized illustration of the changes we observe with the washing of corn root tissue, presented in this fashion to save time and space. Much of the data have been published (Leonard and Hanson, 1972a, b,; Lin and Hanson, 1974b). We study this "Washing effect" because we believe that tracing down the parameters involved in the 2 - 3 fold increase in ion absorption rates may provide clues to the membrane transport mechanisms involved. I will not discuss here the small decline and recovery in respiration, adenine nucleotide or K^+ content; present research indicates that these are injury effects which are associated with, but not necessarily antecedent to, the enhanced ion transport which develops with washing.

The important points here are that washing increases the electrogenic cell potential and decreases the H^+ efflux (Lin and Hanson, 1974b). Electrogenic cell potential is defined as that portion of the total potential collapsed by uncouplers and respiratory inhibitors. If the electrogenic potential is due to H^+ efflux driven by an ATPase, one might expect increased ATPase activity and increased H^+ efflux. Cell resistance does not change with washing, so increases in electrogenic proton or ion flux responsible for the electrogenic potential would have to lie with greater ion current; ie: greater turnover of the ATPase or more ATPase. Only a small increase in K^+-stimulated ATPase activity is found, but we are uncertain whether we are measuring the correct ATPase in the correct fashion.

Most puzzling is the decline with washing in net proton efflux. If there is an increase it must be masked by even greater antiporter activity consuming the protons, thus producing the net decrease. In the course of investigating this possibility we discovered that dithioerythritol, or other sulfhydryl-protecting reagents, would cause a net proton influx (Lin and Hanson, 1974). Further study of this phenomenon produced data indicating that root cell membranes have a H^+/K^+ antiporter. Since this bears on the problem of salt and osmotic balance, I will summarize some the data here.

Figure 6 shows that adding 0.5 mM dithioerythritol (DTE) to washed corn root tissue induces a net proton influx. Further

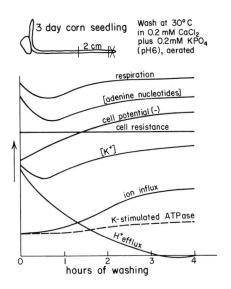

Fig. 5. Generalized graph of changes accompanying the washing of non-growing corn root tissue.

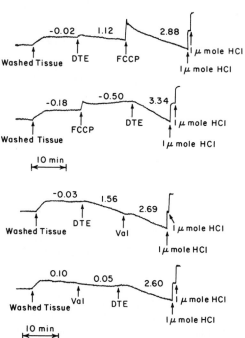

Fig. 6. Induction of H^+ influx in 4-hour washed corn root tissue by dithioerythritol (DTE), and the enhancing effect of uncouplers (FCCP) and ionophores (valinomycin). Medium: 0.2 mM $CaCl_2$ plus 0.2 mM KH_2PO_4, pH 6.0. Additions: 0.5 mM DTE, 10 µM FCCP, 1 µg/mℓ valinomycin. Rates: µmoles H^+/g fr wt. hr. Positive sign indicates H^+ influx. From Lin and Hanson (1976).

addition of an uncoupler, FCCP, greatly accelerates the influx rate. However, FCCP alone does not produce this effect, and hence uncoupling only produces H^+ influx if DTE is present. The same is true for valinomycin, the K^+ ionophore. These results suggest

285

the presence of a H^+/K^+ antiporter like that of the mitochondria. Presumably, DTE activates the enzume by reduction of disulfide bonds or by reversal of some endogenous sulhydryl binding, and access of H^+ and K^+ to the antiporter is through a lipid barrier. The exchange must be passive, driven by electrochemical gradients of H^+ and/or K^+, otherwise uncoupling would have blocked it. Dr. Lin investigated this point in detail, 'and could find no effect of DTE on respiration rate, ATP concentrations in the tissue, or in the activity of the K^+-stimulated ATPase of the microsomal membranes.

Uncoupling does cause reversible depolarization of the cells, reducing the cell potential to the same value in fresh and 4-hour washed tissue (Figure 7). DTE has the opposite effect of hyperpolarizing, and uncoupling will not collapse the hyperpolarization

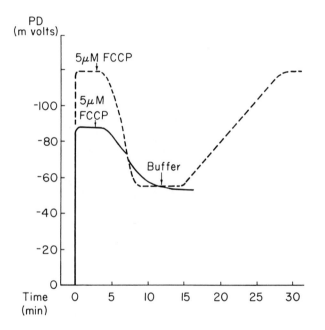

Fig. 7. Reversible depolarization of epidermal cells of fresh (solid line) and washed (dotted line) corn root tissue by uncoupling. Tissue in same solution as Figure 6, with additons and removal of FCCP as indicated. From Lin and Hanson (1974b).

(Figure 8); this is further evidence of the passive nature of the DTE effect.

Table 1 is a summary for washed tissue of the effects produced by DTE. Influx kinetics show a small increase in Km, which is reflected in slightly lower absorption rates from low external concentrations, but Vmax is not significantly affected, again indicating no DTE effect on the energy-linked mechanisms of transport. However, there is a sizable drop in cell resistance due to DTE, and in conjunction with the hyperpolarization this suggests that P_k, the apparent permeability constant for K^+ has been increased by DTE.

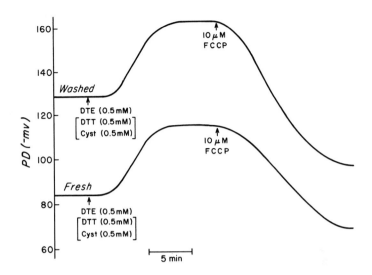

Fig. 8. Hyperpolarization of corn root epidermal cells produced by DTE. Note that the hyperpolarization is in the diffusion potential (cf: action of FCCP, Figure 7). From Lin and Hanson (1976).

Table 1. The effect of dithioerythritol (DTE) on net proton flux, cell potential, cell resistance, and K⁺ absorption kinetics in 4-hour washed corn root segments (see diagram, Figure 5). Data taken from Lin and Hanson (1976). DTE was 0.5 mM. Positive sign indicates proton influx. See original paper for experimental details.

	Proton flux	Cell potential	Cell resistance	K⁺ influx kinetics	
				Km	Vmax
	μ moles/g.hr	mV	megaohms	mM	μ moles/g.hr
- DTE	0.09	-129	12.5	0.022	3.23
+ DTE	1.51	-162	8.0	0.040	3.04

The next question was whether K^+ is effluxing upon DTE treatment. Here we varied pH of the medium, a treatment known to modify H^+ and K^+ fluxes (Poole, 1974). As shown in Table II, the cell potential was unaffected over the range Ph 5 to 7, and hence this component of the elctrochemical potential was constant. I will not deal here with the interesting changes in the control tissue, but simply point out that DTE appears to activate a one-for-one H^+/K^+ exchange.

In other experiments it was determined that the electrochemical gradient of K^+ was directed out of the cells.

Table 2.   The effect of pH on the cell potential and the DTE-
activated H⁺ and K⁺ fluxes in 4-hour washed corn root tissue.   The
medium was 0.2 mM CaCl₂ + 0.2 mM K phosphate with pH adjusted with
KOH.   DTE was 0.5 mM.   See Lin and Hanson (1976) for experimental
details.

pH	Cell potential mV	proton flux* μ moles/g.hr			K⁺ efflux** μ moles/g.hr		
		-DTE	+DTE	Δ	-DTE	+DTE	Δ
5.0	-120	0.33	1.19	0.86	4.27	5.27	1.00
6.0	-115	0.09	1.51	1.42	2.84	4.22	1.38
7.0	-121	-3.48	-1.39	2.09	-0.08	2.05	2.13

*positive sign indicates influx
**extimated efflux from cytoplasm to external solution based on
graphic analysis of the efflux curve of accumulated 86Rb.

Collectively, these show that root cell membranes also have a
H⁺/K⁺ antiporter: in fact, the evidence for this antiporter is
much more conclusive than it is for mitochondria simply because it
can be activated by sulfhydryl protecting agents.

Figure 9 is a chemisosmotic model incorporating what we know
(or think we know) about ion transport in root cells.   It is undoubtedly wrong in detail, but the existance of active and passive
H⁺/K⁺ exchanging systems which cooperate in stabilizing the K⁺
content and the cell potential is likely to be correct.

The evidence for the passive H⁺/K⁺ antiporter has been discussed
above.   It should, of course, operate in both directions transporting K⁺ in at high external pH and K⁺ concentrations, but we have
not yet attempted to determine if it does.   Although not indicated

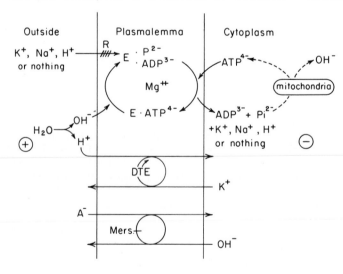

Fig. 9. Chemios-
motic model of
ion transport
in cells with
stable K⁺ con-
centrations
and electrical
potentials.
See text.   From
Lin and Hanson
(1976).

in the figure, we believe that the antiporter has a lipid barrier which must be penetrated by H^+ and K^+, and part of its regulation could lie with the properties of the barrier. Obviously, it could also be regulated by endogenous sulfhydrul-disulfide metabolism. We suspect that high internal K^+ concentrations may make it functional, and we are attempting to find out.

In addition, there must be an active process which develops with washing. Our investigations indicate that an ATPase could provide the energy (Lin and Hanson, 1974a), although there are observations here that are troublesome. Figure 9 assumes an ATPase which produces a proton efflux by means of a polarized alkaline hydrolysis of enzyme-bound ATP. Schematically, this is the simplest way of illustrating the requirements for H^+ efflux plus creation of an extra negative charge for cation transport (of course, the real membrane ATPase may have a more complicated means of achieving this end). Binding of the cation will depend on (a) coulombic forces, (b) properties of the enzyme - ATP complex in binding cations and (c) resistance of the lipid barrier to cation penetration.

The resistance is essential to creating the cell potential. Under conditions of low potential, the ATP hydrolysis cycle can proceed carrying no cation, and yielding an increase in potential. At a certain threshold, the electrophoretic penetration of the barrier by the cation, and cation transport follows. It is observed that Na^+ salts produced higher cell potentials than equivalent K^+ salts (Lin and Hanson, 1976), and this can be explained by greater resistance to penetration of the hydrated Na^+ ion. The model thus has a feedback mechanism for stabilizing the cell potential.

Continuous operation of the ATPase would yield high internal KOH, so there are mechanisms to offset this. If the cell is low in salt, the OH^-/anion antiporter brings in an anion and ejects the hydroxyl, yielding net salt uptake at the expense of the ATP hydrolysis. The anion could be icarbonate (CO_2 penetration would produce the same result as the antiporter) which by the action of phosphoenol pyruvate carboxylase could produce organic salts (eg: K malate).

If the cell is salt saturated, we visualize that the H^+/K^+ antiporter goes into action. The net result here is to waste an ATP, but this might be necessary; allosteric control of the ATPase need not be perfect, and on the evidence probably isn't.

SUMMARY

The regulation of water content in plant cells is an osmotic act governed by salt transport through the cell membranes. It seems probable that the active salt transport and its regulation arises from the functioning of a proton-extruding,cation-carrying ATPase in the membrane, which is balanced in its activity by H^+/cation$^+$ and OH^-/anion$^-$ antiporters, much as proposed by Peter Mitchell.

REFERENCES

Brierley, G.P. (1976) *Mol. Cell. Biochem. 10:* 41.

Chappell, J.B. & Haaroff, K.N.(1967) In *Biochemistry of Mitochondria.* (E. C. Slater et al. (eds.)) p. 75. New York: Academic Press.

Clarkson, D. (1974) *Ion Transport and Cell Structure in Plants.* John Wiley, New York.

Cram, W.J. (1975) In *Ion Transport in Plant Cells and Tissues,* (D.A. Baker & J.L. Hall, eds.), p. 161. New York: American Elsevier.

Glass, A. (1975) *Plant Physiol. 56:* 377.

Hall, J.L. & Baker, D.A. (1975) In *Ion Transport in Plant Cells and Tissues* (D.A. Baker & J.L. Hall, eds.), p. 39. New York: American Elsevier.

Hanson, J.B., Bertagnolli, B.L. & Shepherd, W.D. (1972) *Plant Physiol. 50:* 347.

Hanson, J.B. & Koeppe, D.E. (1975) In *Ion Transport in Plant Cells and Tissue.* (D.A. Baker & J.L. Hall, eds.), p. 79. New York: American Elsevier.

Hensley, J.R. & Hanson, J.B. (1975) *Plant Physiol. 56:* 13.

Hodges, T.K. (1973) *Adv. Agronomy. 25:* 163.

Jeschke, W.D. (1973) In *Ion Transport* (W.P. Anderson, ed.), p. 285. New York: Academic Press.

Leonard, R.T. & Hanson, J.B. (1972) *Plant Physiol. 49:* 430.

Leonard, R.T. & Hanson, J.B. (1972) *Plant Physiol. 39:* 436.

Lin, W. & Hanson, J.B. (1974a) *Plant Physiol. 54:* 250.

Lin, W. & Hanson, J.B. (1974b) *Plant Physiol. 54:* 799.

Lin, W. & Hanson, J.B. (1976) *Plant Physiol.* (in press).

Mitchell, P. (1966) *Chemiosmotic Coupling in Oxidative and Photosynthetic Phosphorylation.* Glynn Research Ltd., Bodmin, Cornwall, England.

Nobel, P.S. (1975) In *Ion Transport in Plant Cells and Tissues* (D.A. Baker and J.L. Hall, eds.), p. 101. New York: American Elsevier.

Pitman, M.G., Mertz, S.M. Jr., Graves, J.S., Pierce, W.S. & Higenbotham, N. (1971) *Plant Physiol. 47:* 76.

Pitman, M.G., Schaefer, N., & Wildes, R.A. (1975) *Plant Science Letters 4:* 323.

Poole, R.J. (1974) *Can. J. Botany 52:* 1023.

Smith, F.A. (1970) *New Phytol. 69:* 903.

Stoner, C.D. & Hanson, J.B. (1966) *Plant Physiol. 41:* 255.

CELL VOLUME CONTROL

Floyd M. Kregenow

National Heart and Lung Institute

Introduction

Cell size is largely determined by factors that control cell water, since in most cells water constitutes 70-90% of cell volume (Dick, 1966). Cell size and its physiological control have been studied with genetic and physicochemical approaches.

Boveri (1905, 1925) pioneered the genetic approach by showing that the sizes of diploid and tetraploid cells from sea urchin embryos were respectively twice and 4 times that of haploid cells. This study established the concept that a relationship exists between cell size and DNA content, which we shall call the DNA-cytoplasmic ratio. When homologous cells in a similar functional state from animals (Mirsky and Ris, 1951; Szarski, 1970; Pedersen, 1971; Bachmann, Goin and Goin, 1972) or plants (Rees, 1972) of different genera are compared, it is found that the DNA-cytoplasmic ratio is remarkably uniform despite major differences in DNA content. Figure 1 shows a collation of data on erythrocytes of different vertebrates taken from the work of several laboratories (Vendrely and Vendrely, 1949; Mirsky and Ris, 1951; Gerzeli, Casati and Meneghetti-Gennaro, 1956; Pederson, 1971). We have compared DNA content in picograms/cell to cell volume in cubic microns. An increase in DNA content is generally associated with an increase in cell volume.

It is unclear how this information about cell size is genetically transmitted. Transmission does not have the characteristics of simple Mendelian inheritance. Furthermore, it is not the total quantity of DNA but the nature of individual chromosomes that is important (Dobzhansky, 1929; Longwell and Svihla, 1960; Commoner, 1964; Schmidtke, Zenzes, Dittes and Engel, 1975). Finally, during the cell cycle, cell growth is not necessarily coupled to the DNA division cycle (Mitchinson, 1971; Fantes, Grant, Pritchard, Sudbery

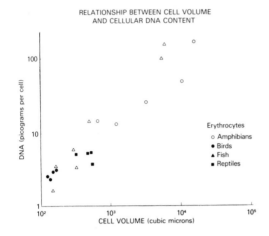

RELATIONSHIP BETWEEN CELL VOLUME
AND CELLULAR DNA CONTENT

Erythrocytes
○ Amphibians
● Birds
▲ Fish
■ Reptiles

Fig. 1. Relationship between cell size of erythrocytes of various vertebrate classes and the organisms' characteristic cellular DNA content. Data from Vendrely & Vendrely, 1949; Mirsky & Ris, 1951; Gerzeli, Casati and Meneghetti-Gennaro, 1956; Pedersen, 1971.

and Wheals, 1975).

The results of the physicochemical approach are more familiar. A membrane solved a fundamental problem for the cell by preventing the dispersion of essential macromolecules, but at the same time its presence created another problem. If a membrane is highly permeable to water, freely permeable to ions like Na, K, and Cl, and encloses a system like cytoplasm which contains not only charged macromolecules but also ions, then the volume of the system is unstable (Overbeck, 1956; Tosteson, 1963).

Cells evolved many different solutions to this osmotic problem. For instance, the membrane of the fertilized trout egg has become virtually water impermeable (Gray, 1932; Krogh and Ussing, 1937; Potts and Rudy, 1969). Some protozoa, on the other hand, have developed a semirigid pellicle and a water excreting system, the contractile vacuole. In cells with rigid walls, volume stability has been achieved by simply allowing hydrostatic pressure to develop within the cell so that differences in pressure and the activity of water nearly balance each other. In such cells, cell wall synthesis controls cell volume. For naked cells, estimates of internal pressure (Rand and Burton, 1964) and observations on the intracellular and extracellular activity of water (Robinson, 1960) have confirmed the previous suspicion of a weak membrane and indicate that hydrostatic forces are not normally important in volume control. How these cells achieve volume control was anticipated by early workers. Jacobs (1931), for instance, suggested that the membrane of such cells could solve the osmotic problem 1) by being impermeable to cations but not to anions, or 2) by being permeable to cations but not to anions. A third scheme was proposed based on the work of Hober (Höber and Höber, 1928; Höber and Hoffman, 1928). It depended on the development of separate regions in the membrane, some of which were permeable to cations but not to anions and others which were permeable to anions but not cations. In each instance the forces determining electroneutrality within the cell would prevent net solute uptake and colloidosmotic swelling.

Observations during the period 1930-1950 demonstrated, however, that animal cells were permeable to both cations and anions. This made it necessary to revise the concept of absolute impermeability to one of relative impermeability. In 1960, Tosteson and Hoffman (1960) combined a revised version of Jacob's first scheme and the pump-leak hypothesis (Shaw, 1955) to formulate an explicit model for the steady-state control of cell volume in sheep red cells. In Figure 2, similar properties are attributed to a duck red cell.

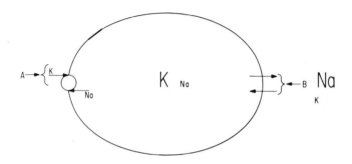

Fig. 2. Schematic drawing of a duck erythrocyte maintaining a steady-state volume and cation concentration. Part A represents the sodium-potassium pump; Part B represents diffusional leaks.

The cell is shown as having a high K and low Na concentration while incubating in a medium with a high Na and low K concentration. According to the model, the cell's cation composition and steady-state volume result from the continued action of a cation pump, labeled A in the drawing, balancing diffusional leaks for Na and K, labeled B in the drawing. The cation pump is pictured as a coupled mechanism, moving Na out of and K into cells against their electrochemical gradients.

VOLUME REGULATION IN DUCK ERYTHROCYTES

Avian erythrocytes, depicted by Figure 2, have proved to be a valuable model for studying the regulation of cell volume. When duck erythrocytes are added to a hypotonic solution, they swell initially; when added to a hypertonic solution, they shrink. These initial changes in volume are expected and result from osmosis. Upon continued incubation, however, the duck red cells return to their original isotonic volume even though they remain in the anisotonic media (Kregenow, 1971a,b). This response to continued incubation in anisotonic media represents volume regulation. It is caused by changes in cation transport that lead to shifts in cell water. We shall see that the cation transport

FLOYD M. KREGENOW

mechanisms responsible for volume regulation are independent of the
pump but have a transport capacity and precision of control nearly
comparable to that of the pump.

The ability of duck red cells to regulate their volume in anisotonic media resembles a form of volume regulation common to euryha-
line invertebrates (Schoffeniels, 1967). In these invertebrate
cells, shifts in cell water are brought about by altering the number of osmotic particles within cells, so that cells change volume
while remaining isosmolar with their surroundings. The adjustable
cellular constituents are primarily organic acids which are usually
controlled metabolically.

In duck red cells the regulatory process involves the controlled
movement of potassium into or out of the cell. The potassium in
turn is accompanied by diffusable anion, primarily chloride, and
finally osmotically obligated water. We have labelled the cellular
elements which regulate the total number of osmotic particles a
Volume Controlling Mechanism (Kregenow, 1971a,b). Although ouabain
inhibited the sodium and potassium pump in duck erythrocytes and
caused sodium and potassium concentrations to change reciprocally,
it did not inhibit the Volume Controlling Mechanism (Kregenow,
1971a,b, 1973). Thus the Volume Controlling Mechanism is functionally separate from the pump.

The process that causes cells to shrink and lose K, Cell Shrinkage, differs from the process that causes enlargement and gain of
K, Cell Enlargement. For this reason, the two phenomena will be
discussed separately.

CELL SHRINKAGE

Figure 3 illustrates what happens when previously enlarged cells

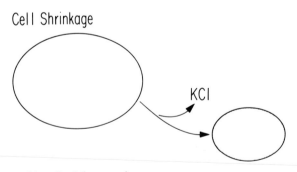

Cell Shrinkage

Fig. 3. Schematic
representation of
Cell Shrinkage.

KCl

Volume Regulation occurs in:
(1) Hypotonic Media
(2) An Isotonic Medium

shrink (Kregenow, 1971a, 1973). The small dark ellipse represents
a cell with a normal volume, and the larger ellipse, a swollen cell.

Cell Shrinkage occurs not only in hypotonic media but also in isotonic media, provided the cells are previously swollen by a procedure that increases their cation content. The swollen state triggers
Cell Shrinkage. The nature of the triggering mechanism is unknown.
The increase in cell size which starts the response can be as small
as 4% or as large as 50%. Cells shrink rapidly, returning to near
their original isotonic volume within 15-45 minutes. Controlled
loss of potassium characterizes the response. Normally, chloride
follows the potassium so that electroneutrality is maintained.
Sodium content and permeability do not change significantly.

Figure 4 shows the fine control of K loss. The cells were
initially swollen 25% by introduction to a hypotonic medium. This

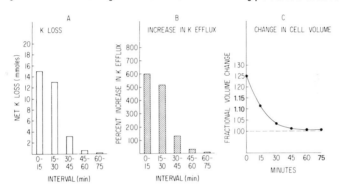

Fig. 4. *The temporal changes in potassium content, potassium efflux
and cell volume as duck erythrocytes regulate their volume in a
hypotonic medium. The measured changes in K content are expressed
in millimoles; the changes in K efflux are presented as the percent
increase above the value of control cells incubating in an isotonic
medium. Changes in cell volume are expressed as fractional volume
-- isotonic control cells have a value of 1.00 on this scale.*

figure compares the net changes in K loss (Part A) with the corresponding changes in K efflux (Part B) and cell volume (Part C) during
successive 15 minute intervals. As cells shrink they lose K. The
loss is the result of a temporary increase in K efflux. Note that
K efflux was initially very high in the swollen cells, but returned
to a normal value as cell K content decreased and cell volume was
corrected. We infer from these findings that the change in K
efflux is constantly determined by the degree of cell swelling.

The increase in K efflux is presumably an increase in simple
diffusion. The following evidence supports this hypothesis (Kregenow, 1971a, 1973). First, the net movement of K is downhill. Eliminating the K electrochemical gradient by raising the extracellular
K concentration blocks both net K loss and volume changes. Second,
the fluxes with or without a K gradient are consistent with a potassium leak hypothesis. Finally, ouabain, which specifically
inhibits the pump, does not alter volume changes, net K loss, or

the increase in K efflux.

To investigate the role played by cellular Na and K, we enlarged cells by selectively loading them with salts (Kregenow, 1974). Figure 5 shows what happened when these experimentally enlarged cells were incubated in isotonic media. Four populations of cells

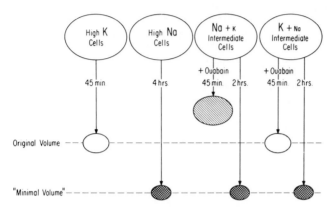

Fig. 5. Schematic representation of the response of experimentally enlarged (cation loaded) cells to incubation in isotonic media. Four kinds of enlarged cells are shown; all contained twice the normal quantity of cations.

were produced: high K cells, high Na cells, and two kinds of intermediate cells. Each of the four populations contained twice the normal quantity of cation; the intermediate cells differed in that they contained both K and Na rather than all K or Na. Intermediate cells containing more Na than K have been labelled Na + K cells, whereas cells containing more K than Na have been labelled K + Na cells.

Consider the high K cells first. Enlarged high K cells reacquired and then maintained their original volume, represented by the smaller cell. The Volume Controlling Mechanism was responsible for this regulation.

High Na cells also shrank, but the process differed. First, they took 5 times longer to reach their original volume and then continued to shrink to 4/5 of that volume before stabilizing at that smaller size. This latter volume has been labelled the "minimal volume" and is represented by the smallest cell. The cation pump, not the Volume Controlling Mechanism, was responsible for this shrinkage. Inhibiting the pump prevented Na and water loss.

In intermediate cells the pump and the Volume Controlling Mechanism operate simultaneously. The pump removes excess sodium and the Volume Controlling Mechanism excess potassium. When intermediate cells are incubated in the presence of ouabain, which prevents the pump from removing Na, only the K + Na cells regulate to their original volume. Na + K cells, in which the quantity of K is limiting, shrink rapidly at first, but stop before reaching their original volume.

Without ouabain, on the other hand, the pump remains operative and both types of intermediate cells act as high Na cells and shrink until they reach the "minimal volume". Since both mechanisms operate, both sodium and potassium are lost and shrinkage is faster than with high Na cells.

Using intermediate cells, we determined that the pathway taken by a potassium ion as it left the cell through the Volume Controlling Mechanism is different from that taken by a sodium ion expelled by the pump (Kregenow, 1974). Eliminating cation loss through either mechanism does not affect the rate of cation loss through the other. If both mechanisms share and compete for the same route, a difference would be noted.

The role played by anions is different. In contrast to the specific requirement for K, there is no specific anion requirement. Cells containing bromide or sulfate instead of chloride still regulate their volume, despite the fact that a different anion accompanies the exiting potassium. However, the response when sulfate is present is somewhat slower.

CELL ENLARGEMENT

Figure 6 portrays the other aspect of volume regulation, Cell Enlargement. This occurs not only in hypertonic media where the

Cell Enlargement

Volume Regulation occurs in:
(1) Hypertonic Media, containing an elevated $[K]_o$ (2.5 ⟶ 15mM)

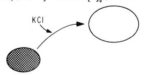

(2) Isotonic Media, containing both norepinephrine and an elevated $[K]_o$ (2.5 ⟶ 15mM)

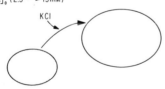

Fig. 6. Schematic representation of Cell Enlargement.

initial osmotic shrinkage seems to trigger the response (Kregenow, 1971b), but also in isotonic media to which norepinephrine is added (Riddick, Kregenow, and Orloff; Kregenow, 1973). (Thus, enlargement can be initiated hormonally.) Cells enlarge 4-15% in volume by accumulating potassium, chloride, and H_2O. The potassium accumulates against an electrochemical gradient. In order for the cells to swell, media K concentration $[K]_o$, has to be greater than 2.5 mM. This effect of $[K]_o$ on cell enlargement plateaus at 15 mM. Typical permeability changes occur, however, at all extracellular K concentrations between 2.5 and 15 mM.

Part of the gain in cell K results from the following sequence.

Permeability to Na increases, bringing sodium into the cell down
its electrochemical gradient. The pump, stimulated by the additional sodium, exchanges Na for K in a one to one fashion. This process accounts for nearly half the K taken up by cells. The rest
of the uphill accumulation of K, and in fact most of the accumulation in the early stages of the response, occurs by a process independent of the pump. If the pump is inhibited K still accumulates
in the cells in addition to sodium. Both processes are associated
with a dramatic increase in permeability which augments all sodium
and potassium fluxes (Kregenow, 1971b, 1973). Figure 7 demonstrates
how great the changes in cation permeability are. Influx and efflux

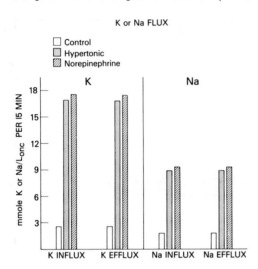

Fig. 7. The increase in
sodium and potassium fluxes
produced by norephinephrine
or hypertonicity. Flux
measurements were determined
for a 15 minute interval
beginning with either 1)
the introduction of $10^{-6}M$
norepinephrine to a suspension of cells in an isotonic
medium or 2) the addition
of cells to a hypertonic
medium (475 milliosmoles/
liter). All media had a
$[K]_o$ of 2.5 mM.

of K and Na in the control condition (open column) are compared
to those after the addition of cells to either hypertonic media
(solid column) or an isotonic solution containing norepinephrine
(slashed column). The ordinate is the flux measurement during
15 minutes. Since the external potassium concentration was only
2.5 mM, the change in permeability occurred without net cation
movement or swelling of cells. Both hypertonicity and norepinephrine increased all four fluxes above control values by a factor
of 5 to 10.
   The following evidence indicates that this generalized increase
in permeability is an intrinsic part of Cell Enlargement (Kregenow,
1971b; Kregenow, Robbie and Orloff, 1976). Identical increases
in permeability develop irrespective of whether the stimulus is
hypertonicity or catecholamines. In addition, almost all of the
increase in permeability is ouabain-insensitive, as in Cell Enlargement. Furthermore, the increase in permeability gradually subsides
as the cells approach their final enlarged volume. In contrast,
if the external potassium is only 2.5 mM and the cells fail to enlarge, the increase in permeability continues. Finally, agents or

experimental procedures which inhibit the increase in permeability invariably inhibit Cell Enlargement as well.

The following picture of Cell Enlargement emerges. Both stimuli, norepinephrine and hypertonicity, activate a ouabain-insensitive transport process(s) whose salient feature is the rapid movement of sodium and potassium in and out of the cell. When the $[K]_0$ is greater than 2.5 mM, Na and K influx exceed effluxes, resulting in the accumulation of both Na and K. This additional Na and K, followed by anions and water, is responsible for Cell Enlargement. Any newly acquired Na is ordinarily extruded by the pump in exchange for extracellular K. The final result is an enlarged cell, containing more K, Cl, and water. Thus, potassium is actively transported into the cell through two mechanisms. One is an unknown mechanism associated with the rapid bidirectional K fluxes, and the other is the Na-K exchange pump. Ouabain inhibits only the pump.

The nature of the mechanism responsible for the increase in permeability to Na and K is unclear, especially as this increase only leads to cation accumulation when the extracellular K concentration is between 2.5 and 15 mM.

Three of the fluxes, K influx and efflux and Na influx require the presence of both extracellular Na and K. If either is absent, movement ceases. These responses saturate, and the shapes of the curves describing them sometimes approximate Michaelis-Menten kinetics. Thus, under some conditions there may be coupling of the fluxes as in co-transport. Na and K would serve as the co-transported species. There are other conditions, however, where the findings are not compatible with co-transport.

HORMONAL EFFECTS

Interest in avian erythrocytes as a preparation for studying how hormones act began with the pioneering studies of Orskov (1956) and Sutherland. Sutherland and co-workers showed that catecholamines produce an elevation of cyclic AMP in pigeon erythrocytes by stimulating an adenyl cyclase system located in the membrane (Klainer, Chi, Freidberg, Rall and Sutherland, 1962; Davoren and Sutherland, 1963a,b; Oye and Sutherland, 1966; Sutherland and Robinson, 1966). This finding served as part of the basis for the general hypothesis that a variety of hormones exert their characteristic effect in receptor cells via the intermediacy of cyclic AMP (Sutherland and Robinson, 1966). Because dibutyryl cyclic AMP mimicked norepinephrine, in that it also activated the volume regulatory system, we suggested that cyclic AMP was responsible for the peremability changes. We have since demonstrated that increases in hormone concentration, up to the point at which transport saturates, cause proportional increases in K influx and cellular cyclic AMP levels (Kregenow, Robbie and Orloff, 1976).

In contrast, the volume regulatory response to hypertonicity is not mediated by cyclic AMP. Cyclic AMP levels were not elevated in the hypertonicity-stimulated cells despite a similar increase in K influx (Kregenow, Robbie and Orloff, 1976). Thus, although

both norepinephrine and hypertonicity have the same effect on transport, cyclic AMP is not an intermediate when the stimulus is hypertonicity. One can picture the two stimuli as initiating events
which are at first dissimilar, but which eventually converge upon
a common pathway leading to cation translocation. Consequently,
the hypertonicity-stimulated transport process becomes a useful
system for examing, independent of cyclic AMP effects, events that
develop beyond the point at which cyclic AMP acts.

CONCLUSION

   We presume that the Volume Controlling Mechanism contains
both   a receptor and effector. Very little is known about the
receptor, which probably senses differences in cell size. We
believe the effector, the portion of the Volume Controlling Mechanism responsible for transport, is composed of more than one element.
This idea is expressed in Figure 8, which shows a cell regulating
its volume. Under these unsteady-state conditions, unlike the
steady-state conditions of Figure 2, membrane pathways C and D become apparent in duck erythrocytes. It is through pathway C, represented by the single arrow, that the Volume Controlling Mechanism,
labelled VCM, controls K loss during Cell Shrinkage. Pathway
D, the double-headed arrow, repesents the pathway or pathways
through which bidirectional cation movements take place. As shown,
transport by the Volume Controlling Mechanism is separate from
transport by the pump. In this model, the Volume Controlling
Mechanism fixes the total Na + K content, while the pump determines
the absolute Na and K concentrations.

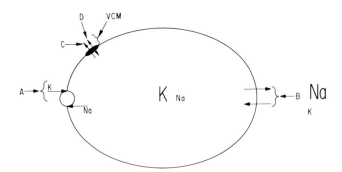

*Fig. 8. Schematic drawing demonstrating the distinction between
the Volume Controlling Mechanism and the sodium-potassium pump in
a duck erythrocyte incubating in anisotonic media. VCM represents
the Volume Controlling Mechanism; A represents the sodium-potassium
pump; B represents the diffusional leaks; C represents the pathway
for K loss during Cell Shrinkage; D represents the pathway(s) for
bidirectional cation movement during Cell Enlargement.*

Many important questions are unanswered. Does the Volume Controlling Mechanism contribute to the control of the steady-state volume of cells? Is it involved in size changes that take place during cell division and differentiation? One speculation is that the emergence of the Volume Controlling Mechanism or a similar system may have represented the crucial evolutionary step that allowed dynamic regulation of volume. This capability was probably a prerequisite for development of plastic cell surfaces and the elimination of the rigid cell wall.

SUMMARY

Physiological control of cell size has long been the subject of genetic and physiocochemical inquiry. Genetic approaches have associated DNA content with cell volume. Physicochemical approaches have concentrated on membrane characteristics and the relationship with water. Avian red cells have proved valuable in such studies. Duck erythrocytes, via cellular elements we have labelled a Volume Controlling Mechanism (VCM), return to their original isotonic volume after being enlarged or shrunken in anisotonic media. They also swell in isotonic media, where the response is hormone dependent. The VCM raises or lowers the total cation content (primarily K); anions follow and the resulting shifts in osmotically obligated water produce the changes in cell size. Such cation movement utilizes transport processes which are functionally separate from the classical cation exchange pump. Dynamic volume control may have permitted cells to lose rigid cell walls and develop plastic cell surfaces.

REFERENCES

Bachmann, K., Goin, O.B. & Goin, C.J. (1972) In *Evolution of Genetic Systems* (Smith, H.H. Ed.) p. 419, New York: Gordon and Breach.

Boveri, T. (1905) *Jena. Z. Naturw. 39:* 445.

Boveri, T. (1925) Cited by E.B. Wilson, in *The Cell*, p. 727, New York: Macmillan Co.

Commoner, B. (1964) *Nature 202:* 960.

Davoren, P.R. & Sutherland, E.W. (1963a) *J. Biol. Chem. 238:* 3009.

Davoren, P.R. & Sutherland, E.W. (1963b) *J. Biol. Chem. 238:* 3016.

Dick, D.A.T. (1966) *Cell Water,* Butterworth, London.

Dobzhansky, T. (1929) *Arch. Entw.Mech.Org. 115:* 363.

Fantes, P.H., Grant, W.D., Pritchard, R.H., Sudbery, P.E. & Wheals, A.E. (1975) *J. Theor. Biol. 50:* 213.

Gerzeli, G., Casati, C. & Meneghetti-Gennaro, A. (1956) *Riv. Istochim. Norm. Patol. 2:* 149.

Gray, J. (1932) *J. Exp. Biol. 9:* 277.

Höber, R. & Höber, J. (1928) *Pfluegers Arch. 219:* 260.

Höber, R. & Hoffman, F. (1928) *Pfluegers Arch. 220:* 558.

Jacobs, M.H. (1931) *Ergbn. Biol. 7:* 1.
Klainer, L.M., Chi, Y.M., Friedberg, S.L., Rall, T.W. & Sutherland, E.W. (1962) *J. Biol. Chem. 237:* 1239.
Kregenow, F.M. (1971a) *J. Gen. Physiol. 58:* 372.
Kregenow, F.M. (1971b) *J. Gen. Physiol. 58:* 396.
Kregenow, F.M. (1973) *J. Gen. Physiol. 61:* 509.
Kregenow, F.M. (1974) *J. Gen. Physiol. 64:* 393.
Kregenow, F.M., Robbie, D.E. & Orloff, J. (1976) *Am. J. Physiol. 321:* 306.
Krogh, A. & Ussing, H.H. (1937) *J. Exp. Biol. 14:* 35.
Longwell, A.C. & Svihla, G. (1960) *Exp. Cell. Res. 20:* 294.
Mirsky, A.E. & Ris, H. (1951) *J. Gen. Physiol. 34:* 451.
Mitchinson, J.M. (1971) *The Biology of the Cell Cycle p. 244:* Cambridge: Cambridge University Press.
Orskov, S.L. (1956) *Acta Physiol. Scand. 37:* 299.
Overbeck, J. Th. G. (1956) *Prog. Biophys. biophys. Chem. 6:* 57.
Oye, I. & Sutherland, E.W. (1966) *Biochim. Biophys. Acta. 127:* 347.
Pedersen, R.A. (1971) *J. Exp. Zool. 177:* 65.
Potts, W.T.W. & Rudy, P.P. (1969) *J. Exp. Biol. 50:* 223.
Rand, R.P. & Burton, A.C. (1964) *Biophys. J. 4:* 115.
Rees, H. (1972) In *Evolution of Genetic Systems* (Smith, H.H.,Ed.) p. 394. New York: Gordon and Breach.
Riddick, D.H., Kregenow, F.M. & Orloff, J. (1971) *J. Gen. Physiol. 57:* 752.
Robinson, J.R. (1960) *Physiol. Rev. 40:* 112.
Schmidtke, J., Zenzes, M.T., Dittes, H. & Engel, W. (1975) *Nature 254:* 426.
Schoffeniels, E. (1967) In *Cellular Aspects of Membrane Permeability p. 157:* Oxford: Pergamon Press.
Shaw, T.I. (1955) *J. Physiol., Lond. 129:* 464.
Sutherland, E.W. Robinson, G.A. (1966) *Pharmac. Rev. 18:* 145.
Szarski, H. (1970) *Nature 226:* 651.
Tosteson, D.C. (1963) In *Cellular Functions of Membrane Transport* (Hoffman, J.F., Ed.) p. 3, Englewood Cliff: Prentice Hall.
Tosteson, D.C. & Hoffman, J.F. (1960) *J. Gen. Physiol. 44:* 169.
Vendrely, R. & Vendrely, C. (1949) *Experentia 5:* 327.

HORMONAL ACTION IN THE CONTROL OF FLUID AND SALT TRANSPORTING EPITHELIA

S.H.P. Maddrell

Cambridge University

INTRODUCTION

While the route and the mechanism by which water crosses fluid secreting epithelia is far from clear, it is relatively well established that ion transport, both that which drives fluid secretion and that involved in systems where fluid flow is minimal, involves ion movements into cells across one area of membrane and out across another. We have seen in the earlier papers just how these ion movements are brought about in a wide range of epithelia.

In many systems the rate of ion transport is controlled by one or more hormones. Since it is supposed that the major barriers to ion movements across epithelia are the cell membranes it comes as no surprise that hormonal control of ion transport is thought to involve changes in the permeability of cell membranes to ions and/ or changes in the rates at which ion pumps sited on the cell membranes operate. These changes can occur very rapidly with a time scale of the order of seconds or the change may take several minutes or even hours to achieve. In the latter case it is thought that new proteins have first to be synthesized and then incorporated into the membranes before any effects are observed. Probably the best known example of this is the effect of aldosterone in increasing the reabsorption of sodium by mammalian kidney tubules. Among insect tissues there is circumstantial evidence to suggest that the ability of some insect Malpighian tubules to transport sulphate ions (Maddrell and Phillips, 1975) might be greatly increased by a slow acting hormone. However, little is known in detail about how these relatively slow changes are brought about by their controlling hormones. They seem at the moment not to involve second messengers within the cell and it is supposed that the hormones themselves enter the cell and may interact directly with the chromosomes to initiate the production of new mRNA which in turn leads to the synthesis of proteins responsible for ion transport.

About the faster acting hormones, more is known. In these cases there is insufficient time for the hormones to enter the cell and there is a transduction of information at the cell surface with the production of so-called second messengers within the cell. Two sorts of internal signals are known to be involved, changes in calcium ion concentration and changes in the levels of cyclic nucleotides (cyclic AMP and cyclic GMP). Much attention has been focused on the cyclic nucleotides and their role in controlling a wide range of cellular activities. In other systems changes in intracellular calcium concentration have been implicated as the main controlling factor. Recently it has become clear that these second messengers do not operate in isolation from one another but rather interact in a cooperative manner to regulate activity. The very widespread nature of this interaction has been the subject of an important recent review (Berridge, 1975). In the present paper I have space only to take a few examples mostly from transporting epithelia of insects to show that they too are controlled in a basically similar fashion to other tissues. A problem peculiar to transporting epithelia is that they are subjected to a flow of substances much of which goes on through their cells; we shall see how they are able to accomodate this.

This paper also describes some recent experiments on insect epithelia which shed some light both on their control mechanism and on the way they carry out transport.

INSECT SALIVARY GLAND

The first example to be considered is the salivary gland of the blowfly, *Calliphora*. In many ways this is an ideal system for study; it survives well in vitro, it can be stimulated by 5-hydroxytryptamine (5-HT) and it is composed of a single homogeneous layer of cells. The tubular glands secrete an isosmotic enzyme-containing fluid largely composed of potassium chloride solution (Berridge and Prince, 1972). The fluid secretion is thought to be driven by an apical potassium pump with chloride ions following passively; potassium and chloride ions enter the cell passively across the basal plasma membrane to replace those transported into the lumen at the other side of the cell (Berridge, Lindley and Prince, 1976). This ion transport leads to an accompanying flow of water in amounts which makes the transported fluid isosmotic. How this coupling is achieved is not clear.

The action of 5-HT is greatly to increase the rate of fluid secretion. That changes in the intracellular level of cyclic AMP is involved is indicated by the following findings (i) in stimulated glands there is a 2-3 fold increase in the intracellular level of cyclic AMP (Prince, Berridge and Rasmussen, 1972), (ii) cyclic AMP applied to the glands causes an increase in fluid secretion (Berridge, 1970), (iii) theophylline can both stimulate secretion by itself and potentiate the action of 5-HT or cyclic AMP in the bathing medium (Berridge, 1970).

Although cyclic AMP can stimulate fluid secretion, electrical measurements show that the stimulated gland is not in the same condition as it is when stimulated by 5-HT. Specifically the transepithelial potential is such that the lumen is at a much higher positive potential than normal. The explanation turns out to be that the permeability of the cell membranes to chloride ions has not increased as it does under 5-HT stimulation (Berridge, Lindley and Prince, 1975). These changes are thought to be mediated by intracellular calcium.

Normal fast secretion can occur in the absence of calcium ions for about ten minutes but then fails. During this time the cells are presumed to use calcium ions from an intracellular store but once this is depleted they require calcium in the bathing medium.

Treatment with the calcium ionophore A 23187 is instructive (Prince, Rasmussen and Berridge, 1973). If it is added to glands in a calcium-free medium, there is a short-lived burst of secretion. Since this occurs even in the absence of 5-HT, it suggests that calcium ions can also increase the activity of the potassium pump.

Whatever modifications of the model future research brings, it seems certain that hormonal control involves changes in intracellular concentration of both calcium ions and cyclic AMP and together these second messengers control secretion by causing changes in both membrane permeability and the rate of action of the potassium pump. Finally it is worth pointing out that hormone action involves changes at both the apical and basal membrane. In discussing mechanisms of epithelial transport there is sometimes a tendency to forget the effects that rapid transport might have on the cells responsible. If the intracellular milieu is not to be dramatically changed during hormonal stimulation, effects on say the apical membrane of the cell must be balanced by others at the basal surface. One should expect therefore hormonal effects to involve changes at both cell surfaces. Indeed this requirement may have been important in leading to the evolution of intracellular second messengers which can virtually simultaneously affect all parts of the cell in response to a hormone which itself is unable, within a sufficiently short time, to reach and affect more than one part of the cell surface.

ION-SELECTIVE MICROELECTRODES AND THE INSECT SALIVARY GLAND

I am most grateful to Dr. M.J. Berridge for permission to quote from his unpublished results using intracellular K^+ selective electrodes. Recent experiments using intracellular K^+-selective microelectrodes have thrown additional light on the behavior of the cells of the blowfly salivary gland during stimulation. The ability to follow the intracellular activity of potassium allows a better understanding particularly of the potential changes occurring at the basal cell membrane which is known to be selectively permeable to potassium. When placed in solutions containing low concentrations

of potassium, the basal cell membrane of an unstimulated gland
rapidly hyperpolarises and then slowly depolarises. The depolari-
sation is readily explained by the finding that there is a steady
loss of potassium from within the cell. When such a potassium
depleted gland is treated with 5-HT there is a recovery of the
large potential across the basal membrane and this is attributable
to a rapid increase in the intracellular potassium activity. One
previously unsuspected effect of 5-HT is thus a stimulation of
potassium uptake across the basal cell membrane. How this is brought about is not yet known but the finding highlights the ability
of the stimulated cell to accelerate potassium trasnport across
both cell membranes, so that the intracellular level is kept within
limits.

In glands stimulated while bathed in more normal bathing media,
the intracellular potassium activity changes scarcely at all in
spite of the very large changes in the rate at which potassium-rich
fluid is produced. This neatly shows how very well the potassium
fluxes across the basal and apical cell walls are normally balanced
and makes the point again that hormonal stimulation leads to coordinated changes at the two cell surfaces.

MAMMALIAN SALIVARY GLAND

As with the insect salivary gland those of mammals secrete not
only fluid but also enzymes and as before the same cells are responsible for both activities. In this case, however, it is known
that the two activities are controlled differently. Stimulation
by acetylcholine released from endings of nerves of the parasympath-
etic system controls fluid secretion, while noradrenaline from
sympathetic nerves causes mainly an accelerated release of amylase
into the saliva.

It is proposed that the following elements are involved in the
stimulation of fluid transport. Acetylcholine causes an influx of
sodium into the cell as a result of a rise in sodium permeability
at the basal surface of the cell. At the same time calcium ions
are thought to enter and they then trigger two further events.
First there is an increase in potassium permeability of the basal
cell membrane which causes a marked hyperpolarization (Petersen,
1970), and a transient loss of potassium (Burgen, 1956). The
potential increase must accelerate sodium entry which of course
makes sense in that the stimulated salivary gland secretes large
quantitites of sodium-rich fluid. It is conceivable that the loss
of potassium ion from the cell may allow its replacement with sodium
without there being any osmotic swelling of the cell.

The other change thought to be triggered by the influx of calcium ions is an increase in level of cyclic GMP (Schultz, Hardman,
Schultz, Baird and Sutherland, 1973) and it is supposed that this
activates apical pumps involved in the production of saliva.

Cyclic AMP is not thought to play a role in fluid secretion but
to regulate the release of amylase-containing granules from the
apical surface.

So in this case again the current hypothesis proposes that changes in the intracellular levels of calcium ions and of a cyclic nucleotide (cyclic GMP) are both involved in causing both an increase in permeability to ions and an activation of ion pumps, and these changes involve both the apical and basal cell membranes. The loss of potassium from the cell compensating for the uptake of sodium, which the cell needs for the secretion of saliva, may be an example of the changes needed to maintain the transporting cell in a viable physiological state during its activity.

MALPIGHIAN TUBULES OF A BLOOD-SUCKING INSECT

Insect Malpighian tubules secrete fluid at a rate which in many cases is known to be under hormonal control (Maddrell, 1971). In most insects the fluid secreted is largely isosmotic potassium chloride solution but those of blood sucking insects, which need to excrete sodium chloride from the ingested blood, have a more pronounced ability to transport sodium ions (Maddrell, 1977). In addition the fluid intake of blood suckers is irregular so that they need to be able to make large changes in the rate of fluid excretion. The hormonal effects on their Malpighian tubules are correspondingly more dramatic for example, the rate of fluid secretion by tubules from *Rhodnius* may increase by a factor of one thousand during hormone stimulation to a rate approaching $5\mu l.min^{-1}.cm^{-2}$ (Maddrell, 1969). In *Rhodnius* that cyclic AMP is involved in the tubule response to hormone stimulation is shown by the facts (i) that the intracellular levels of cyclic AMP increase initially by about four times and are then maintained at a level about twice that of the resting tubule (Aston, 1975) and (ii) that the inclusion of more than about $4 \times 10^{-5}M$ cyclic AMP in the bathing solution causes a marked accleration in the rate of fluid secretion (Maddrell, Pilcher and Gardiner, 1971). One puzzling finding is that theophylline causes no accleration of fluid secretion (Maddrell, Pilcher and Gardner, 1971) although the tubules contain phosphodiesterase able to break down cyclic AMP (Aston, 1975). A possible explanation may lie in the fact that the Malpighian tubules of *Rhodnius* have active transport systems for a variety of organic compounds both anions (Maddrell, Gardiner, Pilcher and Reynolds, 1974) and cations (Maddrell and Gardiner, 1976). Theophylline as a base might well be readily removed from the cells by the very active transport system for such basic compounds that *Rhodnius* tubules have.

The intracellular changes in cyclic AMP have an interesting parallel in the behaviour of the transepithelial potential (Fig. 1), which is most lumen positive at the time when the cyclic AMP level is at its highest. The tubules of *Rhodius* are thought to have an apical cation pump (Fig. 2) and intracellular recordings (Prince and Maddrell, unpublished observations) show that the changes in transepithelial potential are almost entirely accounted for by changes across the apical surface. It seems probable that cyclic AMP has as one of its effects the stimulation of the apical cation pump.

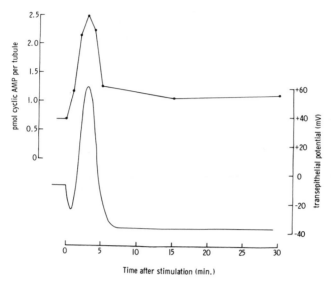

Fig. 1. Changes in the intracellular level of cyclic AMP and trans-epithelial potential following hormonal stimulation of Malpighian tubules of Rhodnius at 22°C. The upper trace shows changes in cyclic AMP (redrawn from Aston, 1975) and the lower line, changes in trans-epithelial potential (lumen with respect to the bathing (medium).

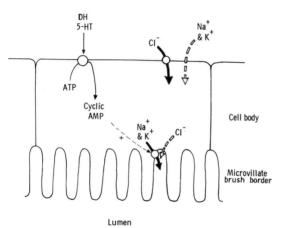

Fig. 2. Hypothetical model of ion transport across the Malpighian tubule of Rhodnius. The possible role of cyclic AMP in stimulating the apical cation pump is indicated. The role of calcium ions, if any, remains unknown.

Very little is known of the role of calcium ions in the control of secretion largely because the tubules seem to have a more pronounced ability to secrete in calcium free media than has for example the blowfly salivary gland; secretion persists for more than 1 h in such media. A possible explanation for this comes from recent work with the electron microprobe (see section: *Possible intracellular store of calcium in Malpighian tubules.*)

It should not be thought, however, that stimulation merely involves an increase in the intracellular level of cyclic AMP. Treatment with tryptamine which binds to the 5-HT receptor site prevents any stimulatory action by cyclic AMP (Maddrell, Pilcher and Gardiner,

1971). Probably one of the effects that cyclic AMP must have if it is to induce secretion depends on the state of the 5-HT receptor site and if this is occupied by a substance such as tryptamine, the effect does not occur. It remains to be seen whether this has anything to do with the intracellular level of calcium ions.

ELECTRON PROBE X-RAY MICROANALYSIS OF INSECT MALPIGHIAN TUBULE CELLS

One of the new techniques which have become available for the study of epithelia is the analysis of X-rays emitted by excitation with an electron microprobe. In this technique a tissue is suddenly deep frozen to about -180°C so as to immobilize the ions and water in it. While still frozen and hydrated, thin sections (1 μm thick) are cut and transferred to a modified scanning electron microscope. A focussed beam of electrons is used both to visualize the tissue and to excite the atoms in a selected area of the frozen section so that they emit X-rays characteristic of the elements present. By analysing the X-rays emitted, the concentrations of such bio- logically important elements as sodium, potassium, chlorine, cal- cium, phosphorus etc. can be determined in particular areas of the tissue as small as .07 μm². Details of the technique are described more fully in a recent paper (Gupta, Hall and Moreton, 1977). A limitation of the method is that it provides information only as to the concentration of various elements and the probability that the activities of the ions may be somewhat lower must be borne in mind.

We have recently used this technique on the fluid secreting parts of the Malpighian tubules of *Rhodnius* (Gupta, Hall, Maddrell and Moreton, 1976). There are three findings which are worth mentioning in the present context.

The first of these is that after stimulation of rapid fluid secretion by treatment with 5-hydroxytryptamine (5-HT) (Maddrell, Pilcher and Gardiner, 1971), the microprobe shows that the general intracellular concentration of sodium increases from about 10mM to around 50mM. The fluid secreted by such stimulated tubules is known to be rich in sodium ions, while the fluid in the lumen of unstimulated tubules contains very much lower concentrations of sodium ions. The situation is summarized in Fig. 3. It is known that in stimulated tubules that there is a strong correlation between the relative rates of transport of sodium and potassium ions and the intracellular concentrations ions as measured by a tracer technique (Maddrell, 1977); this has led to the hypothesis that the tubules may possess a cation pump on the luminal (apical) membrane capable of transporting either sodium or potassium ions but with a higher affinity for sodium ions. Stimulation of fluid secretion by 5-HT may therefore, cause a switch from secretion (at a very low rate) of a sodium poor fluid to (fast) secretion of a fluid much richer in sodium by increasing the relative rate at which sodium enters the cell. This might be simply due to an increase in permeability to sodium ions, but this will require further investigation. As with mammalian salivary glands it will

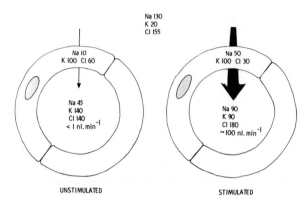

Fig. 3. *Intracellular ion concentrations in stimulated and unstimulated Malpighian tubules of* Rhodnius *as revealed by electron probe X-ray microanalysis (Modified after Gupta, Hall, Maddrell and Moreton, 1976).*

be interesting to know how the cells cope osmotically with an increase in their sodium content - is there a compensating loss of potassium?

Two other observations have implications for an understanding of the mode of operation of these tubules. The first of these is that the microprobe shows clearly that the basal lamina surrounding the Malpighian tubule is not as inert and may be less permeable than had previously been supposed. The concentration of potassium associated with it is much higher than that of the bathing medium while the concentrations of sodium and chloride are depressed to below the levels in the surrounding fluid (Fig. 4). So although the basal lamina at least in some Malpighian tubules is permeable

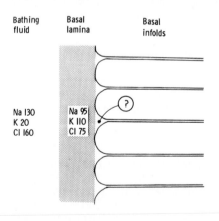

Fig. 4. *Electron probe measurements of ion concentrations in the basal lamina of a Malpighian tubule of* Rhodnius *(Modified after Gupta, Hall, Maddrell and Moreton, 1976). When stimulated, fluid is drawn rapidly through the basal lamina and extracellular space of the basal infolds. Because of the differential affinity of the basal lamina for anions and cations, and for sodium and potassium, the composition of the fluid reaching the cell surface may differ significantly from that of the general bathing medium.*

to such large compounds as horseradish peroxidase (Kessel, 1970), the present findings strongly suggest that the lamina in *Rhodnius* is more permeable to potassium ions than to sodium and chloride ions. If the effect is a large one, it may well be that the fluid actually in contact with the basal cell membrane differs significantly in

composition from that in the bulk of the bathing medium. Any such
effect will of course be magnified during the fast fluid secretion
initiated by hormonal action.

Finally we have been able to make determinations of concentrations of sodium, potassium and chloride in and between the apical
microvilli. The results show (Fig. 5) that there do not appear
to be the gradients of these ions in the extracellular spaces between the microvilli which would give grounds for supposing that

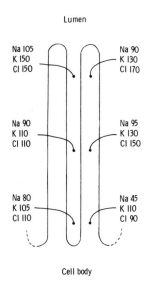

Fig. 5. *Estimations of ionic concentrations in the apical microvilli and intervening extracellular
spaces of stimulated Malpighian
tubules of Rhodnius determined
by electron probe x-ray microanalysis
(Modified after Gupta, Hall, Maddrell
and Moreton, 1976).*

ion and fluid flow might be coupled by a standing osmotic gradient
(Diamond and Bossert, 1967). If anything the results suggest
that the osmotic concentration of the fluid in the extracellular
spaces increases towards the lumen.

POSSIBLE INTRACELLULAR STORE OF CALCIUM IN MALPIGHIAN TUBULES

The use of the microprobe has suggested a possible explanation
for the finding that *Rhodnius* Malpighian tubules will secrete for
prolonged periods in calcium-free media. The cells of insect
Malpighian tubules unlike salivary glands contain large numbers
of granular structures termed dense bodies or mineralised granules
(Sohal, 1974). It now transpires that these structures contain
high concentrations of divalent and trivalent metal ions, among
them calcium. The calcium concentration may reach 30 mM (B.L.
Gupta, personal communication). It seems very probable that these
structures could act as a calcium store additional to those found
in other cells and this may well allow Malpighian tubules to
operate for extended periods under apparently calcium-free conditions.

CONCLUSIONS

From this brief survey of transporting epithelia two general points emerge. The first is that, as in other tissues, hormonal stimulation leads to changes in the level of intracellular controlling factors, both calcium and ions and cyclic nucleotides. These agents act in a coordinated way to accelerate transport by causing changes both in passive membrane permeability and in the activity of ion pumps.

The second point worth emphasizing is that hormonal action is in a few cases now known to lead to changes at both surfaces of transporting cells which are so well coordinated that the intracellular milieu is relatively little disturbed by what may be very large changes in ion traffic through it. Where fast transport of sodium rich fluids is involved, however, the intracellular sodium concentration may rise but in at least one case there may be compensating losses of potassium perhaps directed at avoiding osmotic swelling. Future studies of transporting epithelia may well find a fruitful area in considering how the intracellular environment is maintained or, if it is changed, how possibly damaging consequences are minimized. One way of coordinating transport across both cell membranes is for activity at one membrane to affect transport at the other. A particularly neat example of this is thought to occur in the gills of fresh water fish (Maetz, 1971). The operation of apical sodium pumps leads, as in any active aerobic tissue, to the production of carbon dioxide, and, through the action of carbonic anhydrase to the appearance in the cells of hydrogen and bicarbonate ions. These ions then leave the opposite side of the cell on carriers which exchange them for extracellular sodium and chloride ions. In fact the stoichiometry of the system is such that the production of hydrogen and bicarbonate ions is not nearly sufficient to match the rate at which sodium and chloride ions are transported out of the cell. However, the system does accelerate the uptake of sodium and chloride ions and their uptake is geared to some extent to the rate at which they are pumped out of the cell, for the faster the pumps operate, the more bicarbonate and hydrogen ions are produced and the faster sodium and chloride ions can be taken up. Very similar systems may be involved in salt glands of birds (Peaker, 1971) and in frog skin (Erlij, 1971). All these cases are of epithelia engaged in ion transport, where accompanying water movements would be counter-productive; the exchange of intracellular and extracellular ions is, of course, particularly suited to such transport in that it would not produce any osmotic concentration gradient. It will be of considerable interest to see how coordination of ion transport across the cell membranes of fluid-transporting epithelia is achieved.

ACKNOWLEDGEMENTS

I thank Dr. Charlotte Mangum for some helpful comments. The electron probe X-ray microanalysis was carried out in the Biological Microprobe Laboratory of the Department of Zoology, Cambridge and was supported by a grant from the Science Research Council to (the late) Professor T. Weis-Fogh and Drs. P. Echlin, B.L. Gupta, T.A. Hall and R.B. Moreton.

REFERENCES

Aston, R.J. (1975) *J. Insect Physiol. 21:* 1873.

Berridge, M.J. (1970) *J. Exp. Biol. 53:* 171.

Berridge, M.J. (1975) *Adv. in Cyclic Nucleotide Research. 6:* 1.

Berridge, M.J., Lindley, B.D. & Prince, W.T. (1975) *J. Physiol. 244:* 540.

Berridge, J.H., Lindley, B.D. & Prince, W.T. (1976) *J. Exp. Biol. 64:* 311.

Berridge, M.J. & Prince, W.T. (1972) *Adv. Insect. Physiol. 9:* 1.

Burgen, A.S.V. (1956) *J. Physiol. 132:* 20.

Diamond, J.M. and Bossert, W.H. (1967) *J. Gen. Physiol. 50:* 2061-2083.

Eerlij, D. (1971) *Phil. Trans. Roy. Soc. Lond. B. 262:* 153.

Gupta, B.L., Hall, T.A., Maddrell, S.H.P. & Moreton, R.B. (1976) In preparation.

Gupta, B.L., Hall, T.A. & Moreton, R.B. (1977) In *Transport of ions and water in animals.* (eds. B.L. Gupta, R.B. Moreton, J.L. Oschman & Wall, B.J.). London: Academic Press.

Kessel, R.G. (1970) *J. Cell. Biol. 47:* 299.

Maddrell, S.H.P. (1969) *J. Exp. Biol. 51:* 71.

Maddrell, S.H.P. (1971) *Adv. Insect Physiol. 8:* 199.

Maddrell, S.H.P. (1977) In *Transport of ions and water in animals.* (eds. B.L. Gupta, R.B. Moreton, J.L. Oschman & Wall, B.J.). London: Academic Press.

Maddrell, S.H.P., Gardiner, B.O.C., Pilcher, D.E.M. & Reynolds, S.E. (1974) *J. Exp. Biol. 61:* 357.

Maddrell, S.H.P. & Gardiner, B.O.C. (1976) *J. Exp. Biol. 64:* 267.

Maddrell, S.H.P. & Phillips, J.E. (1975) *J. Exp. Biol. 62:* 367.

Maddrell, S.H.P., Pilcher, D.E.M. & Gardiner, B.O.C. (1971) *J. Exp. Biol. 54:* 779.

Maetz, J. (1971) *Phil. Trans. Roy. Soc. Lond. B. 262:* 209.

Peaker, M. (1971) *Phil. Trans. Roy. Soc. Lond. B. 262:* 262.

Petersen, O.H. (1970) *J. Physiol. 210:* 205.

Prince. W.T., Berridge, M.J. & Rasmussen, H. (1972) *Proc. Natn. Acad. Sci., U.S.A., 69:* 553.

Prince, W.T., Rasmussen, H. & Berridge, M.J. (1973) *Biochem. Biophys. Acta. 329:* 98.

Schultz, G., Hardman, J.G., Schultz, K., Baird, C.E. & Sutherland, E.W. (1973) *Proc. Natn. Acad. Sci., U.S.A., 70:* 3889.

Sohal, R.S. (1974) *Tissue and Cell, 6:* 719.

NON-EQUILIBRIUM THERMODYNAMICS OF
WATER MOVEMENT

Robert A. Spangler

State University of New York at Buffalo

INTRODUCTION

Non-equilibrium thermodynamics, like its classical progenitor, provides the logical framework and formalism to analyze a system essentially independent of the specific details of the system's structure. Being concerned with processes occurring within the system as it naturally evolves toward equilibrium, and the relationships between these processes and the forces driving them, non-equilibrium thermodynamic theory provides a useful and satisfying approach to the unsolved problems of biological transport. Demanding no mechanistic details, the application of non-equilibrium thermodynamics correspondingly yields little or no mechanistic information about the system. Its utility lies rather in circumscribing the domain of the theoretically possible, and serving as a logical guide for specific efforts aimed at unraveling the mechanisms involved.

This paper will briefly review the major threads of the theory of non-equilibrium thermodynamics, with particular emphasis upon its salient features concerning biological transport, and especially the movement of water. At the onset, however, it must be noted that water transport is inextricably entwined with the transport of other molecules; hence the title of this communication is not meant in a restrictive sense. Indeed, the heart of non-equilibrium thermodynamics lies in the interplay between the various processes simultaneously occurring.

NON-EQUILIBRIUM THERMODYNAMICS

Upon the foundation laid by Onsager, (1931a) the subsequent work of numerous investigators, including that of Staverman (1952), and Kedem and Katchalsky (1958, 1960, 1963a,b,c) developed the thermodynamics of irreversible processes to the stage of practical application to the complex systems of biology.

The cardinal principle of non-equilibrium thermodynamics is a generalization of the second law of classical thermodynamics: processes in the natural evolution of a system must be accompanied by an increase in the entropy of the universe.

In terms of the entropy of a finite system, this principle is frequently expressed:

$$\frac{ds}{dt} = \frac{d_e s}{dt} + \frac{d_i s}{dt} \qquad \frac{d_i s}{dt} \geq 0 \qquad (1)$$

with the equality holding only at equilibrium. $\frac{d_e s}{dt}$ is the exchange of entropy with the environment, and is of indefinite sign. $\frac{d_i s}{dt}$, the internal production of entropy, is the quantity of interest, positive in any non-equilibrium process.

The generalized expression devised by Onsager provides a good starting point for the application of this principle. Choosing a set of variables α_j, sufficient to define the state of a system relative to its equilibrium, with all $\alpha_j = 0$ at equilibrium, the entropy of the system can be expanded as a Taylor's series:

$$S - S_o = \Delta S = \frac{1}{2} \frac{\partial^2 S}{\partial \alpha_j \partial \alpha_k} \bigg._o \alpha_j \alpha_k + \cdots \qquad (2)$$

with the first order term vanishing since S is an extremum at the reference state. Taking the time derivative:

$$\frac{ds}{dt} = \sum_{jk} \frac{\partial^2 S}{\partial \alpha_j \partial \alpha_k} \alpha_k \frac{d \alpha_j}{dt} = \sum_j \frac{\partial S}{\partial \alpha_j} \bigg._\alpha \frac{d \alpha_j}{dt} \qquad (3)$$

The $\frac{d \alpha_j}{dt}$, time derivatives of state variables, represent the consequence of a process occurring within the system; they are the generalized *flows* or *fluxes*, commonly assigned the symbol J_j. $\frac{\partial S}{\partial \alpha_j} \bigg._\alpha$, being the first order term in the expansion about equilibrium, are the generalized *forces*, X_j. Thus one arrives at the familiar form

$$\frac{ds}{dt} = \sum_j X_j J_j \qquad (4)$$

It is worth noting that, in equation (3), while S and α are generally extensive variables, the generalized forces, being of the dimensions of their ratio, are intensive. It is also important to observe that at equilibrium, with $\frac{ds}{dt} = 0$, both the X's and the J's vanish.

The couplets of forces and fluxes satisfying equation (4) are known as conjugate pairs; or more precisely, for each independent flux there exists a conjugate force such that (4) is satisfied.

It is also noteworthy that only the summation need meet the inequality of (4), and not each force-flux product independently.

316

It is this fact, of course, which permits the flow of one process to proceed counter to its conjugate force, so long as other processes, simultaneously occurring, maintain the entropy production positive.

The fluxes may be represented by means of a Taylor's expansion in the forces (a set sufficient to define the state of the system relative to the reference equilibrium). Thus

$$J_i = \sum_j \frac{\partial J_i}{\partial X_j} X_j + \frac{1}{2} \sum_{jk} \frac{\partial^2 J_i}{\partial X_j \partial X_k} X_j X_k + \cdots \quad (5)$$

where the partial derivatives are evaluated at the reference equilibrium. The linear phenomenological laws of non-equilibrium thermodynamics are nothing more than the leading term of this expansion:

$$J_i = \sum_j L_{ij} X_j \quad (6)$$

with

$$L_{ij} = \frac{\partial J_i}{\partial X_j}\bigg|_o \quad (7)$$

The validity of (6) is thus restricted to a "small" region in the vicinity of equilibrium. The limits of this domain of linearity depend upon (mathematically) the magnitude of the higher order expansion terms in (5), and (physico-chemically) upon the process involved. The concept of linearity implies that the L_{jk} coefficients are not functions of the X's, although they may be - and generally are - functions of the reference state at which they are defined, as expressed by (7).

Onsanger's monumental contribution to the theory of non-equilibrium thermodynamics is his demonstration, quite generally, of the generally, of the reciprocal relations commonly bearing his name (Onsager, 1931a,b). For a properly chosen set of forces and fluxes, such that the entropy production is given by the sum of force-flux products, Onsager showed that the L_{kj} coefficient matrix must be symmetrical. That is, $L_{kj} = L_{jk}$ for any choice of K and j. This result of Onsager, based upon fluctuation theory, reflects what is apparently a fundamental property of the interaction of physical processes.

The existence of these reciprocal relations is of considerable practical, as well as theoretical, interest and value. In a system in which n independent processes can occur, for example, reciprocity reduces the number of transport parameters which must be determined experimentally, from n^2 to $\frac{n(n+1)}{2}$.

By combining equations (4) and (6) we obtain the well-known bilineal form of entropy production:

$$\frac{ds}{dt} = \sum_{jk} X_j L_{jk} X_k \geq 0 \quad (8)$$

In order to insure that the inequality is satisfied for any arbitrary choice of values for the forces, the coefficient matrix L_{ij} must be a positive definite matrix. This condition requires all diagonal ("straight") coefficients, L_{ii}, to be positive. In addition, restrictions are placed upon the relative magnitudes of the off-diagonal ("cross coupling") coefficients. In a two flow system, for example, these restrictions assume the form

$$L_{11} L_{22} > L_{12}^{2} \tag{9}$$

Analogously, in a more complex system, the general inequality

$$L_{ii} L_{jj} > L_{ij}^{2} \tag{10}$$

must necessarily hold for any two processes, taken pairwise, if the L matrix is positive definite: satisfaction of (10) for all possible pairs is not sufficient, however, to establish the positive definite character of the coefficient matrix.

In connection with the general properties of the Onsager reciprocal relations, it is of interest to examine the persistence of this symmetry beyond the region of linear dependence of fluxes upon the forces. This question was recently elucidated by Sauer (1973). Combining equations (5) and (7), the flows can be expressed generally as

$$J_j = \sum_k L_{jk} X_k + \frac{1}{2} \sum_{ki} \frac{\partial^2 J_j}{\partial X_k \partial X_i} X_k X_i + \cdots \tag{11}$$

Reciprocal relations in the non-linear regime is taken to mean symmetry of the λ coefficient in the psuedo-linear expression

$$J_j = \sum_k \lambda_{jk}(X) X_k \tag{12}$$

in which the λ coefficients are functions of the forces:

$$\lambda_{jk} = L_{jk} + \sum_i \frac{\partial \lambda_{jk}}{\partial X_i} X_i + \cdots \tag{13}$$

Putting (13) into (12), and comparing equivalent terms in the X's with those of (11), we obtain the set of relations

$$\frac{\partial \lambda_{ij}}{\partial X_k} + \frac{\partial \lambda_{ik}}{\partial X_j} = \frac{\partial^2 J_i}{\partial X_j \partial X_k} \tag{14}$$

in the second order terms. Since the partial derivatives of the right hand side are symmetrical with respect to j and k, there are only $\frac{n^2(n+1)}{2}$ such equations, while there exist n^3 different derivatives of the Λ's. Thus equation set (14) is insufficient to uniquely determine the $\partial \Lambda_{ij}/\partial X_k$'s and values satisfying these relations can be assigned in an infinite number of ways. Sauer has shown (Appendix I) that it is always possible to find at least one set of values which will ensure that $\Lambda_{ij} = \Lambda_{ji}$, to the second order in the X's. Similar arguments can be applied to the higher order

318

terms. From this point of view, then, only symmetry of the linear L_{ij} coefficients embodies physical significance; preservation of symmetry relations into the non-linear domain becomes a matter of mathematical convenience.

It must also be borne in mind that this definition of symmetry differs from that based upon differential transport coefficients, ℓ_{ij}, given by

$$\ell_{ij} = \frac{\partial J_i}{\partial X_j} \qquad X \neq o \qquad (15)$$

The distinction between these coefficients, and the L_{ij}, being that the derivatives are evaluated at the non-equilibrium state, rather than at the reference equilibrium. The differential coefficients, (15), outside the region of linearity, are in general not symmetrically related.

Loss of symmetry among the differential coefficients is a necessary condition for the appearance of certain non-linear dynamic phenomena, such as Prigogine's dissipative structures (Glansdorf and Prigogine, 1971), which are excluded in the linear region where reciprocal relations hold.

The unity of this theoretical structure hinges upon the derivation of explicit expressions for the conjugate flows and forces, in terms of measureable system vairables. Based upon the assumed validity of Gibbs' equation in a non-equilibrium state,

$$Tds = d\upsilon + PdV - \sum_i \mu_i \, dn_i \qquad (16)$$

such explicit forms for the driving forces have been worked out. In what might be called the conventional approach, the system is viewed as a continuum, and both forces and fluxes are cast as continuous space variables. From this treatment one can obtain the expression (DeGroot and Mazur, 1962; Katchalsky and Curran, 1965)

$$T = - \sum_i J_i \cdot \nabla \tilde{\mu}_i + \sum_r J_r A_r \qquad (17)$$

for isothermal processes in which bulk flow and viscosity are neglected. σ is the local rate of entropy production, per unit volume, and $\tilde{\mu}_j$ the electrochemical potential of the j-th species. J_r and A_r are scalar quantities, the rate and affinity, respectively, of the r-th chemical reaction as defined by DeDonder (DeDonder and Van Rysselberghe,1936). Specifically

$$A_r = - \sum_i \nu_{ir} \mu_i \qquad J_r = \frac{1}{\nu_{ir}} \frac{d_r C_j}{dt} \qquad (18)$$

where ν_{jr} is the stoichiometric number of species j in the r-th reaction, and $\frac{d_r C_j}{dt}$ is the time rate of change of the concentration of j resulting from the reaction process alone.

The conjugate force-flux pairs can be readily identified in this local form of entropy production, and the corresponding phenomenological equations written in terms of local coefficients. Because of their dependence upon the local reference state, however, these coefficients are not necessarily uniform through space, although Onsager relations should hold among the linear coefficients.

WATER TRANSPORT

In a specific application, such as that of water transport, however, direct utilization of a local field equation is virtually impossible since the space distribution information required is inaccessible to experimental measurement. For the most part, the investigator must be content with the characterization of two phases, separated by the transporting layer: a membrane, epithelial sheet, or more complex structure, together with the possible existence of unstirred fluid layers adjacent to the physical structure. In order to achieve explicit relations between the "macroscopic" flows and forces, those defined in terms of quantities observable external to the membrane, one can resort to the integration of the local equations over the region of the barrier (Goldman, 1943; Kedem and Katchalsky, 1958; Patlak, Goldstein and Hoffman, 1963). This procedure frequently requires assumptions concerning the behavior of parameters over the interval of integration. The equation derived by Goldman (1943) based upon the assumption of a constant electric field, is a well-known example of this approach.

Alternatively, the entire system may be treated as a "black box", which the entropy production and related phenomenological equations cast from the onset in terms of variables observable in the external homogeneous phases. This approach has been termed the non-equilibrium thermodynamics of discontinuous systems (Sauer, 1973).

In either case, the transition to global equations is not completely trivial (Mason, Wendt and Bresler, 1972), and introduces certain new features.

Since the thermodynamic state of the membrane must be time invariant, the resultant theory is restricted to the steady state. This constraint imposes relationships upon the local fluxes, in the case of the flow of species i, for example, of the form

$$\frac{\partial C_i}{\partial t} = - \nabla \cdot J_i + \sum_r \nu_{ir} J_r = 0 \qquad (19)$$

These relationships, required only by steady state conditions and not the basic physical processes, can in turn introduce coupling in the integrated phenomenological equations, between processes which are not coupled in the microscopic description. This point is perhaps most clearly illustrated by an analogous case, the chemical reaction sequence:

$$B \quad \overset{1}{} \quad D \quad \overset{2}{\underset{3}{}} \quad \overset{D}{\underset{E}{}} \qquad (20)$$

"Locally" each of the elemental reaction processes is described as an uncoupled process, with its rate dependent only upon its conjugate affinity. Thus

$$J_r = L_{rr}A_r \qquad r = 1,2,3 \qquad (21)$$

$$T\frac{ds}{dt} = \sum_{r=1}^{3} A_r J_r = \sum_{r=1}^{3} L_{rr}A_r^2$$

Placed within a black box so that only changes in B, D and E can be observed, the system can be uniquely described, in the steady state, by the two processes, B → D and B → E. The rates of these processes are given by the appearance of D and E, respectively, or just the original J_2 and J_3. Steady state conditions require $J_1 = J_2 + J_3$. The conjugate forces of the reduced two flow system, A_2^* and A_3^*, are then given by

$$A_2^* = A_2 + A_1$$
$$A_3^* = A_3 + A_1 \qquad (22)$$

since these affinities retain the entropy production

$$J_2 A_2^* + J_3 A_3^* = \sum_{r=1}^{3} J_r A_r = T\frac{ds}{dt} \qquad (23)$$

Without much difficulty, it can be shown that the transformed phenomenological equations become

$$J_2 = (L_{22} - \frac{L_{22}^2}{\sum_r L_{rr}}) A_2^* - \frac{L_{22}L_{33}}{\sum_r L_{rr}} A_3^* \qquad (24)$$

and

$$J_3 = - \frac{L_{22}L_{33}}{\sum_r L_{rr}} X_2^* + (L_{33} - \frac{L_{33}^2}{\sum_r L_{rr}}) X_3^* \qquad (25)$$

It is interesting to observe, in this simple system, that re-formulation in terms of fewer degrees of freedom, by virtue of imposing the steady state constraint, has resulted in the appearance of coupling between the processes quite apart from the molecular mechanisms of each process.

In this illustration, symmetry of the coefficient matrix resulted. One may well question, however, whether reciprocal relations can be expected generally in the over-all coefficient matrices of the discontinuous treatment. Sauer (1968) has shown, however, that under very general conditions, symmetry and linearity of the local transport coefficients will be reflected by symmetry in integrated steady-state relations.

The process of integrating the local equations over the membrane can often result in non-linear force-flux relations, even though the local laws are themselves linear. The linearized form of the integrated relationships, that approached in the limit of vanishingly small forces, should then exhibit reciprocity, and can be identified with the linear phenomenological coefficients introduced in the discontinuous treatment.

By the same token, composite systems of series and/or parallel arranged barriers, each described by linear integrated relations, can result in non-linear properties of the composite system (Kedem and Katchalsky, 1963a,b,c).

Taking the approach of the discontinuous system development, entropy production in a membrane system can be written in terms of only those quantities observable in the media on the two sides of the barrier. With reference to Figure 1 all translational fluxes are taken as positive in a left-to-right direction.

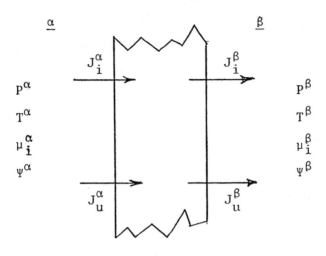

Figure 1

And the fluxes are defined by the rate of appearance (or disappearance) of the i-th species on the two sides of the membrane. The entropy production (not written in terms of conjugate flows and forces), through Gibbs' equation, can be shown to be:

$$\frac{ds}{dt} = J_\mu^\alpha \left(\frac{1}{T^\beta} - \frac{1}{T^\alpha}\right) + \sum_i \left(J_i^\alpha \frac{\tilde{\mu}_i^\alpha}{T^\alpha} - J_i^\beta \frac{\tilde{\mu}_i^\beta}{T^\beta}\right) \qquad (26)$$

where $\tilde{\mu}_i$ is the electrochemical potential, and the summation is taken over all species in the bathing solutions. With steady state conditions in the membrane region, the flow of internal energy is constant through the membrane, $J_u^\beta = J_u^\alpha$. The same is true for the flux of those species which are not stoichiometrically involved in chemical reaction within the membrane. For species which undergo chemical reaction the conservation relation

$$J_j^\alpha + \sum_r \nu_{jr} J_r = J_j^\beta \qquad (27)$$

must be used. Since the stoichiometric numbers, ν_{jr}, vanish for those species not involved in the reaction, equation (27) is valid for all molecular fluxes. From (27) and (26) we get

$$\frac{ds}{dt} = J_u^{\alpha}\left(\frac{1}{T\beta} - \frac{1}{T\alpha}\right) + \Sigma_j J_j^{\alpha}\left(\frac{\tilde{\mu}_j^{\alpha}}{T\alpha} - \frac{\mu_j^{\beta}}{T\beta}\right) - \Sigma_j \frac{\mu_j^{\beta}}{T^{\beta}} \Sigma_r \nu_{jr} J_r \qquad (28)$$

Inverting the order of summation in the final term, and noting that the affinity of the r-th reaction can be expressed

$$A_r = - \Sigma_i \nu ir \, \tilde{\mu}_i \qquad (29)$$

by virtue of charge conservation in the reaction, we obtain

$$\frac{ds}{dt} = - J_u \Delta\left(\frac{1}{T}\right) + \Sigma_i J_j \Delta\left(\frac{\tilde{\mu}_j}{T}\right) + \frac{1}{T\beta} \Sigma_r JrAr^{\beta} \qquad (30)$$

or in the special case of an isothermal system

$$T \frac{ds}{dt} = \Sigma_j J_j \, \Delta\tilde{\mu}_j + \Sigma_r J_r A_r^{\beta} \qquad (31)$$

The Δ operator is defined as the difference, α slide-β side. The β on the affinities indicate that they are taken relative to the β side, while the molecular fluxes are measured at the α side.

Restricting further consideration to the isothermal case, one can now expect phenomenological relations of the form

$$J_j = \Sigma_k L_{jk}\Delta\tilde{\mu}_k + \Sigma_r L_{jr} \, A_r^{\beta}$$
$$J_r = \Sigma_j L_{rj}\Delta\tilde{\mu}_j + \Sigma_s L_{rs}A_s^{\beta} \qquad (32)$$

to hold, with Onsager's reciprocity relations valid.

FACILITATED DIFFUSION

Several points concerning (32) deserve mention. A chemical reaction, for example, with one or more reactants confined to the membrane (e.g. facilitated diffusion) does not appear in these expressions at all (Katchalsky and Spangler, 1968). Steady-state conditions require the net flux (integrated over the membrane) of such a reaction to vanish, and consequently the presence of the reaction process cannot be detected externally.

This choice of convention in defining molecular fluxes and affinities is arbitrary, no more fundamental than its converse, or the infinity of linear combinations possible. Alternative choices, however, may be expected to change the values of the phenomenological coefficients without destroying the matrix symmetry. Caplan and DiSimone (1970) have shown, for example, that in a symmetric membrane of uniform properties, the choice

$$J_i^* = \frac{J_i^{\alpha} + J_i^{\beta}}{2} \qquad , \quad A_r^* = \frac{A_r^{\alpha} + A_r^{\beta}}{2} \qquad (33)$$

leads to the vanishing of the cross-coupling between translational flux and chemical reaction. As noted above, the appearance of a finite coupling coefficient in the integrated equations bears no direct implication of coupling of processes at the molecular level.

ACTIVE TRANSPORT

Active transport of the species, j, can be defined (Sauer, 1973) in the context of phenomenological relations (32), by a non-zero cross-coupling coefficient L_{jr} between the flow of the j-th species and a chemical reaction in which its apparent stoichiometric number is zero. Operationally this would be manifest by non-vanishing flow of species j with all forces other than the affinity of the driving reaction set to zero.

Kedem (1961) on the other hand, proposes a definition of active transport based upon the non-zero value of the cross-coupling co-efficient in the resistance matrix, obtained by inversion of the phenomenological coefficient matrix. If such inversion is carried out, equations (32) are transformed into

$$\Delta \tilde{\mu}_j = \sum_k R_{jk} J_k + \sum_r R_{jr} J_r$$
$$A_r = \sum_s R_{rj} J_j + \sum_s R_{rs} J_s \tag{34}$$

In this frame of reference, j is actively transported if $R_{jr} \neq 0$. One would then observe a non-zero electrochemical potential difference for that species, with all flows stopped with the exception of J_r.

The two definitions are not generally equivalent, although they may coincide in particular cases. With the existence of coupling between the translational flows of j and k ($L_{jk} \neq 0$), for example, the active transport of j alone, in the L-definition, will almost certainly mean that both j and k are actively transported in relation to the R-definition.

Neither criterion of active transport appears to be the more fundamental, and the distinction becomes one of convenience in application. This ambiguity serves to illustrate, however, the strong role played by the nature of constraints placed upon the system when establishing a precise definition of qualitative phenomena in rather complex systems.

Some considerable effort has been devoted to the problem of assessing the thermodynamic properties of the active transport engine in specific tissues, and this remains one of the most challenging applications of the non-equilibrium treatment. Drawing largely upon the work of Essing and Caplan (1968), Schultz (1976) has recently reviewed various techniques of interpreting measurable flows and forces in terms of the affinity and coupling parameters of the metabolic process driving transport.

This task is a difficult one, inasmuch as usually neither the driving reaction affinity, A_r, nor its progression rate, J_r, is accessible to direct evaluation. Oxygen consumption, sometimes taken as proportional to J_r, may not in fact bear a fixed relation to the transporting reaction rate as a result of coupling to various other metabolic processes through common intermediates.

The question of constancy of the reaction affinity under the differing experimental conditions established to decipher the transport properties (25), raises the interesting secondary question of where the affinity is assigned in the biochemical sequence. If, for example, one designates by A_r the affinity of the chemical reaction step directly involved in the transport process, the linear phenomenological relations, with only the one reaction, are

$$J_i = \sum_j L_{ij} X_j + L_{ir} A_r$$
$$J_r = \sum_i L_{ir} X_i + L_r A_r \tag{35}$$

In addition, A_r, involving the chemical potentials of metabolites generated by a lengthy reaction sequence, may be expected to vary with reaction rate. Within the spirit of linear phenomenological laws, this could be represented by letting $A_r = A^0 - RJ_r$, in which A^0 is some ultimate affinity. When this relationship is introduced into (35), however, we recover

$$J_i = \sum_j \left(L_{ij} - \frac{RL_{ir}L_{jr}}{1 + RL_r} \right) X_j + \frac{L_{ir}}{1 \_ RL_r} A^0$$
$$J_r = \sum_i \frac{L_{ir}}{1 + RL_r} X_i + \frac{L_r}{1 + RL_r} A^0 \tag{36}$$

again a linear form with symmetry preserved. Thus the phenomological cal formulation is indifferent to the incorporation of intermediary reaction steps within the black box, provided, of course, that the intervening sequence is not branched and thereby coupled to unknown processes not accounted in the phenomenological equations.

The isothermal entropy production (31) and related phenomenological equations (32) are written in terms of the fluxes of the n species* composing the system, and the conjugate forces. In practical circumstances, this choice of fluxes may not be the most convenient, nor the most revealing of the properties of the system. Through appropriately chosen transformations, the original set of conjugate flows and forces can be converted into an equivalent set, maintaining entropy production invariant and preserving Onsager relations within the new coefficient matrix (Meixner, 1943). In vector notation, the transformations

---

*Generally in a system of n species, only n-1 independent fluxes exist. In this treatment of membrane permeation as a discontinuous system, the membrane structure itslef plays the role of the (n+1) -th component. The flows in (31) and (32) are accordingly taken with reference to the membrane.

$$\bar{J}^* = T\bar{J} \qquad\qquad X^* = (T^{-1})^T X$$

$$L^* - TLT^t \qquad\qquad\qquad (37)$$

$$X^{*T}J^* = X^T T^{-1} \; J^* = X^T J$$

satisfy these conditions.

Coleman and Truesdell (1960) have also pointed out that an infinity of transformations exist which maintain the correct entropy production, but do not preserve reciprocity. These transformations are not of the form (37), however, and generally are dependent upon the phenomenological coefficients characteristic of a given system (See Appendix II).

SOLVENT DRAG

A well known transformation, and certainly one found useful in the consideration of water movement, is the choice of volume flow as a frame of reference developed by Kedem and Katchalsky in their pioneering papers (Kedem and Katchalsky, 1958, 1961). Its utility arises from the fact that one component - in this case water - is present in considerable molecular excess over the remaining constituents. Considering for the moment only non-electrolytes, upon expanding the $\Delta\mu_j$ terms, the dissipation function can be written

$$T\frac{ds}{dt} = J_W (\bar{V}_W \vec{\Delta P} + \Delta\mu_W^C) + \Sigma_j J_i (\bar{V}_j \Delta P + \Delta\mu_j^C) \qquad (38)$$

where the superscript c indicates only the composition dependent part of the chemical potential differences. For convenience, the water terms are separated out with the subscript w; the summation is taken over the remaining components. The $\bar{V}_j$ are the partial molar volumes, and are assumed to be constant.

Introducing the volume flow, defined to be

$$J_V = \bar{V}_W J_W + \Sigma_j \bar{V}_j J_j \qquad (39)$$

and some (yet to be defined) average concentration parameters, $\tilde{C}_j$, we obtain for the dissipation function

$$T\frac{ds}{dt} = J_r \Delta P + \Sigma_j (J_j - \tilde{C}_j J_V)(\Delta\mu_j^C - \frac{\bar{V}_j}{\bar{V}_W} \Delta\mu_W^C)$$

$$+ \frac{J_V}{\bar{V}_W} \{ \Delta\mu_W^C - \Delta\mu_W^C \Sigma_i \bar{V}_i \tilde{C} + \bar{V}_W \Sigma_i \tilde{C}_i \Delta\mu_i^C \} \qquad (40)$$

in which J_W has been eliminated to retain only independent flows. In this form, the awkward coefficient in brackets would appear to be part of force conjugate to volume flow J_V. Noting that the $\tilde{C}_j$ are completely arbitrary, however, the bracketed term can be made to disappear through appropriate choices for these terms. The Biggs-Duhem relation, in isothermal systems, can be put in the form

$$\sum_j C_j d\mu_j^C + C_w d\mu_w^C = 0 \tag{41}$$

If then the parameters $\tilde{C}_j$ are chosen such that

$$\tilde{C}_j = \frac{\int_\alpha^\beta C_j d\mu_j^C}{\int_\alpha^\beta d\,\mu_j^C} \tag{42}$$

with the integration taken between the conditions on the two sides of the membrane system, we achieve the relation

$$\tilde{C}_w \Delta\mu_w^C + \sum_j \tilde{C}_j \,\Delta\mu_j^C = 0 \tag{43}$$

In addition, the scaling factor γ can be chosen to achieve a pseudo-homogeneity condition

$$\tilde{C}_w \overline{V}_w + \sum_j \tilde{C}_j \overline{V}_j = 1 \tag{44}$$

Upon insertion of (43) and (44) into the bracketed term of the dissipation function (40), it vanishes. Consequently, with these definitions, we obtain the transformed set of conjugate forces and fluxes

$$J_V = J_w \overline{V}_w + \sum_j \overline{V}_j J_j, \qquad X_V = \Delta P \tag{45}$$

$$J_j^* = J_j - \tilde{C}_j J_V, \qquad X_j^* = (\Delta\mu_j^C - \frac{\overline{V}_j}{\overline{V}_w} \Delta\mu_w^C$$

and

$$T \frac{ds}{dt} = J_V X_V + \sum_i J_i^* X_i^* \tag{46}$$

A linear phenomenological relation can then be written with some confidence that the coefficient matrix will be symmetric:

$$J_V = L_p \Delta P + \sum_j L_{vj} X_j^*$$

$$J_i^* = (J_i - \tilde{C}_i J_V) = L_{iv} \Delta P + \sum_j L_{ij}^* X_j^* \tag{47}$$

These relations are frequently further rearranged, obtaining

$$J_V = L_p (\Delta P - \sum_i \,_i \tilde{C}_i X_i^*) \tag{48}$$

$$J_i = \tilde{C}_i (1 - \,_i) J_V + \sum_j (L_{ij}^* - \frac{L_{iv} L_{vj}}{L_p}) X_j^*$$

in which Staverman's reflection coefficients (Staverman, 1952) defined here to be

$$i = \frac{-Liv}{\tilde{C}_i L_p} \tag{49}$$

have been used. Equations (48) exhibit in particularly lucid form the coupling between volume flow and solute flow ("solvent drag"), and the interrelationship this phenomenon bears to the osmotic effectiveness of the particular solute.

In this regard, it is interesting to note that, despite the attractive intuitive interpretation of their significance, the concentration parameters $\tilde{C}_j$ have their origin in the thermodynamic properties of the bathing solutions, and are only tenuously related to the actual concentrations averaged over the membrane in any particular instance.

For this reason, the set of $\tilde{C}_j$ need not correspond to any physically attainable solution; electroneutrality, for example, is not necessarily exhibited by the $\tilde{C}_j$. In the vicinity of equilibrium, however, the average concentrations defined in this way have been shown to approach the simple arithmetic mean of the two bounding solutions, and hence converge to a possible equilibrium state.

In the case of electrolytes, to the extent the electroneutrality condition

$$\sum_j z_j \tilde{C}_j = 0 \tag{50}$$

is met, the electrical potential difference becomes a part of the force conjugate to J^* in (47) and (48). Carrying over the coupling to the reactions of active transport, and making the approximation

$$X_j^* \cong RT \left(\frac{\Delta C_i}{\tilde{C}_i} + \frac{z_i F}{RT} \Delta\Psi \right) \tag{51}$$

good in sufficiently dilute solution, we arrive at the more familiar form

$$J_v = L_p \left[\Delta P - RT \sum_j (\Delta C_j + \tilde{C}_j \frac{z_j F}{RT} \Delta\Psi) \right] + L_{vr} A_r$$

$$J_i = \tilde{C}_i (1 - i) J_v + RT \sum_j W_{ij} (\Delta C_j + \tilde{C}_j \frac{z_j F}{RT} \Delta\Psi) + L_{ir} A_r \tag{52}$$

The coefficients $\omega_{ij} = \frac{1}{\tilde{C}_j} (L_{ij}^* - \frac{L_{iv} L_{jv}}{L_p})$ are not necessarily symmetric.

As an illustration, greatly simplified, of the application of non-equilibrium thermodynamics to water transport, consider the specific question: under what conditions can isotonic fluid flow be expected in a series arrangement of two membranes? This question is, of course, inspired by the observation of isotonic fluid resorption in the proximal tubule of the kidney, a phenomenon rather poorly understood. In a recent model study (Sackin and Boulpaep, 1975), efforts to achieve strictly isotonic flow in the Diamond-Bossert model of water transport (Diamond and Bossert, 1967) were unsuccessful.

The system examined here is essentially similar to the model proposed by Curran (1960) and extensively analyzed by Patlak, Goldstein and Hoffman (1963). Thus the restricted results obtained in this treatment are contained with the more general conclusions of Patlak, Goldstein, and Hoffman.
The model is shown in figure 2.

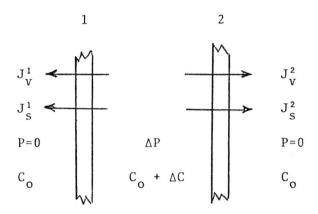

Figure 2

For simplicity, only a single solute, a non-electrolyte, is considered. The two bounding solutions are of identical composition, and no pressure difference exists across the composite structure.

From (52) we write, for either membrane

$$J_v{}^i = L_p{}^i (\Delta P - {}^i RT\Delta C) \qquad i = 1,2$$
$$J_s{}^i = \hat{c} (1 - {}^i)J_v{}^i + w^i RT\Delta C + J_a{}^i \qquad i = 1,2 \tag{53}$$

in which active transport of the solute, J_a, is allowed across both structures.

Steady state conditions require

$$J_v{}^1 + J_v{}^2 = 0 \tag{54}$$

readily yielding

$$-J_v{}^1 = J_v{}^2 = \frac{L_p{}^1 L_p{}^2}{L_p{}^1 L_p{}^2} ({}^1 - {}^2)RT\Delta C \tag{55}$$

For isotonic flow, we have

$$\frac{J_s{}^1}{J_v{}^1} = \frac{J_s{}^2}{J_v{}^2} = C_0 \tag{56}$$

Letting $\hat{c} \approx C_0 + \frac{\Delta C}{2}$, and combining equations (53-60) leads to the two equalities

$$J_v^i = \frac{(\sigma^1 - \sigma^2) J_a^1}{w^1 (\frac{1}{L_p^1} + \frac{1}{L_p^2}) + (\sigma^1 - \sigma^2)(\bar{C}^1 - \frac{\Delta C}{2})} \tag{57}$$

$$\frac{(\sigma^1 - \sigma^2) J_a^2}{w^2 (\frac{1}{L_p^1} + \frac{1}{L_p^2}) - (\sigma^1 - \sigma^2)(\bar{C}^2 - \frac{\Delta C}{2})}$$

As expected, volume flow depends upon active transport, and a difference in the reflection coefficients of the two barrier layers. We shall assume $(\sigma^1 - \sigma^2) > 0$. Under these circumstances, the denominator of the first expression is always positive (until ΔC becomes large) and the direction and magnitude of volume flow is fixed by the active transport of membrane 1.

Reference to the second equality of (57) would suggest active transport at the second barrier is also required to achieve isotonic flow. This is not necessarily the case, however, since the denominator in this ratio can vanish for particular relationships among the coefficients, in which case $J_a^2 = 0$ for finite volume flow. This situation can be achieved for sufficiently small values of ω^2 and a non-zero reflection coefficient. The Curran model, and its refinement in the Diamond-Bossert model, visualize the second barrier as a non-selective diffusion barrier, with $\sigma^2 = 0$ and a corresponding large value of the permeability coefficient. This brief treatment would suggest, then, that isotonic volume flow may not be achievable in models of that configuration.

SUMMARY

The theoretical structure of non-equilibrium thermodynamics provides a general and plastic framework upon which the analysis of biological phenomena, such as water movement, can be elaborated. Applied to particular systems, it can aid in circumscribing the domain of the theoretically possible, and serving as a guide to the experimental characterization of the system. It

1. Provides constraints upon the model parameters to ensure positive entropy production;
2. Reduces the number of independent transport parameters requiring experimental evaluation through the Onsager reciprocal relations; and
3. Serves as a prescription for the transformation to more convenient forces and fluxes.

Application to a biological system frequently requires a series of approximations; viewed within the context of the theory, the nature and significance of the error arising from approximation procedures can be assessed.

APPENDIX I

$$J_i = \sum_j L_{ij} X_j + \frac{1}{2} \sum_{jk} a_{ijk} X_j X_k + \cdots \qquad a_{ijk} \cong \frac{\partial^2 J_i}{\partial X_j \partial X_k} \tag{A1}$$

In pseudo-linear form:

$$J_i = \sum_j \text{ij} X_j = \sum_j (L_{ij} + \sum_k \lambda_{ijk} X_k + \cdots) X_j \quad \lambda_{ijk} \equiv \frac{\partial \text{ij}}{\partial X_k} \quad \text{(A2)}$$

Equating powers of X in the two series expansions yields

$$\lambda_{ijk} + \lambda_{ikj} = a_{ijk} = a_{ikj} \quad \text{(A3)}$$

Symmetry in Λ_{ij} requires

$$\lambda_{ijk} = \lambda_{jik} \quad \text{(A4)}$$

Letting

$$\lambda_{ijk} = \frac{1}{2} (a_{ijk} + a_{jik} - a_{kji}) \quad \text{(A5)}$$

satisfies both conditions A3 and A4, and is the needed relation.

APPENDIX II

Given the initial set of conjugate fluxes and forces such that $J = LX$, L symmetric, any set of fluxes J^* such that $J^* = (L+A)X$, where A is an arbitrary antisymmetric matrix, will yield the correct entropy production, since $X^T J^* = X^T L X + X^T A X$. The second term vanishes since A is antisymmetric. Thus $X^T J^* = X^T J$, the original entropy production.

In order to find the transform M necessary to express J^* in the original fluxes, such that $J^* = MJ$, we have

$$J^* = (L+A)X = MJ = MLX$$

$$ML = L+A, \quad M = (\underline{1} + AL^{-1})$$

where $\underline{1}$ is the identity matrix. Thus the transform necessary to generate J^* will be dependent upon the phenomenological coefficients of the particular system.

REFERENCES

Caplan, S.R. & DeSimone, J.A. (1970) *Chem. Eng. Progr. Symp. 66:* 43.

Coleman, B.D. & Truesdell, C. (1960) *J. Chem. Phys. 33:* 28.

Curran, P.F. (1960) *J. Gen. Physiol. 43:* 1137.

DeDonder, D. & Van Rysselberghe, P. (1936) *Thermodynamic Theory of Affinity,* Stanford Univ. Press.

DeGroot, S.R. & Mazur, P. (1962) *Non-Equilibirium Thermodynamics* Amsterdam, North Holland.

Diamond, J.M. & Bossert, W.H. (1967) *J. Gen. Physiol. 50:* 2061.

Essing, A. & Caplan, S.R. (1968) *Biophysical J. 8:* 1434.

Glansdorf, P. & Prigogine, I. (1971) *Thermodynamic Theory of Structure, Stability and Fluctuations,* New York, Wiley-Interscience.

Goldman, D.E.   (1943)   *J. Gen. Physiol. 27:* 37.
Katchalsky, A.   & Curran, P.F.   (1965)   *Nonequilibrium Thermodynamics in Biophysics,* Cambridge, Harvard Univ. Press.
Katchalsky, A.   &   Spangler, R.A.   (1968)   *Quart. Rev. Biophys. 1:* 127.
Kedem, O.   (1961)   In *Membrane Transport and Metabolism,* ed. A. Kleinzeller & A. Kotyk, New York: Academic Press.
Kedem, O. & Katchalsky, A.(1958)*Biochim. Biophys. Acta. 27:* 229.
Kedem, O. &   Katchalsky, A.   (1961)   *J. Gen. Physiol. 45:* 143.
Kedem, O. &   Katchalsky, A.   (1963a)   *Trans. Faraday Soc. 59:* 1918.
Kedem, O. & Katchalsky, A.   (1963b)   *Trans. Faraday Soc. 59:* 1931.
Kedem, O. & Katchalsky, A.   (1963c)   *Trans. Faraday Soc. 59:* 1940.
Mason, E.A., Wendt, R.P.   & Bresler, E.H.   (1972)   *J. Chem. Soc., Faraday Trans. II, 68:* 1938.
Meixner, J.   (1943)   *Ann. Physik 43:* 244.
Onsager, L.   (1931a)   *Phys. Rev. 37:* 405.
Onsager, L.   (1931b)   *Phys. Rev. 38:* 2265.
Patlak, C.S., Goldstein, D.A. & Hoffman, J.F.   (1963)   *J. Theor. Biol. 5:* 426.
Sackin, H. &   Boulpaep, E.L.   (1975)   *J. Gen. Physiol. 66:* 671.
Sauer, F., Stationarer Stofftransport durch Membranen von makroscopischer Dicke, Unpublished Memo, Max Planck Institut für Biophysik, Frankfurt, a/M, 1968.
Sauer, F.   (1973)   In *Handbook of Physiology, Renal Physiology,* ed, J. Orloff & R.W. Berliner, Washington, Ann. Physiol. Soc.
Schultz, S.G.   (1976)   *Kidney Internat. 9:* 65.
Staverman, A.J.   (1952)   *Trans. Faraday Soc. 48:* 176.

PROBLEMS OF WATER TRANSPORT IN INSECTS

John E. Phillips

University of British Columbia

INTRODUCTION

Isosmotic transport of fluid by vertebrate epithelia has been intensively studied in recent years. This has led to the formulation of detailed mathematical models which describe the transport activity in terms of cell geometry and the properties of plasma membranes and junctional complexes (Curran, 1960; Diamond and Bossert, 1967; Diamond, 1971; Hill, 1975a, b; Skadhauge, 1977). In particular the 'Standing-Gradient Osmotic Flow' model of Diamond and Bossert (1967) has gained such widespread respectability with biologists over the last ten years (Berridge and Oschman, 1972) that it has become commonplace for investigators to propose and readily accept functional models for less well studied epithelia on the basis of ultrastructural organization, often without substantial and convincing physiological corroboration.

The Standing-Gradient model has been satisfactorily applied to these insect epithelia which transfer isomotic fluid (Maddrell, 1971; Berridge & Oschman, 1972). However, there are other unusual cases of fluid transport which do not involve isosmotic transfer and which are concerned with water conservation in insects. For these systems, the application of this hypothesis has been acceptable only after major modification and in some cases has not seemed appropriate at all (Phillips, 1970: Maddrell, 1971; Ramsey, 1971; Wall, 1971; Noble-Nesbitt, 1973; Stobbart and Shaw, 1974; Wall, 1975; Phillips, 1977). In some insects the ultrastructural organization of the major cells responsible for osmoregulation is not readily explained by the 'Standing-Gradient' hypothesis and the significance of the membrane arrangements must remain a matter for speculation. This article summarizes the attempts of insect physiologists to reconcile their observations on these unusual insect systems with the recent dogma for vertebrate epithelia. This is perhaps an appropriate time to ponder such exceptional cases of fluid transport since the Diamond and Bossert model is presently under critical reappraisal. Indeed, Hill finds various osmosis models wanting (Hill, 1975) and has recently revived the 'radical' alternative of electro-osmosis (Hill, 1975b).

GENERAL NATURE OF THE INSECT EXCRETORY PROCESS

The success of insects in the terrestrial environment can be attributed in large measure to the evolution of efficient mechanisms for conservation of body water. These mechanisms include an integumentary cuticle of very low permeability, direct absorption of water vapour from the atmosphere in some species, and the ability of the excretory system to produce very concentrated urine or dry excreta which equals or greatly exceeds in osmolarity that from the best mammalian kidneys (Phillips, 1970; Maddrell, 1971; Wall, 1971; Stobbart and Shaw, 1974). Since excretory systems of most insects lack any anatomical organization analogous to the counter-current system of the mammalian kidney, clearly alternative mechanisms have evolved in insects for concentrating the excreta. In fact, at least four distinct mechanisms have evolved amongst various insects, in spite of a superficial similarity in organization of most insect excretory systems at the anatomical level (Wall and Oschman, 1975; Phillips, 1977).

A typical insect exretory system (Fig. 1) consists of a series of blind-ended Malpighian tubules, a single cell-layer thick, lying free in the hemocoel, and discharging their content into the gut. Here the fluid, mixed with material periodically released from the midgut, moves back through the hindgut into the enlarged terminal section, the rectum, from which excretion occurs through the anus.

As in most animals, excretion in terrestrial insects is a two-step process (Maddrell, 1971; Stobbart and Shaw, 1974). The Malpighian tubules secrete an isosmotic primary urine containing most blood sonstituents, and so have a function analogous to the vergebrate glomerulus. Because this fluid secretion is commonly driven by K^+ transport, the composition of this fluid is unlike that of the hemolymph and would alone radically upset hemolymph composition were it not for a second step, selective reabsorption in the hindgut (Fig. 1). In fact, most of the reabsorption and osmotic regulation occurs in the rectum (Phillips, 1964a,b,c; 1965, 1969, 1970; Stobbart and Shaw, 1974). Perhaps only 0.5% of the water secreted by the Malpighian tubules is actually excreted during antidiuresis, while the remaining fluid and most of its essential constituents are recycled. This serves to clear the hemolymph of metabolic wastes, which are concentrated in the rectum due in part to the extracellular cuticle, which acts as a molecular sieve (Phillips and Dockrill, 1968; Phillips, 1977).

Since a close coupling of water movement to solute transport has been widely observed for epithelia, a brief summary of solute absorption in the rectum is in order. Phillips (1964a,b,c) studied

334

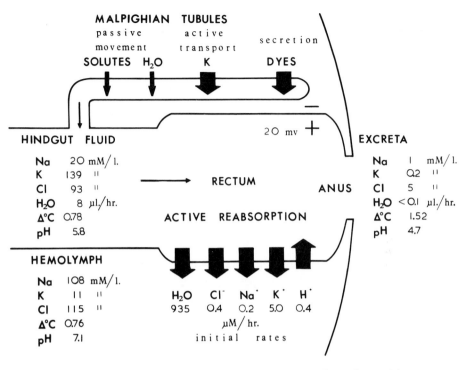

Fig. 1. Diagrammatic summary of secretory and reabsorptive processes occurring in the excretory system of the desert locust (Phillips,
1970). The rates and concentrations are for hydrated animals. The
reabsorptive rates for individual ions are those observed in the
isolated rectum in vivo when the initial concentration of the fluid
introduced into the rectum is the same as the normal hindgut fluid.
The value for chloride is for absorption from a NaCl solution of
low potassium concentration.

changes in the volume and concentration of the rectal content in
vivo, following introduction of test solutions through the anus
into recta which had been ligated at their junction with the colon.
When water absorption is prevented by the addition of sucrose to
the lumen fluid (Fig. 1), Na^+, K^+, and Cl^- are all actively transported against 10 to 100-fold concentration differences and under
various experimental circumstances, against electropotential differences (Phillips, 1964b; Goh, 1971). Indeed, transport of each
of these three monovalent ions can occur in the absence of substantial absorption of the other two ions (Table 1). We have
concluded that all three ions are actively transported, as are five
neutral amino acids normally present in the hemolymph (Balshin and
Phillips, 1971; Balshin, 1972).

335

Table 1. *The rate of fluid transport to the hemocoel and absorbate concentrations when everted rectal sacs are bathed on the lumen side with Ringer solutions containing only one of the major mono-valent ions* (Na^+, K^+, Cl^-) *normally present, with the foreign counter-ion shown in brackets. The control Ringer contained equal amounts of NaCl and KCl, a ratio which was found to support the highest rates of fluid absorption. Concentrations of other con-stituents were the same in all the Ringers. All the absorbate was collected directly from rectal sacs at hourly intervals. Mean + S.E. of values for the 2nd. to 5th. h (Steady-state condition) for 8 - 17 animals (Phillips, 1977).*

RINGER (200 mM, 375 mOsm)

	Control[*]	K(NO₃)	Na(NO₃)	Cl(choline)
Fluid transport	6.5	5.2	4.5	2.4
($\mu l\ h^{-1}\ rectum^{-1}$)	± 0.6	± 0.1	± 0.2	± 1.1
Absorbate Concentration (mM)				
K+	77	175	0.1	11
	± 2	± 4	± 0.5	± 1
Na^+	69	5	156	12
	± 3	± 1	± 5	± 1
Cl^-	109	3	6	101
	± 4	± 1	± 1	± 3
Osmolality	299	357	317	351
	± 19	± 4	± 4	± 3
(as % of Ringer)	80%	97%	87%	92%

*Control Ringer contained 100 mM NaCl and 100 mM KCl.

Acidification of the rectal contents, which has the possible function of regulating hemolymph pH and precipitating the principal nitrogenous waste (urates) is caused by active secretion of H^+ into the lumen and/or HCO_3^- into the hemocoel. Recent experiments in our laboratory using the short-circuit current method suggest that this particular transport process is of greater magnitude than that of other ions and is stimulated 3-fold by the substitution of acetate for chloride in the bathing Ringer (Williams, 1976). Since analysis of absorbate from the locust rectum indicates an anion deficit (Table 1), substantial HCO_3^- transfer to the hemocoel side is a strong possibility. The dehydrated locusts, the ionic and osmotic concen-trations of the excreta are increased (eg. to 1900 mOsm, 600 mM Na^+) by a reduction in the maximum rate of ion transport and an increase in water absorption (Phillips, 1964a,b,c; Wall, 1967), probably under hormonal control.

WATER ABSORPTION IN THE PAPILLATE RECTUM

Early experiments (Phillips, 1964a) using ligated locust recta in vivo conclusively demonstrated how many insects form hyperosomotic

excreta in the absence of a counter-current system. Water is absorbed
from isosmotic or hyperosmotic solutions introduced into the locust
rectum against osmotic concentration differences which actually increase during absorption (eg. to 1 Osm) because a proportional net
uptake of solute from the lumen does not occur. That is, the absorbate is hyposmotic to the bathing Ringer (Table 1) rather than
isosmotic, as in the case for the vertebrate gall bladder, ileum
or proximal convoluted tubule. To over-state the case, like a de-
salination plant, the rectum extracts fresh water from salt, thereby
concentrating the lumen contents. The insect rectum appears to be
the only known case of hyposmotic fluid transport across a simple
epithelium. Except for this difference in osmolality, a comparison
of various other transport parameters indicates that the locust
rectum is quantitatively rather similar to the toad bladder and frog
skin (Table 1), and can be considered to have 'tight' junctions
between the cells.

Herein lies the first problem, because all local osmosis hypotheses which involve an open-ended channel predict an isosmotic
or hyperosmotic absorbate; eg. Diamond and Bossert (1967), and
Sackin and Boulpaep (1975). Related models which are not presently
favoured, such as Curran's 'double-membrane theory' (Curran, 1960),
which has been extended by Patlak et al. (1963), involve a hyperosmotic compartment located between two varriers with different reflection coefficients. These models do allow for the transport of
hyposmotic absorbate under specific conditions. The same is true
for the electro-osmosis models recently proposed by Hill (1975b).

A second difficulty in applying these various models is that
they predict a strict stoichiometric dependence of new water transfer on the rate of solute transport across the epithelium when
solutions of equal osmotic concentration are present on both sides
of the membrane (Phillips, 1970). That is, water absorption should
cease when solute transport is abolished by various experimental
perturbations. This is the operational basis which Kedem (1965)
has used to distinguish between secondary and primary transport of
water. Experimental demonstration of such a close correlation between net water and solute movement across epithelia is perhaps
the strongest evidence in favour of local osmosis models. Several
early studies on insect recta failed to demonstrate such a coupling
of solute and water flow. The rate of water absorption from ligated
locust recta in vivo seemed virtually independent of net solute
absorption (Phillips, 1964a), since uptake continued when a hyperosmotic solution of an impermanent sugar was introduced into the
lumen. Similar results were obtained by Stobbart for the desert
locust (Stobbart, 1968), Wall (1967) for the cockroach and Ramsay
for the mealworm (Grimstone, Mullinger and Ramsay, 1968). A weak-
ness of these experiments was that transient versus steady-state
conditions could not be distinguished because transport activity
was not followed with time (Phillips, 1964a; Grimstone, Mullinger
and Ramsay, 1968; Stobbart, 1968) or was found to fall rapidly
with time (Wall, 1967). Rates were estimated on samples removed

at a set time, usually short, following introduction of known
solutions into the rectum lumen. It was also clear for studies on
the locust rectum (Phillips, 1964a,b) that the low levels of monovalent ions (5-10 mM) which were maintained in the lumen contents
during rectal absorption from sugar solutions were the result of a
dynamic equilibrium between active absorption and passive back-diffusion of ions. That is, some ion recycling was occuring across
the rectal wall in the absence of net ion absorption. While not
excluding the possibility of a primary water pump, which is thermodynamically possible for the locust rectum, Phillips (1965) pointed
out that these observations cold be explained by a solute-linked
water transport followed by back-diffusion of solute (i.e. solute
recycling) so that a net absorption of solute is not observed. A
structural basis for proposing specific models of this type emerged
from ultrastructural studies of rectal papillae of the blowfly by
Berridge and Gupta (1967) and rectal pads of the cockroach by
Oschman and Wall (1969).

The rectal pads of the cockroach are structurally complex, although they are similar in general design to other epithelia which
transport fluid (Fig. 2). The pads consist of a single layer of
tall columnar cells. The lateral plasma membranes between these
cells are highly folded into finger-like interdigitations, each containing a mitochondrion. The resulting intercellular channels of
200Å width are blocked by junctional complexes at both apical and
basal borders but are in continuity with dilated intercellular sinuses. Only limited exit to the hemocoel is possible from these
spaces, at those points where trachea penetrate the epithelium into
a sinus which exists between the epithelium and muscle layers. Wall
and Oschman (1970) propose that solute-linked water transport occurs
by local osmosis into the 200 Å spaces and that the solute is then
reabsorbed in the tracheal channel and across the flat basal plasma
membrane to yield a hyposmotic absorbate. It is the latter activity
which distinguishes the rectum from various vertebrate epithelia.
This model is reduced in Fig. 3 to a structurally simple form which
ignores postulated local intracellular osmotic gradients (Wall, 1971)
and which is modified to emphasize recent observations by Phillips
(1975), Schmidt-Nielsen and Goh, (1971) on solute-water coupling.
In support of this type of proposal, Wall, Oschman and Schmidt-Nielsen (1970) have obtained samples of fluid by micropuncture from the
dilated intercellular sinuses in dehydrated cockroaches and found
that this fluid (691 mOsm) is more concentrated than the lumen contents (562 mOsm) by 120 mOsm, i.e. a favourable gradient for local
osmosis is present. This is the only direct verification of local
hyperosmosity in lateral spaces of any epithelium. Electrolytes
account for only half of the solute in such spaces (Goh, 1971).
Fluid from the sub-epithelial sinus had an osmolality one-third lower
than the lumen content (Goh, 1971). These observations provide strong
circumstantial evidence for the model but do not prove solute-water
coupling. Moveover, there is no direct evidence that most of the
water crossing the rectal wall moves through the type of space which

338

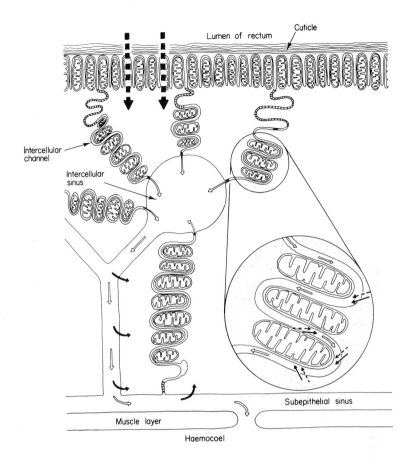

Fig. 2. A hypothetical scheme to explain water absorption from the rectum. The drawing is based on that of Oschman and Wall (1969) for Periplaneta, but the general scheme would apply as well to Calliphora and Schistocerca. The solid continuous arrows refer to active movements of solutes, the solid broken arrows to passive movements of water and the open continuous arrows to fluid movements in the extracellular spaces (from Maddrell, 1971).

is large enough to be micropunctured, or that the fluid becomes diluted on the lateral channels due to solute reabsorption rather than solvent entry into these spaces.

Other evidence for this model is less conclusive. Berridge and Gupta (1967) report distention of the lateral intercellular spaces, as seen in electron-micrographs, of Calliphora recta previously exposed to conditions which should favour enhanced fluid absorption. Wall, Oschman and Schmidt-Nielsen (1970) were able to observe a similar distension of spaces in vivo at the onset of water absorttion in the cockroach. This is in spite of their failure to

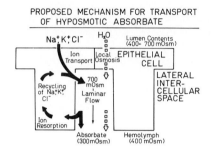

Fig. 3. *In this diagram by Phillips (1977), epithelial structure of Fig. 2 is oversimplified and emphasis is given to summarizing specific aspects of solute-water coupling, which are suggested by recent experiments with everted rectal sacs of the desert locust (Goh, 1971; Phillips, 1975, 1977). Prolonged absorption of water requires a supply of at least one monovalent ion on the lumen side, but ions from the hemolymph cannot sustain fluid transport. Tissue ions can maintain fluid absorption until they are depleted by loss to the hemocoel. Concentrations of all three monovalent ions are lower in the absorbate than in the lumen contents (Table 2). This is consistent with ion reabsorption and recycling before the fluid leaves the rectum.*

detect a similar correlation by electron microscopy (Oschman and Wall, 1969). Finally, Berridge and Gupta (1968) demonstrated histochemically the localization of Mg-activated ATPase specifically on the lateral membrane stacks of *Calliphora*; however, this activity was not stimulated by Na^+ or K^+ and therefore has not been identified as a 'pump' ATPase.

We have recently used everted rectal sacs of the desert locust, which exhibit relatively constant transport rates for several hours, to obtain direct confirmation that water transport is dependent on ion absorption and recycling within the rectal wall (Goh, 1971, Phillips, 1975). The following predictions of the model (Fig. 3) have been confirmed:

a) Water absorption occurs from an isosmotic solution of an impermeant sugar but this eventually ceasea after 1 hour because tissue ions are continually lost (Fig. 4).

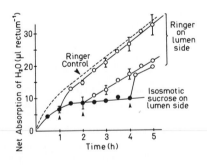

Fig. 4. *The volume of fluid obsorbed by everted rectal sacs of the desert locust with time after exposure to various bathing media (Goh, 1971; Phillips, 1975, 1977). In all cases Ringer was placed on the hemocoel side and either Ringer (control or isosmotic sucrose solution (closed circles) were present on the lumen side. Fluid absorption stops after 1 hr. in the latter situation, but this can be restored at various times by placing Ringer on the lumen side (open circles). Absorption was determined gravimetrically with corrections made for any changes in tissue weight so that the amount of fluid transferred to the hemocoel is indicated.*

b) During such experiments (a), the hyposmotic absorbate collected during the first hour contains substantial amounts of Na^+, K^+, Cl^-. This is expected if recycling within the epithelium is not 100% efficient (Goh, 1971; Phillips, 1975).

c) Ions are not recycled to the cell from the hemolymph directly, because fluid absorption does not continue from sucrose solution after 1 hour (a) even though Ringer is present in the hemocoel side.

d) After water absorption from isomotic sucrose solution has ceased (a), it can be at least partially restored by adding any one of Na^+, K^+, or Cl^- on the lumen side (Table 1), but all other minor constituents of the Ringer together have no restorative effect (Goh, 1971; Phillips, 1975). Therefore, prolonged abosrption of water requires a supply of at least one of these monovalent ions from the rectal lumen.

e) The absorbate is most hyposmotic when all three monovalent ions are present, which is consistent with reabosrption of all these ions. It should be noted that the absorbate concentration of each ion is lower than that in the Ringer (Table 1). If transport of each ion drives water movement by creating a favorable osmotic gradient in the tissue, it follows that their lower concentration in the absorbate must be the consequence of their reabsorption.

f) Only low concentrations (15 mM) of sodium and potassium chloride are required on the lumen side to sustain fluid transport, and absorbate composition changes relatively little as external ion levels are increased (Fig. 5). These low levels of ions are comparable to those reported to be present in the rectal lumen during absorption from 'initially pure' sugar solutions and are the consequence of ion recycling (see above). Possibly back diffusion of ions to the lumen can support additional absorption of water from sugar solutions in vivo. This potential route of recycling is eliminated in experiments with overted recta because the very large volume of circulated fluid on the lumen side maintains ion levels in this compartment at the μM level; i.e. ions which move to the lumen are in effect lost to the tissue. Moreover, the osmotic permeability in the hemocoel to lumen direct is only one-third that in the reverse direction (Goh, 1971). This asymmetry means that osmotic gradients which are initially established across the rectal wall during absorption from sugar solutions will subsequently decline slowly. These two processes may largely explain the apparent discrepancy between earlier in vivo experiments and the results reported here for in vitro rectal sacs.

Two problems with this model deserve mention (Phillips, 1970; Wall, 1971). Analysis of micropuncture samples from intercellular spaces by Wall (1971) indicate that recycling of unknown organic solutes may contribute substantially to water transport in vivo. Secondly, the model assumes that water moves by osmosis across the apical border of the rectal epithelium. Since the osmolality of excreta may reach 1600 mOsm, even higher osmolalities must be postulated within the epithelium cells. This has not been demonstrated. The total concentration of inorganic ions, amino acids and urea per kg. tissue H_2O in the rectal wall of partially dehydrated desert locusts only equals that of the hemolymph (Phillips, 1970; Balshin, 1972). The presence of other solutes or reduction of water activity by macro-molecules must be postulated according to this model. An alternative which is suggested by recent observations on paracellular shunts (Hill, 1975a; Sackin and Boulpaep,

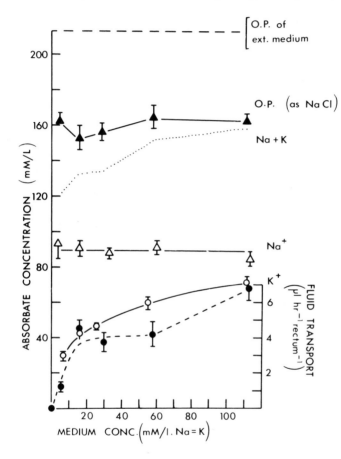

Fig. 5. The dependence of fluid transport rate (broken line) and the ionic composition of the absorbate (solid line) collected from the hemocoel side on the concentration of monovalent ions placed on the lumen side of everted rectal sacs of desert locusts (Phillips and Meredith, unpublished data). The ionic concentration of the external medium was varied by diluting a bathing medium containing equal levels of NaC1 and KC1 with an isosmotic solution of sucrose so that osmolarity was constant (expressed as the equivalent NaC1 solution; upper broken horizontal line). No solution was placed initially inside the sac on the hemocoel side, and all absorbate was collected from the sacs at hourly intervals for analysis. Each point represents the mean (+ S.E.) for determinations made over the 2nd to 5th. h, when steady-state absorption is approximated, on 8-17 individuals, substantial fluid absorption occurs when external NaC1 and KC1 concentrations are well below 30 mM. The composition of the absorbate is relatively independent of ion concentration in the bathing medium. The absorbate remains hyposmotic to the bathing medium by 20%.

1975) is that water moves from the lumen to hyperosmotic lateral channels by way of the apical junctional complexes, rather than across the apical plasma membrane. The electrical thightness of the rectal apithelium (Table 2) would require that the apical junctional complexes be very semipermeable or the basal junctional complexes might be the source of the high electrical resistance.

While the recycling model which has been proposed for the papil-

Table 2. A comparison of various physiological parameters for the locust rectum (Phillips, 1964a,b,c; Williams, 1976) with values for selected vertebrate epithelia with 'tight' and 'leaky' cell junctions (Fromter and Diamond, 1972). Except for osmolarity of the absorbate, the locust rectum is quantitatively similar to such 'tight' epithelia as the toad bladder and frog skin. Correlates of junctional tightness (Fromter and Diamond, 1972) include transepithelial resistance R, osmotic water permeability P_{osm}, potential difference (PD) set up by active ion transport between identical solutions, rate of active ion transport between identical solutions, maximum (limiting) concentration ratio against which salt transport can occur, and the ratio of the osmolarity of the transported fluid (absorbate) to that of the bathing medium.

Epithelium	R (ohm cm^{-2})	P_{osm} (ml cm^{-2}s^{-1}) osmolar^{-1}	Transport PD (mV)	Ion Transport rate (μmol cm^{-2} h^{-1})	Limiting concentration ratio	Osmolarity ratio (Absorbate/ Bathing med.)
Frog skin	2,000	1x10^{-5} (no ADH)	100	1.5	10,000	>>1
Toad urinary bladder	800	9x10^{-6} (no ADH)	35	1.6	>30	>>1
Frog stomach	500	2x10^{-5}	30	4	1,000,000	~1
Rat submandibular salivary duct	400	?	71	12	70	>>1
Locust rectum	5,000	5.5x10^{-6}	35	1.0	>30	<1
LEAKY JUNCTIONS						
Rat gall bladder	28	4x10^{-5}	0	13	12	1.0
Rat proximal tubules	6	4x10^{-3}	0	55	1.3	1.0

late rectum (Figs. 2,3) is reasonably consistant with available data, it should be recognized that it has been preferred over variations of the 'double-membrane' model (Phillips, 1970), or the electro-osmosis model recently revived by Hill (1975b), largely because of the widespread acceptance of the Diamond and Bossert hypothesis for various vertebrate epithelia. In the author's view it is too early to reject outright these alternative and presently less attractive hypotheses (Phillips, 1970) particularly when the Standing-Gradient hypothesis is at present the object of serious

criticism (Hill, 1975a; Sackin and Boulpaep, 1975; Skadhauge, 1977).
At least some other insect recta clearly operate by a somewhat different mechanism than that proposed for the papillate rectum. These
other types of recta will now be considered.

RECTAL ABSORPTION OF WATER VAPOUR

   Other terrestrial insects form powder-dry faces in the terminal
hindgut by water absorption from the gaseous phase. This has been
demonstrated by Ramsay (1971) for the cryptonephridial system of the
mealworm (also Machin, 1975) and by Noble-Nesbitt (1973, 1975) for
the anal sacs of the fire-brat. Only some relevant points will be
considered. The rectal epithelium of these insects lack elaborate
foldings of the lateral plasma membrane and so the mechanism of
water absorption proposed for the cockroach, locust and blowfly
(Fig. 3) appears inoperable. Since the gut of both insects can
absorb water vapour, either directly from sub-saturated atmospheres
surrounding fecal material or from air taken through the anus, solute uptake from the lumen is clearly not essential for water transport. As a consequence, dehydrated mealworms produce feces which
are in equilibrium with an atmosphere of 88% relative humidity
(R.H.), equivalent to a 6 Osm solution, while for some individuals
the value may reach 75% R.H.
   The mealworms, like other larval Coleoptera and Lepidoptera,
possess a crytonephridial complex (Fig. 6). This is believed to
act somewhat like the counter-current system of the mammalian kidney
to build up local, high osmotic concentrations in the perirectal
space and thereby reduce the osmotic difference against which the
rectum must transport water. KCl is actively secreted into the
tissues of the complex from the hemolymph by the terminal ends of
the Malpighian tubules, which are closely applied to the rectum
and are enclosed within a water-impermeable, myelin-like sheath
except at the window-like leptophragmata. The latter are the probable sites of KCl transport, or perhaps they simply provide a
pathway by which K^+ reaches the large cells of the Malpighian tubules which seem better equpped for high rates of K^+ transport.
In effect, a recycling of KCl occurs between the tissues of the
complex and the hemolymph by way of the Malpighian tubules and this
helps draw water from the lumen. However, less than 50% of the
osmotic pressure of the perirectal fluid can be accounted for by
ions. Ramsay found that this fluid contains a material of molecular
weight 10,000 which possesses unusual properties during freezing
and this may be responsible for additional lowering of water activity. There remains the problem of how water moves across the posterior rectal epithelium against a local concentration difference of
possibly 3000 mOsm (Grimstone, Mullinger and Ramsay, 1968). There
is an element of doubt as to the reality of this osmotic concentration difference because it is not certain that samples of perirectal
fluid which have been obtained by micropuncture include only fluid
from the very terminal region of the perirectal space. This element

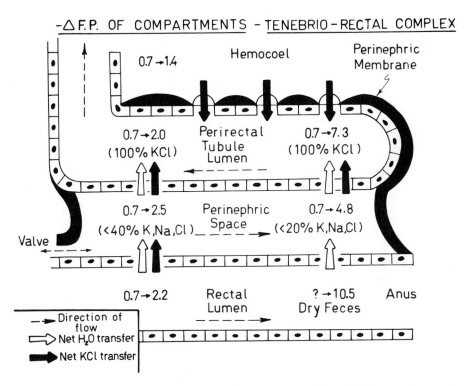

$-\triangle$ F.P. OF COMPARTMENTS – TENEBRIO-RECTAL COMPLEX

0.7→1.4   Hemocoel   Perinephric Membrane

0.7→2.0   Perirectal   0.7→7.3
(100% KCl)   Tubule   (100% KCl)
             Lumen

0.7→2.5   Perinephric   0.7→4.8
(<40% K,Na,Cl)   Space   (<20% K,Na,Cl)

Valve

0.7→2.2   Rectal   ?→10.5   Anus
           Lumen   Dry Feces

- → Direction of flow
⇨ Net H₂O transfer
➡ Net KCl transfer

Fig. 6. Diagram of the cryptonephridial complex of the mealworm in longtiduinal section to illustrate the major membranous barriers and compartments (Phillips, 1971). The unbracketed figures in each compartment indicate measured osmotic concentrations expressed as freezing point depression (-Δ°C). The first figure indicates the average value for hydrated larvae, while the figure after the arrow indicates the average meximum value reached in dehydrated larvae. Contributions of individual ions to the total osmotic pressure are shown in brackets in some cases. Postulates sites and direction of potassium transport and osmotic flow of water across membranes are indicated by large arrows. Movements of fluid within compartments are indicated by broken lines with small arrows. (Based on data from Ramsay, 1964, and Grimstone, Mullinger, and Ramsay, 1968).

of uncertainty concerning the osmotic concentration across the rectal epithelium does not exist in the case of the fire-brat.
    The ability of some arthropods to absorb water directly from the air has been well-established by numerous workers since the initial observation over forty years ago. At one time, this was thought to occur through the integument, but Noble-Nesbitt (1973, 1975; see also Machin, 1975) has shown that only the posterior end of desiccated fire-brats is capable of absorbing atmospheric water (Fig. 7) and this ability is abolished by blocking the anus. The

345

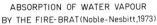

ABSORPTION OF WATER VAPOUR
BY THE FIRE-BRAT(Noble-Nesbitt,1973)

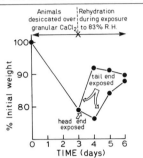

Fig. 7. Effects of exposing head end or tail end alone to high humidity on subsequent rehydration of previously desiccated fire-brats. Experiments conducted at 37°C; dessication over granular calcium chloride; rehydration in 83% R.H. Mean values for the weights of the insects are plotted and are expressed as per cent initial weight against time in days. The lines joining the mean values indicate the time-course of weight change.

anal sacs of this insect can absorb water from atmospheres of 45% relative humidity. The anal values exhibit cyclical opening and closing only when desiccated animals are subsequently re-exposed to hydrating conditions. These movements decrease again as the insect regains body water. Calculations based on the estimated volume of the rectal chambers indicate that these cyclical movements are frequent enough to deliver sufficient air by diffusion to account for the observed rates of water absorption, provided uptake occurs in the rectal chambers only (Maddrell, 1971). The insect, once it has regained its original body-water content, ceases to absorb water so that the process is truly a regulatory one.

The anal sacs, where the absorption of water is presumed to occur, consists only of a simple epithelium containing one cell type, covered on the lumen side with cuticle. These possess highly folded apical membranes which extend almost the whole width of the cell and which are associated with numerous mitochondria to give an extraordinarily close-packed hexagonal array of these two organelles (Fig. 8). According to the current interpretation of such ultrastructural organization, these long narrow channels might wrongly be thought to be involved in isosmotic fluid transport. To appreciate the enormity of the task these cells must perform, a saturated solution of NaCl is in equilibrium with an atmosphere of 75% R.H. and would be concentrated to dryness of 45% R.H. Osmosis would be very inefficient, if indeed a fluid could still exist under such conditions, because the number of water molecules which might accompany transfer of small solute molecules would approach 1 to 1 at 45% R.H. (Wall, 1971). It is difficult to make any definite suggestion as to how these cells might reduce the activity of the water within themselves by more than half so that water could move thermodynamically downhill into them from air at 45% R.H. This problem might be avoided if the low activity were created in the extracellular channels between the apical infoldings and the cuticle. Noble-Nesbitt (1973) has suggested that organic macromolecules of the type found in the perirectal space of the mealworm (Grimstone, Mullinger and Ramsay, 1968), presumably with a high proportion of hydrophilic groups, may be secreted between the apical infoldings

346

where they absorb water of hydration. In support of this view, electron micrographs of this epithelium show that the apical extracellular channels are much wider in desiccated than in hydrated firebrats. This expansion appears to result from swelling during fixation. The osmotic nature of this welling was confirmed by increasing the osmolarity of the faxative, which reduces the width of the extracellular spaces (Noble-Nesbitt, 1973). Moreover, during moulting the apical complex of infoldings and mitochondria temporarily disappears and the ability of the insect to take up water is also lost during this period (Noble-Nesbitt, 1973). The high density of mitochondria immediately adjacent to the apical membrane may be an important feature of the uptake mechanism. Noble-Nesbitt (1973) has suggested that the deep narrow infoldings may permit a standing gradient of water potential to be set up from the apical to basal side of the cell by differential activity along the infoldings. There remains the problem of releasing the water so acquired and making it available to the hemocoel. Perhaps hydrates solute is secreted into the lumen of the anal sacs and is then moved forward by peristalsis to the rectum proper, where it is absorbed by cells which are ultrastructurally similar to those of the papillate rectum.

A second possibility (Maddrell, 1971; Noble-Nesbitt, 1973) is that the apical membrane of the anal chambers contain hydrophilic macromolecules. When hydrated these might undergo a conformational change which presents adhering water to the cell interior where it might then be released because of the different intracellular conditions (eg. ionic of pH). Such an energy-requiring system would constitute a primary water pump, a possibility long discarded by most but not all physiologists (Phillips, 1970; Ramsay, 1971; Noble-Nesbitt, 1973). At least this possibility appears to be thermodynamically feasible (Maddrell, 1971).

THE SECRETING RECTUM OF SALINE WATER MOSQUITO LARVAE

It is not only those insects capable of absorbing water vapour which raise difficulties for physiologists attempting to relate ultra-structure to cell function on the basis of generalizations from studies of vertebrate epithelia.

Until recently it has been assumed that saline-water (S.W.) insects form hyperosmotic urine in the same manner as terrestrial insects, by absorbing water without proportional amounts of solute from the rectum. However, Meredith and Phillips (1973) questioned this assumption on the basis of ultrastructural studies of the rectum of a saline-water mosquito larva, A. campestris. Formation of hyperosmotic urine in S.W. larvae is associated with a posterior rectal segment absent in strictly freshwater species (Fig. 9). This segment consists of a single layer of cells of one type which lack the elaborately developed lateral intercellular spaces of the papillate rectum. Rather, the apical membrane infoldings, which are regular and tightly packed, extend 60% of the distance across the cell. Most of the mitochondria, which are very abundant, are

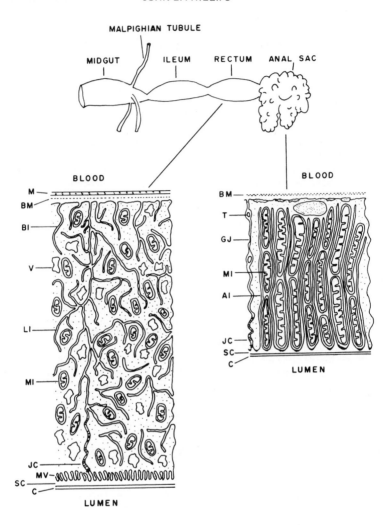

Fig. 8. Hindgut of the fire-brat, Thermobia (Thysanura), based on electron microscopy by Noirot and Noirot-Timothee (1971a,b). Hindgut is comprised of ileum, rectum, and air sac. The air sac is structurally more elaborate than the rectum, as most of the cell volume consists of highly folded apical infolds and closely adhering mitochondria. It is thought that the air sac is not a passive air bladder, but is the site of water uptake from the atmosphere (From Wall and Oschman, 1975). Abbreviations: BM, basement membrane; BI, basal infold; MI, mitochondrion; AI, apical infold; SD, septate junction; SC, subcuticular space; C, cuticle.

associated with these infoldings. The arrangement is more reminiscent of the anal sacs found in the fire-brat. Since S.W. larvae ingest large quantities of slaine waters (Bradely and Phillips,

1975), these workers concluded that these animals did not have to conserve water but need only rid themselves of the excess ions so ingested. They proposed that this was achieved by active secretion of hyperosmotic fluid across the apical plasma membrane into the lumen of the posterior rectum (Meredith and Phillips, 1973).

Indeed, if the gut is ligated anterior and posterior to the rectum, it swells with secretion and the osmolality of the lumen content increases from a value isosmotic to the hemolymph to that of the seawater in which the larvae are reared (Bradley and Phillips, 1975, 1976). As expected if the rectum is the principal site of regulation, the ionic composition of the secretion resembles that of sea water except that potassium levels are an order of magnitude higher. Secretion of Na^+, K^+, Mg^{++}, and Cl^- all occur against large concentration differences of 2 to 18-fold. In fact, recta of larvae living in 200% seawater can secrete against much larger concentration differences (Bradley and Phillips, 1976). The

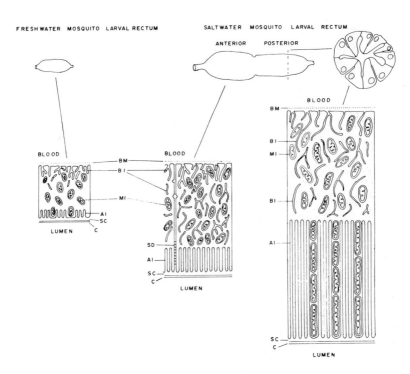

Fig. 9. Morphology of rectum of fresh-water mosquito (Aedes aeghpti) and salt-water mosquito (Aedes campestris). Based on electron micrographs published by Meredith and Phillips (1973). (From Wall and Oschman, 1975). Drawn approximately to scale. Abbreviations: BM, basement membrane; BI, basal infold; MI, mitochondrion; AI, apical infold; SD, septate junction; SC, subcuticular space; C, cuticle.

average trans-epithelial potential difference across the posterior rectum decreases from 10 to 6 mV (lumen positive) during the course of secretion by ligated racta (Bradley and Phillips, 1976). It is therefore necessary to postulate some form of active transport of all four ions to the lumen side. Secretion is only turned on in the posterior rectum when larvae are in hyperosmotic waters (Bradley and Phillips, 1976). A model (Fig. 10) for the arrangement of transport processes in this epithelium has been proposed by Bradley and Phillips based on the influence of different hemolymph ion ratios on electro-potential differences across the apical border and whole rectal wall (Bradley and Phillips, 1976). The posterior rectum of saline-water mosquito larvae plays a role analogous to the gills of marine teleosts, the rectal glands of elasmobranchs, and the salt glands of birds and reptiles. Indeed the electro-potential profile across the posterior rectum of mosquito larvae (Fig. 10) and secreting salt glands of birds (Peaker and Linzell, 1975) are rather

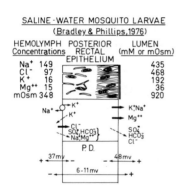

Fig. 10. A model by Bradley and Phillips (1976) for the cellular location of ion transport processes in the posterior rectum of saline-water mosquito larvae. Solid arrows indicate active transport and broken arrows depict passive movements. This model is based in part on measured ion concentration differences (Bradley and Phillips 1975) developed across this simple epithelium as a result of hyperosmotic secretion to the lumen side (figures shown in upper part of diagram), and on the observed electro-potential differences (Bradley and Phillips, 1976) across the two plasma membranes (shown in lower part of the diagram).

similar. The cell interior is negative to both the blood and the lumen so that electrogenic transport of cations is postulated to occur across the apical plasma membrane in both epithelia. The apical rather than the basal plasma membrane of these secreting cells is thought to be the site across which large concentration-differences are developed (Peaker and Linzell, 1975). The argument has been repeatedly made for these invertebrate epithelia that the relative folding of the plasma membranes is consistent with such an hypothesis. The baso-lateral membrane is in general highly folded to allow isosmotic exchange between the cell and blood while the apical membrane is relatively flat so that less water will accompany ions actively secreted into the lumen. In essence, lack of infoldings on the apical side allows transfer of very hyperosmotic fluid across the cell border.

This interpretation clearly does not hold for the mosquito posterior rectum where the predominant cell feature is the elaborate folding of the apical border in association with numerous mitochondria. The salt-secreting cells (dark cells) of the brine shrimp (Artemia salina) exhibit a similar organization (Copeland, 1967) so this arrangement may be common amongst arthropods. A comparison of channel dimensions (Table 3) indicates that the apical infoldings are much narrower and longer than those in epithelia where isosmotic transport is known to occur. Ironically the channels in the mosquito rectum most closely approach the dimensions which Hill calculates are necessary for isosmotic fluid transport, assuming the Standing-Gradient model of Diamond and Bossert (Hill, 1975a). Not only are the extracellular channels very narrow but the cyto-

Table 3. The dimensions of membrane-bound channels associated with solute-coupled water transfer for selected epithelia. All values are from Hill (1975a) except for the apical foldings of the mosquito posterior rectum (Meredith and Phillips, 1973). Assuming typical permeability values for animal membranes, Hill (1975a) has calculated that the geometric function L^2/R must exceed 50 cm for the absorbate to achieve isotonicity with the bathing medium. The mosquito posterior rectum most closely approaches this value but secretes a very hyperosmotic fluid.

SYSTEM	10^4 Length (cm)	10^4 Radius (cm)	L^2/R	SECRETION
Rabbit gall bladder	30	1.5	0.06	isotonic
Rat small intestine	20	0.7	0.06	isotonic
Rat proximal tubule	8	0.05	0.12	isotonic
Avian salt gland	7	0.13	0.04	hyperosmotic
Insect Malpighian tubule	5	0.01	0.28	isotonic
Locust rectal pad	-	-	1.0	hyposmotic
Mosquito posterior rectum	37	0.015	9.0	hyperosmotic
Required for isosmotic transport (Hill, 1975)	-	-	50	

Values from Hill (1975), except for mosquito (Meredith and Phillips, 1973).

plasmic processes are of similar dimension so that the assumption of good mixing on the intracellular side of the membrane, which is made by various local osmosis hypotheses (Diamond and Bossert, 1967; Diamond, 1971; Sackin and Boulpaep, 1975) is very improbable. Berridge and Oschman (1972) have argued that this arrangement might aid in isosmotic transport by allowing parallel decreasing gradients to develop along the length of alternating backward and forward channels. This would seem inappropriate in the case of the mosquito posterior rectum.

Possibly the absence of mixing on the cytoplasmic side and the 'sweeping in effect' which should accompany fluid entry into cytoplasmic processes during secretion might lead to local accumulation of major intracellular solutes which are not transported. In turn this would reduce the amount of water following ions secreted into the channels. Moreover, ions which leak back from the lumen into the cell must follow a long path down these narrow cytoplasmic processes where they are available to ion pumps for re-secretion into extracellular channels. Such active recycling of Na$^+$ might establish a concentration gradient of this ion along the channels to augment the one postulated above. In essence, an increasing osmotic gradient might be established along the length of the infoldings from hemolymph to lumen side rather than a decreasing one as implied by the Diamond and Bossert model. Clearly there is need to investigate the feasibility of this idea using the approach of mathematical modeling. As it stands, the ultrastructural specialization observed in the posterior rectum of the mosquito larvae is difficult to reconcile with the Standing-Gradient model of Diamond and Bossert.

SUMMARY

1) Some insects conserve water by absorbing a hyposmotic fluid against large osmotic concentration differences in a papillate rectum. This may involve solute-linked water movement by local osmosis into lateral intercellular spaces and subsequent recovery of solute (i.e. recycling) within the epithelium. Some direct evidence for such coupling has recently been obtained.

2) Other insect recta absorb water from sub-saturated air. Since water may be absorbed from atmospheres with relative humidities much lower than those in equilibrium with saturated salt solutions, models involving osmotic flow seem inappropriate. The elaborate infoldings of the apical plasma membrane may contain or secrete molecules which absorb water of hydration from the rectal lumen of such insects.

3) Saline-water mosquito larvae form hyperosmotic excreta in the rectum by secreting into the lumen a fluid several times more concentrated than the hemolymph. Unlike various vertebrate organs which secrete hyperosmotic fluids, these recta exhibit exceptional foldings of the apical rather than the baso-lateral plasma membrane. This appears contrary to the widely held view that very long narrow channels represent specialization of plasma membranes for isosmotic transport of fluid across epithelia.

REFERENCES

Balshin, M. (1972) Ph.D. Thesis, Univeristy of British Columbia.
Balshin, M. & Phillips, J.E. (1971) *Nature N.B. 233:* 53.
Berridge, M.J. & Gupta, B.L. (1967) *J. Cell Sci. 2:* 89.
Berridge, M.J. & Gupta, B.L. (1968) *J. Cell Sci. 3:* 17.
Berridge, M.J. & Oschman, J.L. (1972) *Transporting Epithelia.*
    New York: Academic Press.

Bradley, T.J. & Phillips, J.E. (1975) *J. Exp. Biol. 63:* 331.
Bradley, T.J. & Phillips, J.E. (1976) *J. Exp. Biol.* In press.
Copeland, E.C. (1967) *Protoplasma 63:* 363.
Curran, P.F. (1960) *J. Gen. Physiol. 43:* 1137.
Diamond, J.M. (1971) *Federation Proc. 30:* 6.
Diamond, J.M. & Bossert, W.H. (1967) *J. Gen. Physiol. 50:* 2061.
Frömter, E. & Diamond, J. (1972) *Nature N.B. 235:* 9.
Grimstone, A.V., Mullinger, A.M. & Ramsay, J.A. (1968) *Phil. Trans. R. Soc. Lond, 253:* 343.
Goh, S. (1971) M. Sc. Thesis, University of British Columbia.
Hill, A.E. (1975a) *Proc. R. Soc. Lond. B. 190:* 99.
Hill, A.E. (1975b) *Proc. R. Soc. Lond. B. 190:* 115.
Kedem, O. (1965) *Symp. Soc. Exp. Biol. 19:* 61.
Machin, J. (1975) *J. Comp. Physiol. 101:* 121.
Maddrell, S.H.P. (1971) *Adv. Insect Physiol. 8:* 199.
Meredith, J. & Phillips, J.E. (1973) *Z. Zellforsch. 138:* 1.
Noble-Nesbitt, J. (1973) *Comparative Physiology,* Edited by L. Bolis, K. Schmidt-Nielsen and S.H.P. Maddrell, London:North-Holland.
Noble-Nesbitt, J. (1975) *J. Exp. Biol. 62:* 657.
Oschman, J.L. & Wall, B.J. (1969) *J. Morphol. 127:* 457.
Patlak, C.S., Goldstein, D.A. & Hoffman, J.F. (1963) *J. Theor. Biol. 5:* 426.
Peaker, M. & Linzell, J.L. (1975) *Salt Glands in Birds and Reptiles.* Cambridge: Cambridge University Press.
Phillips, J.E. (1964a) *J. Exp. Biol. 41:* 15.
Phillips, J.E. (1964b) *J. Exp. Biol. 41:* 39.
Phillips, J.E. (1964c) *J. Exp. Biol. 41:* 67.
Phillips, J.E. (1965) *Trans. Roy. Soc. Can. Sect. V. 3:* 237.
Phillips, J.E. (1969) *Can. J. Zool. 47:* 85.
Phillips, J.E. (1970) *Am. Zool. 10:* 413.
Phillips, J.E. (1975) *Am. Zool. 15:* 794.
Phillips, J.E. (1977) *Federation Proc.,* In press.
Phillips, J.E. & Dockrill, A.A. (1968) *J. Exp. Biol. 48:* 521.
Ramsay, J.A. (1971) *Phil. Trans. R. Soc. Lond. B. 262:* 251.
Sackin, H. & Boulpaep, E.L. (1975) *J. Gen. Physiol. 66:* 671.
Skadhauge, E. (1977) In *Transport of Ions and Water in Animals,* (Gupta, B., Moreton, R., Wall, B., Oschman, J.L. Editors) London: Academic Press, In press.
Stobbart, R.H. (1968) *J. Insect Physiol. 14:* 269.
Stobbart, R.H. & Shaw, J. (1974) *The Physiology of Insecta,* 2nd Ed. Vol. V. (M. Rockstein, Ed) P. 362 New York: Academic Press.
Wall, B.J. (1967) *J. Insect Physiol. 13:* 565.
Wall, B.J. (1971) *Federation Proc. 30:* 42.
Wall, B.J. & Oschman, J.L. (1970) *Am. J. Physiol. 218:* 193.
Wall, B.J. Oschman, J.L. & Schmidt-Nielsen, B. (1970) *Science 167:* 3924.
Williams, D. (1976) M.Sc. Thesis, University of British Columbia.

THE ROLE OF HYDROSTATIC AND COLLOID-OSMOTIC PRESSURE ON THE
PARACELLULAR PATHWAY AND ON SOLUTE AND WATER ABSORPTION

Emile L. Boulpaep

Yale University

INTRODUCTION

The available morphological and electrophysiological evidence suggests that the functional organization of the proximal tubule epithelium can be represented by two parallel pathways, one transcellular the other paracellular (Boulpaep, 1970; Whittenbury, Rawlins and Boulpaep, 1973). As shown in Figure 1, the transcellular path includes the apical or luminal cell membrane, the cellular cytoplasm and the peritubular or basolateral cell membrane.

peritubular capillary

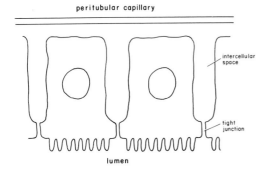

intercellular space

tight junction

lumen

Fig. 1. *Schematic representation of proximal tubular epithelium.*

Substances crossing the epithelium this way may further use either the lateral intercellular spaces or the basal labyrinth before traversing the basement membrane and capillary. The paracellular path is composed of the zonula occludens or tight junction and the lateral intercellular space. The importance of the existence of

355

this path is twofold: it adds additional permeation barriers to account for in the analysis of transepithelial fluxes, and it provides an additional compartment or pool for solute and or solvent exchange.

This paper is concerned with the effects of hydrostatic and colloid osmotic pressure terms on the paracellular compartment or its boundaries and finally how these affect the combined fluxes of ions and water.

THE PARACELLULAR PATH IS HIGHLY PERMEABLE

The proximal tubule of the kidney has a paracellular pathway which is very leaky to ions, small non-electrolytes and most likely water.

Evidence that tight junctions may become sites of ion permeation was suggested already by the work of Ussing and Windhager (1964) on the frog skin indicating that in the presence of hypertonic urea the tight junction may become a shunt for solute permeation. The proximal tubule was the first epithelium where a quantitatively significant and dominant paracellular ion permeability was described (Windhager, Boulpaep and Giebisch, 1967; Boulpaep, 1967). The finding of extensive interaction between the peritubular cell membrane potential and the luminal cell membrane potential difference, (after correcting for changes in chemical potential difference at one side of the epithelium) led Boulpaep (1967) to propose the model diagramed in figure 2. The diagram represents the epithelium by an equivalent circuit that corresponds to the structure shown in Figure 2 for *Necturus* proximal tubule. Across the peritubular or serosal cell membrane the potential difference is V_1 of opposite sign and larger than V_2 across the luminal of mucosal cell membrane; over the entire epithelium transepithelial potential difference, lumen negative V_3 is measured obviously equal to the sum of V_1 and V_2. The diagram represents the epithelium as a set of three barriers, 1) peritubular cell membrane, 2) luminal cell membrane, 3) paracellular path, each endowed with its particular set of electromotive forces E and ionic resistances R. Although more elaborate diagrams have been proposed, each membrane is here represented by the equivalent single comprehensive e.m.f. E_1, E_2, and E_3, which represent for each barrier the sum of all the ionic electromotive forces due to diffusional pathways. Similarly R_1, R_2, R_3 each represent the equivalent resistance of all ionic diffusive pathways within the ionic batteries together with any series resistance or undefined parallel leak within a single barrier. The findings of electrical interaction between contralateral cell membranes could be adequately explained if R_3 is very samll compared to R_1 and R_2 (Boulpaep, 1967). Direct evidence from cable analysis demonstrated that the sum of R_1 and R_2 was 2 orders of magnitude larger than R_3 (Windhager, Boulpaep and Giebisch, 1967; Boulpaep, 1972). Hence overall transepithelial resistance is almost solely determined by the paracellular resistance R_3. Indeed transepithelial resistance of proximal tubule is very low (Hegel, Frömter and Wick, 1967; Boulpaep and

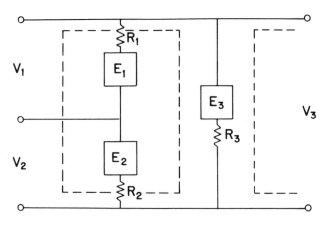

Fig. 2. Model of the equivalent electrical circuit for the proximal tubule cell and overall epithelium. The dashed lines indicate the cell borders V_1 = potential difference across the peritubular membrane of the cell; E_1 = electromotive force of the peritubular membrane which may be of either polarity and is equivalent to the combined electromotive forces of all diffusional pathways; R_1 = resistance of the peritubular (and lateral) membrane of the diffusional pathway. The same elements are represented for the luminal membrane with subscript 2. V_3 = transepithelial potential difference. E_3 = paracellular electromotive force (of either polarity) resulting from dissipative leaks. R_3 = paracellular resistance (modified after Boulpaep, 1967).

Seely, 1971; Boulpaep, 1972).

The permeability to small non-electrolytes of either the tight junction or the paracellular route was established both from net flux and unidirectional flux measurements. The method for the determination of the apparent transepithelial permeability coeffic- ient of (e.g.) raffinose is illustrated in Figure 3. A split-drop is prepared within the tubular lumen as a fluid droplet of isosmotic raffinose confined by two oil blocks. The peritubular compartment is blood or normal electrolyte Ringer solution. Volume changes of the droplet will ensue as the result of two opposing tendencies: that of peritubular electrolytes mainly NaCl ot enter the split- drop and that of the intrauminal non-electrolyte raffinose to leave the split-drop. A biphasic volume change develops with time as shown in Figure 3. It was possible to calculate an apparent trans- epithelial P_{NACl} of $3 \ 10^{-6}$ cm.S^{-1} and $P_{raffinose}$ of $0.5 \ 10^{-6}$ cm. S^{-1}. From the initial rising phase of the curve and with approp- riate corrections the shrinking phase. In this way paracellular permeability for non-electrolytes such as raffinose may be evaluated (Boulpaep, 1972) since it has been established that raffinose does not permeate across the cell membranes proper (Whittenbury, Sugino and Soloman, 1960, 1961). Another method is the measurement of undirectional non-electrolyte fluxes using the compartmental analysis illustrated in figure 4. Providing the solute under study does not enter the cell, a three-compartmental approach is valid which includes 1) the lumen 2) the lateral

357

RELATIVE VOLUME CHANGES IN SPLIT-DROPS
OF ISOTONIC RAFFINOSE

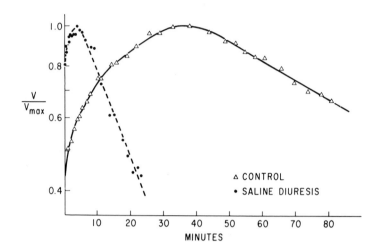

Fig. 3. Method for the determination of P_{NACl} and $P_{raffinose}$ from intraluminal split-drop containing isotonic raffinose at time = 0. Changes in volume are expressed as fraction of the maximal volume reached. Triangles and continuous lines: control condition. Dots and interrupted lines: volume expansion (from Boulpaep, 1972).

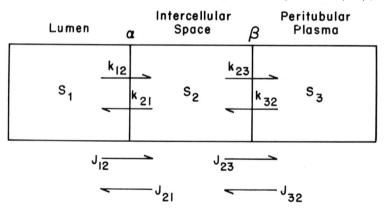

Fig. 4. Schematic illustration of 3-compartment series system consisting of tubular lumen (1), intercellular space (2), and peritubular fluid (3) compartments. S_1, S_2 and S_3 denote amount of solute in individual compartments k_{12}, k_{21}, and k_{32} are rate coefficients across tight junction (A) and effective intercellular space membrane (B), respectively. J_{ij} are unidirectional fluxes (from Berry and Boulpaep, 1975).

intercellular space 3) the peritubular capillaries. Unidirectional rate constants k and fluxes were determined for sucrose in the absence of net volume or net solute flux of sucrose or any other species (Berry and Boulpaep, 1975). The true transepithelial permeability coefficient for sucrose amounts to 1 x 10^{-6} cm. S^{-1}. In addition the latter approach allowed an estimate of the relative restriction to sucrose flow provided by the tight junction (barrier α) and subsequent lateral intercellular space (barrier β in figure 4). The paracellular permeability to sucrose is controlled by both structures, with about 70% of the non-electrolyte resistance contributed by the tight junction. It is noteworthy that the transepithelial permeability coefficients for electrolytes such as NaCl and non-electrolytes such as raffinose or sucrose span only a narrow range. Moreover the ratio of the permeability coefficients does not depart widely from that expected in free solution thus suggesting that these solutes permeate via a watery channel which bypasses the cells.

Two general arguments in favor of water being able to cross the tight junction of *Necturus* proximal tubule have also been presented. First a comparison of transepithelial (Grandchamp and Boulpaep, 1974) and cell membrane (Whittenbury, Sugino and Solomon, 1960) hydraulic conductivity (filtration coefficient) indicates a discrepancy, which suggests that the paracellular pathway constitutes an additional path for water movement (Boulpaep, 1970; Whittenbury, Rawlins and Boulpaep, 1973). Second, the finding of solute entrainment across the paracellular pathway [presumably caused by the tight junction] indicates that when solvent flow is directed from lumen to plasma, a significant fraction of fluid crosses the epithelium directly via the lateral intercellular spaces rather than through the cell (Berry and Boulpaep, 1975). Whereas we believe that solvent movement has both a transcellular and paracellular component (Sackin and Boulpaep, 1975), others have taken the extreme view of modeling the epithelium as if all solute and solvent flow occurs via the tight junction (Schafer, Patlak and Andreoli, 1975).

EFFECTS OF HYDROSTATIC AND COLLOID OSMOTIC PRESSURE

Let us assume that the net transport of salt and water is the result of two paths acting in parallel. It would then be reasonable to assess whether fluxes in the two paths are oriented in the same direction, or whether (e.g.) the paracellular route spills in a backward direction some of the solute which is transported forward by the transcellular path. Since hydrostatic pressure gradients or capillary colloid osmotic pressure may determine the size of the lateral intercellular spaces it may be expected that these factors could affect the balance between transcellular and paracellular fluxes. Indeed, it is well known that net flux of salt and water across the amphibian proximal tubule is greatly influenced by isotonic extra-cellular volume expansion (Boulpaep, 1972). Furthermore, this effect occurs in the absence of changes in glomerular filtration rate, hormone levels or renal innervation. A number of "physical factors" are altered during volume expansion, as is indicated by the white open arrows in figure 5. Extracellular volume expansion was found to be associated with an increase in luminal and peritubular hydrostatic pressure and a decrease in capillary colloid osmotic pressure (Grandchamp and Boulpaep, 1974).

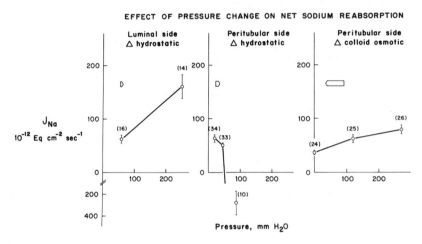

Fig. 5. Effects of pressure change on net sodium reabsorption. Left panel, change in luminal hydrostatic pressure; middle panel, change in peritubular hydrostatic pressure; right panel, change in peritubular capillary colloid osmotic pressure. For comparison, the spontaneous variations observed during volume expansion are indicated by the open arrows (Grandchamp and Boulpaep, 1974).

The mechanism of action of volume expansion on net solute movement was investigated. The separate effects of increased luminal and peritubular hydrostatic pressure were tested as was that of peritubular osmotic pressure. Figure 5 illustrates the dependence of net sodium transport on each of these physical factors (Grandchamp and Boulpaep, 1974). Among the three parameters tested, changes in colloid osmotic pressure seem the most influential in volume expansion. In addition it was shown that the apparent unstirred layer for diffusion on the peritubular side is also a function of colloid-osmotic pressure (Asterita and Windhager, 1975). If e.g. lowering protein concentration in the renal capillaries reduces intrinsic reabsorptive capacity of Na, then this may be the result of either a decrease in active sodium transport or a change in the passive components of solute movement across the paracellular shunt.

Both electrophysiological and flux measurements were performed in order to distinguish between the two alternatives. Cell membrane and overall transepithelial resistances were measured under conditions in which colloid osmotic pressure was modified in the peritubular capillaries. Cellular resistance shown in figure 6A was not influenced significantly by colloid osmotic pressure. The transcellular path thus remains unaltered. However, as shown in figure 6B and C, transepithelial input resistance and specific

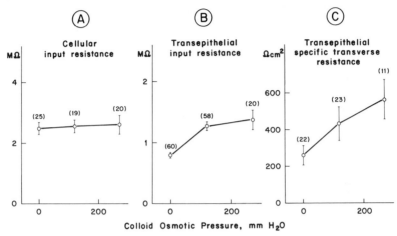

Fig. 6. Resistance changes produced by an alteration in peritubular capillary osmotic pressure. A indicates the resistance change, mainly reflecting the transcellular permeation path. B and C estimate the variations in resistance of the paracellular shunt (from Grandchamp and Boulpaep, 1974).

transepithelial resistance decreased in a protein-free environment and increased with an increase of colloid osmotic pressure (Grandchamp and Boulpaep, 1974). Since it is known that in control conditions the cell membrane resistance is about two orders of magnitude larger than that of the paracellular pathway (Windhager, Boulpaep and Giebisch, 1967; Boulpaep, 1972), the preceding observations indicate that paracellular ionic conductance varies with colloid osmotic pressure changes. Similar observations and conclusions were reported with respect to isotonic extracellular volume expansion (Boulpaep, 1972) where an increased extracellular volume was found to correlate with an enhanced paracellular conductance.

The apparent permeability coefficients across the entire epithelium were also determined as a function of colloid-osmotic pressure (Grandchamp and Boulpaep, 1974). Changes in NaCl and raffinose permeability coefficients at three different colloid osmotic pressures, are illustrated in Figure 7 using the approach shown in figure 3. The permeability coefficient P_{NaCl} was significantly different between 0 and 272 mm H_2O of colloid osmotic pressure. The alterations in $P_{raffinose}$ followed a similar but not statistically significant decline with rising oncotic pressure (Grandchamp and Boulpaep, 1974). Experiments determining P_{NaCl} and $P_{raffinose}$ in control and volume expanded animals showed a significant increase of both permeability coefficients after saline loading (Boulpaep, 1972). This was also found with enhanced peritubular pressure alone (Hayslett, 1973).

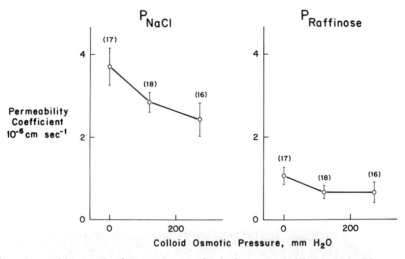

Fig. 7. Effect of change in peritubular capillary colloid osmotic pressure on sodium and raffinose permeability coefficients (from Grandchamp and Boulpaep, 1974).

Passive unidirectional fluxes and the active transport flux of sodium were calculated using the electrochemical potential gradient across the tight junctions, the transepithelial permeability coefficient for sodium and the net sodium flux known from the measured volume absorption rate. Using the constant field equations, passive unidirectional fluxes were estimated as shown by the thick dark arrows in Fig. 8. The net passive component is represented by the thin arrows as the difference, while the thin arrows crossing the capillary endothelium depict net absorption rates measured directly. By subtraction the large open arrows or active transport terms were calculated in Figure 8. The figure indicates the changes in flux components as from left to right the colloid osmotic pressure is raised. The net increase or reabsorption from 37 to 79 x 10^{-12} Eq.cm.$^{-2}$ s^{-1} is not the result of an increase in active transport but caused by a fall in the passive permeability of the shunt and hence a reduction in backflux of sodium ions from 139 to 48 10^{-12} Eq.cm.$^{-2}$ s^{-1} (Grandchamp and Boulpaep, 1974). When the same analysis was applied to the condition which altered simultaneously luminal and peritubular hydrostatic pressure in addition to colloid osmotic pressure, i.e. saline loading, similar conclusions were drawn. The passive and active components of sodium flux are depicted in Fig. 9. The fall in net flux from 161

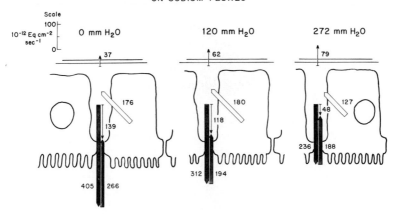

EFFECT OF COLLOID OSMOTIC PRESSURE CHANGE
ON SODIUM FLUXES

Fig. 8. Estimates of sodium flux components in the proximal tubule and influence of change in peritubular capillary colloid osmotic pressure. The length of the arrows together with the figures represent the magnitude of each component. Open arrows: active flux component. Dark arrows: unidirectional passive fluxes. Lower thin arrow: net passive backflux. Upper thin arrow: net reabsorptive flux. Increase in net sodium flux from left to right is due to decrease of passive backflux from left to right (from Grandchamp and Boulpaep, 1974).

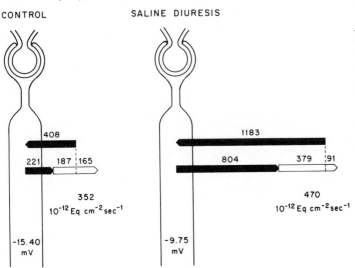

Fig. 9. Necturus proximal tubule. Calculated unidirectional transepithelial passive fluxes of sodium; black arrows. Experimentally determined net sodium flux, and total calculated active sodium flux: white arrows. On left: control condition.

On right: after volume expansion (from Boulpaep, 1972).

to 91×10^{-12} Eq.cm.$^{-2}$s^{-1} is caused by increased passive permeability in saline loading rather than a decrease in active transport. It thus appears that paracellular permeability is greatly determined by the physical factors mentioned above. Further, alterations in these pressure terms may perform a physiological role in the regulation of salt and fluid absorption.

The preceding section emphasized the role of hydrostatic and colloid osmotic pressure on paracellular permeability to small ions which are transported. The right side of figure 7 illustrated the effect on a non-electrolyte which crosses the epithelium by-passing the cells (Boulpaep, 1972), whereas raffinose permeability is increased by volume expansion (Boulpaep, 1972). A more reliable alternative approach is to determine true non-electrolyte permeability coefficients in the absence of solute or solvent movement. This was performed for sucrose in conditions where net fluid absorption, net salt absorption and net sucrose movement was zero (Berry and Boulpaep, 1975). Pressure factors were altered by means of saline loading as described by (Boulpaep, 1972). The permeability coefficient for sucrose in a paired comparison dropped from 1.1×10^{-6}cm. s^{-1} in control to 0.8×10^{-6}cm. s^{-1} in volume expansion (Berry and Boulpaep, 1975). Note that raffinose permeability obtained during net solute and solvent fluxes had shown a rise in permeability with volume expansion (Boulpaep, 1972). How does one reconcile the finding of a fall in paracellular sucrose permeability with a rise in ionic conductance found in saline loading? The apparent contradiction follows from the fact that the raffinose permeability is obtained from net fluxes, while the ionic conductance measurements reflect primarily the response of the zonula occludens. Thus a lesser degree of restriction is offered by the tight junction. In contrast, the sucrose permeability coefficient, depends on both the tight junction and lateral intercellular space. Indeed, if net fluid movement is allowed to occur across the epithelium, then lateral intercellular spaces probably widen sufficiently such that in those conditions the tight junction becomes the only effective barrier to permeation as opposed to the absence of net volume flux. It is thus possible that although the permeability of the tight junction to sucrose might have increased in volume expansion, when measured under zero net volume flow, a simultaneous decrease of lateral intercellular space size may be responsible for the overall fall in sucrose permeability following volume expansion. Additional information can thus be obtained regarding the relative alteration in junctional and interspace properties from non-electrolyte permeation patterns.

The effect of hydrostatic or osmotic pressure differences on the hydraulic conductivity is most complex. Actually all measurements of hydraulic conductivity are probably subject to large errors, since in order to determine a measurable change in water flux, large driving forces are usually applied to the epithelium for rather long time periods. Changes in structure and unstirred layer artifacts are not excluded unless instantaneous flow measurements are made under minute displacements of the appropriate driving force. Considering these reservations, Figure 10 depicts only apparent hydraulic conductivities as these were obtained from displacements of luminal hydrostatic pressure alone. (P_L), peritubular hydrostatic pressure (P_{PT}) and colloid osmotic pressure (π). Whereas from colloid osmotic pressure steps the transepithelial hydraulic conductivity (Lp) would range from 1.1 to 2.3×10^{-8} cm.s^{-1}(cm H$_2$O)$^{-1}$, L_p calculated from ΔP_L would be 5.2×10^{-8} cm.s^{-1} (cm H$_2$O)$^{-1}$ and L_p obtained from small and large steps of ΔP_{PT} ranges from 9.2×10^{-8}cm. s^{-1}(cm H$_2$O)$^{-1}$ to 1.4×10^{-6} cm. s^{-1} (cm H$_2$O)$^{-1}$. It is concluded that physical factors alter the proximal tubular water permeability and most likely the structural changes associated with these pressure changes are associated with the paracellular pathway as shown below.

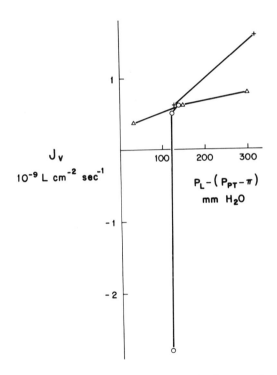

Fig. 10. Transepithelial volume fluxes plotted against net driving force (sum of luminal hydrostatic, peritubular hydrostatic and colloid osmotic pressure difference). Crosses represent experiments in which P_L was selectively altered. Open circles represent experiments in which P_{PT} was selectively modified. Triangles represent experiments in which peritubular was changed (from Grandchamp and Boulpaep, 1974).

J_v

10^{-9} L cm^{-2} sec^{-1}

$P_L - (P_{PT} - \pi)$

mm H$_2$O

Having established that hydrostatic and osmotic pressure gradients alter the paracellular barriers to ions, non-electrolytes and water, it is important to see which structural elements in that path are responsible for the observed changes and what changes occur in extra- or inter-cellular compartments.

The morphological basis of the tight junctional permeability pattern is still poorly understood. Estimates of the width and depth of the tight junction treated as a parallel slit are difficult to estimate from electron microscopy of thin sections. In particular, the width of the junction cannot be resolved accurately, although Bentzel (1972) claimed that the zonula occludens width in *Necturus* proximal tubule increased from 25 A^O in control to 60 A^O in volume expansion, while the zonula adhaerens remained unaltered. Unless the trilaminar structure of the plasma membrane can be resolved in sections exactly perpendicular to the membrane such measurements do not appear very reliable. The fact that lanthanum ions are able to penetrate the tight junction is no indication of the upper size of zonula occludens width (Whittenbury, Rawlins and Boulpaep, 1973). The control *in vivo* transepithelial resistance of 70 ohm. cm.^{2} in *Necturus* proximal tubule (Boulpaep, 1972), if residing entirely in the tight junction (total length of 800 cm.cm.$^{-2}$ epithelium and a depth of 4600 A^O), would predict a width of the junction of 8.2 A^O whereas the perfused kidney with 429 ohm.cm^2 (Grandchamp and Boulpaep, 1974) would require a junctional width of 1.3 A^O. Both these values are possible if in fact focal regions of fusion occur between the apposed plasma membrane leaflets. Replicas of zonulae occludentes in freeze-fractured material have shown a network of strands of fibrils which are the site of contact points between the cells, and the number of these strands have been correlated with the electrical conductance of the epithelium (Claude and Goodenough, 1973). However it is not clear how a single strand if truly occlusive would not give a high electrical resistance to the limiting junction. With respect to described paracellar permeability changes in *Necturus* proximal tubule during volume expansion it was reported that the net work of proximal tubule junctional elements becomes less developed after saline loading in the same species (Humbert, Grandchamp, Pricam, Perrelet and Orci, 1976). In the absence of reliable morphological correlates, a functional slit width of 14.4 A^O has been calculated during conditions of no net solute or solvent flow from non-electrolyte permeability measurements (Berry and Boulpaep, 1975).

Lateral intercellular spaces are known to dilate or shrink in a large variety of circumstances. There exists some ambiguity with respect to the functional changes correlating with either small or large interspaces sizes. Wide lateral intercellular spaces have been associated with high rates of either active transport of solute or net transport of solute and water (Tormey and Diamond, 1967; Schmidt-Nielson and Davis, 1968). On the other hand experimental conditions of low net solute flux and high solute backflux, such as saline loading (Boulpaep, 1972); were also sometimes associated with a widening of lateral intercellular spaces at least in certain limited flow conditions (Bentzel, 1972). Maunsbach and Boulpaep (1975) have recently critically measured the interspace morphology in well controlled fixation procedures and during separate changes of hydrostatic pressure and colloid osmotic pressure during volume expansion. They measured the total volume of the intercellular space per unit length of proximal tubule from composite assemblies of electron micrographs covering an entire tubular cross-section. Intercellular space area was expressed as

a fraction of total tubular cross-section of the epithelium. It
was found that hydrostatic pressure profoundly influences the
geometry of the interspace. Total interspace volume decreased
with elevation of luminal hydrostatic pressure, but increased
dramatically with a decrease of luminal hydrostatic pressure which
was accompanied by a reversal of the normal transepithelial hydrostatic pressure gradient. On the other hand volume expansion in
two different groups of experiments did not lead to an increase
in intercellular spaces (Maunsbach and Boulpaep, 1975). Since
electrical conductance, apparent transepithelial NaCl and raffinose
permeability increases in saline loading it is most likely that
these changes are not related to a modification of the intercellular
spaces but rather of the junctional complexes.

THE EFFECTS OF PRESSURE ON COUPLED SOLUTE-SOLVENT MOVEMENT

The preceding considerations reviewed how physical factors such
as hydrostatic and osmotic pressure differences alter the transport properties of various intraepithelial permeation barriers or
the size of composition of different compartments. However the
question remains how these effects combine to determine overall
coupling between the movements of salt and water across a complex
epithelium. Since no transepithelial driving forces exist for
solute or solvent movement, the ultimate driving force for fluid
absorption is active solute transport into some intra-epithelial
compartment where osmotic exchange must occur. A three-compartmental double-membrane model was proposed by Curran and Mac Intosh
(1962) and accounts for water movement in the absence of a transepithelial osmotic or hydrostatic pressure difference. The
evidence in favor of an important paracellular compartment suggests
that the site for solute-solvent interaction may actually be the
lateral intercellular spaces. A key requirement is the existence
within the epithelium of a region which is hyperosmotic to either
tubular lumen or peritubular capillaries. Hypertonic intercellular spaces have only been detected and directly sampled in insect
rectal pads (Wall, Oschman and Schmidt-Nielsen, 1970). If lateral
intercellular spaces perform such an important role in the exchange
between solutes and water it is likely that alterations of the
barriers lining this space or of the intercellular space composition
may affect the balance between solute and solvent flow.

The proximal tubule represents a special case where solute and
solvent invariably are absorbed in an isosmotic ratio (Windhager,
Whittenbury, Oken, Schatzmann and Solomon, 1969). There is no
indication that interventions such as volume expansion or separate
shifts in one of the three pressure terms discussed above lead to
a departure from isotonic transport. The double membrane model
of Curranand MacIntosh (1962) does not predict an isotonic reabsorbate, if the second barrier at the exit from the lateral intercellular space (e.g. the basement membrane) does not reflect solute.
As an explanation for isotonic transport, Diamond and Bossert (1967)

resorted to a modification of the double membrane model wherein
the intermediate compartment is an inhomogeneous lateral intercellular space with no effective restriction at the exit of the
channel. Active transport is presumed to occur from cell to
lateral intercellular space and osmotic equilibration occurs along
the length of the intercellular channels by inflow of water from
the cell. In order to attain isosmotic transport Diamond and
Bossert confined all active transport sites to the apical end of
the interspace. Several criticisms may be raised against some
of the basic assumptions underlying the Diamond and Bossert model
(Sackin and Boulpaep, 1975). For the special realistic case of
a uniform distribution of solute pumps along the alteral intercellular spaces, Sackin and Boulpaep (1975) provided a general
proof that models of the type as proposed by Diamond and Bossert
would always deliver an hypertonic reabsorbate. Therefore two
new models were developed which are appropriate for proximal
tubular reabsorption in *Necturus* kidney.

The first model is a continuous model similar to Diamond and
Bossert (1967) but with different boundary conditions, uniform
active transport along the lateral cell membranes and permeable
tight junctions (Sackin and Boulpaep, 1975). The model was calculated both for control conditions and after volume expansion.
In the case of a low baseline tight junctional water permeability,
it was found that volume expansion decreases the average concentration in the interspace. Moreover back diffusion of salt through
the tight junction rose from about 1% of net flux to 31% of net
transepithelial flux. The osmolarity of the reaborbate which was
predicted to be 15% hypertonic in control condition was less hyperosmotic (11%) after volume expansion. If a high baseline water
permeability of the tight junction had been postulated, interspaces
were only minimally hypertonic and volume expansion would reduce
the hypertonicity even further, and again the osmolarity of the
reabsorbate would be less hypertonic after saline loading.

The finding of minimal concentration differences predicted
along the lateral intercellular spaces of *Necturus* proximal tubule
prompted the development of a second model where the interspace
is considered a homogeneous compartment (Sackin and Boulpaep, 1975).
A five compartment model was taken consisting of 1) the lumen
2) the cell, 3) the interspace, 4) the peritubular space and 5)
the capillary. The five effective barriers confining the regions
were α) the tight junction, β) the lateral cell membranes   ) the
luminal cell membrane, ∂) the end of the interspace and basement
membrane ϵ) the capillary endothelium. The model is illustrated
in figure 11 with only compartments 1 to 4 and barriers α to δ .
The advantage of this approach is the inclusion of electrical
driving forces, the inclusion of the cell cytoplasm as an additional region and the treatment of individual ion fluxes instead of
neutral salt flows. The fluxes across each barrier were calculated
using the standard Kedem and Katchalsky (1962) equations. Since
all membrane parameters for each barrier were not known, some had

to be treated as unknowns in simultaneous solutions of simple mass
balance equations on volume flow and on ion flow for the different
barriers, combined with equations stating the restriction of
electroneutrality.

The results obtained for transcellular fluxes of Na^+, Cl^- and
water are shown numerically and graphically as dark arrows in figure
11, whereas the transjunctional flux components of Na^+Cl^- and water

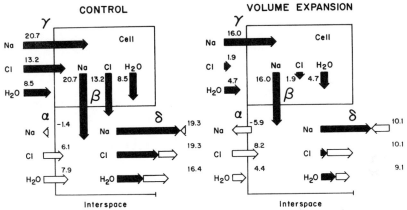

Fig. 11.   Ion and water fluxes for L_p = 6.1 x 10^{-6}cm s^{-1}(cm H_2O)$^{-1}$.
The left side illustrates the control condition; the right side
illustrates volume expansion. Fluxes crossing the cell from lumen
to interspace are indicated by solid arrows. Fluxes bypassing the
cell are indicated by open arrows. The numbers indicate the
magnitude of the net flux across each of the four barriers shown.
Na^+ and Cl^- fluxes are in 10^{-8}mmol s^{-1}/cm^2 epithelium. Water
fluxes are in 10^{-7} ml s^{-1}/cm^2 epithelium (from Sackin and Boulpaep,
1975).

are illustrated by the open arrows. The net results or the magnitude of the emerging fluxes at the level of the basement membrane
are indicated numerically and graphically as the algebraic sum of
the dark and open arrows. Figure 11 depicts the case of a rather
low baseline junctional water permeability. Introducing these
membrane parameter changes known to be associated with volume expansion allowed to compute the predicted flows on the right side
of figure 11. Whereas net volume flow across the tight junction
fell from 7.9 to 4.4×10^{-7}ml s^{-1} cm.$^{-2}$ epithelium, the fractional
contribution of the zonula occludens to volume flow remained unaltered. However the Na^+ flux directed from interspace to lumen
rose appreciably after volume expansion such that backflux was 64%
of the net transepithelial flux as opposed to 9% in control conditions. Moreover, not shown in the figure, volume expansion led
to a fall in the estimated interspace ion concentrations, the
peritubular space ion concentration and in the degree of hypertonicity of the reabsorbate. Finally, it was also possible to pre-

dict some of the pressure parameter modifications subsequent to volume expansion. In combination with the well documented rise in hydrostatic pressure in compartment 1 and 5 and fall in colloid osmotic pressure in compartment 5 (the capillaries) after saline loading the following pressure changes were predicted: a rise in interspace (compartment 3) and peritubular space (compartment 4) hydrostatic pressure together with a drop in colloid osmotic pressure in the peritubular space (compartment 4) (Sackin and Boulpaep, 1975).

It is concluded that a comprehensive model such as that illustrated in figure 11 is required in order to integrate the various membrane parameter changes and compartment property changes induced by physical facotrs. Moreover, the interaction of hydrostatic and colloid osmotic pressure with a complex epithelium such as the proximal tubule may lead to a better understanding of those properties of various barriers and compartments. Finally the studies reported highlight the role of pressure parameters as powerful regulators of overall proximal tubular function.

REFERENCES

Asterita, M.F. & Windhager, E.E. (1975) *Am. J. Physiol.* 228:1393.

Bentzel, C.J. (1972) *Kidney Int. 2:* 324.

Berry, C.A. & Boulpaep, E.L. (1975) *Am. J. Physiol. 228:* 581.

Boulpaep, E.L. (1967) *Transport und Funktion intracellulärer Elektrolyte,* (Kruck, F., Editor) P. 98, Urban and Schwarzenberg, München.

Boulpaep, E.L. (1970) *In* "Electrophysiology of Epithelial Cells", Symp. Med. Hoechst. F. K. Schattauer Verlag. Stuttgart. p. 91.

Boulpaep, E.L. (1972) *Am. J. Physiol. 222:* 517.

Boulpaep, E.L. & Seely, J.F. (1971) *Am. J. Physiol. 221:* 1084.

Claude, P. & Goodenough, D.A. (1973) *J. Cell Biol. 58:* 390.

Curran, P.F. & MacIntosh, J.R. (1962) *Nature (London) 193:* 347.

Diamond, J.M. & Bossert, W.H. (1967) *J. Gen. Physiol. 50:* 2061.

Grandchamp, A. & Boulpaep, E.L. (1974) *J. of Clin. Invest. 54:* 69.

Hayslett, J.P. (1973) *J. Clin. Invest. 52:* 1314.

Hegel, U., Frömter, E. & Wick, T. (1967) *Pflügers Arch. 294:* 274.

Humbert, F., Grandchamp, A., Pricam, C., Perrelet, A. & Orci, L. (1976) *J. Cell Biol. 69:* 90.

Kedem, O. & Katchalsky, A. (1972) *Trans. Farad. Soc. 59:* 1918.

Maunsbach, A.B. & Boulpaep, E.L. (1975) *Abstracts:* 6th Intern. Congr. Nephrol.: 147.

Sackin, H. & Boulpaep, E.L. (1975) *J. Gen. Physiol. 66:* 671.

Schafer, J.A., Patlak, C.S. & Andreoli, T.E. (1975) *J. Gen. Physiol. 66:* 445.

Schmidt-Nielsen, B. & Davis, L.E. (1968) *Science 159:* 1105.

Tormey, J.D. & Diamond, J.M. (1967) *J. Gen. Physiol. 50:* 2031.

Ussing, H.H. & Windhager, E.E. (1964) *Acta Physiol. Scand. 61:* 484.

Wall, B.J., Oschman, J.L. & Schmidt-Nielsen, B. (1970) *Science* *167:* 1497.

Whittembury, G., Rawlins, F.A. & Boulpaep, E.L. (1973) *In*"Transport mechanisms in epithelia" (Ussing, H.H. & Thorn, N.A., Ed.), p. 577. Munksgaard, Copenhagen.

Whittembury, G., Sugino, N. & Solomon, A.K. (1960) *Nature 187:* 699.

Whittembury, G., Sugino, N. & Solomon, A.K. (1961) *J. Gen. Physiol. 44:* 689.

Windhager, E.E., Boulpaep, E.L. & Giebisch, G. (1967) *Proc. Intern. Congr. Nephrol. 3rd Washington. 1:* 35.

Windhager, E., Whittembury, G., Oken, D.E., Schatzmann, H.J. & Solomon, A.K. (1959) *Am. J. Physiol. 197:* 313.

King, G.A. 100,108
Kirk, R.G. 243,248
Kirschner, L.B. 112,114,115,117, 119
Klainer, L.M. 299,302
Kleinzeller, A. 275
Kline, R. 215,216,223,230
Koch, J.H. 129,141
Koefoed-Johnson, V. 184,185
Koppe, D.E. 278,279,290
Kokko, J.P. 250,252,253,272,276
Kostyuk, P.G. 195,197
Kregenow, F.M. 291,293,294,295, 296,297,298,299,302
Kriebel, M. 112,119
Krogh, A. 111,115,119,275,276, 292,302
Kyte, J. 249,271,273

Ladle, R.O. 216,231
Lane, N. 234,235,248
Langendorf, H. 194,198
Leaf, A. 187,188,189,190,197,198
LeBlanc, O.H. Jr. 79,85
Leclercq, J. 91,96
Lee, C.O. 195,197,215,216,217, 218,219,220,221,223,224,229, 230
Lee, D.L. 97,98,109
Lelkes, P.I. 11,14
Leonard, R.T. 284,290
Lev, A.A. 190,198,216,217,219, 221,230
Levitin, H. 250,257,273
Levring, T. 16,29
Lewis, S.A. 201,212
Lichtenstein, N.S. 189,198
Lin, W. 284,285,286,287,288,289, 290
Lindley, B.D. 304,305,313
Ling, G.N. 217,230,275,276
Linzell, J.L. 250,353
Lipton, P. 187,198
Litman, B.J. 11,14
Little, C. 119
Lockau, W. 58,66
Lockhart, J.A. 30,44
Lockwood, A.P.M. 129,141
Loewenstein, W.R. 159,182,194, 197,198

Long, W.S. 70,71,85
Longwell, A.C. 291,302
Lu, C.Y.-H. 70,71,85,86
Lubbers, D.W. 215,230
Luke, R.G. 250,257,273
Luttge, U. 6,14
Lutz, M.D. 249,253,261,267,273
Lyon, T. 223,230

Mache, T.E. 108,109
Machin, J. 344,345,353
MacIntosh, J.R. 184,185,249,272, 367,370
MacKnight, A.D.C. 187,188,189, 190,198
Maclachlan, G.A. 50,52
Maddrell, S.H.P. 90,92,94,96, 303,307,308,309,310,311,313 333,334,339,346,347,353
Maetz, J. 117,119,129,141,146, 159,312,313
Maffly, R.H. 189,197
Maginniss, L.A. 129
Malnic, G. 250,257,273
Malvin, G. 141
Mangos, J.A. 146,159
Mantel, L.H. 129,141
Markscheid-Kaspi, L. 226,228,230
Martin, A.W. 114,119
Martin, P.M. 111,114,115,118
Martin, R.B. 11,14
Mason, E.A. 320,332
Maunsbach, A.B. 366,367,370
Mazur, 319,331
McIver, D.J.L. 188,198
McNutt, N.S. 200,212
Means, G. 57,66
Meixner, J. 325,332
Meneghetti-Gennaro, A. 291,292, 301
Meredith, J. 342,347,349,351, 353
Merz, J.M. 97,103,105,106,109
Mertz, S.M. 283,290
Meves, H. 252,272
Mirsky, A.E. 291,292,302
Mitchel, D.J. 11,14
Mitchell, P. 55,66,77,84,85,278, 289,290
Mitchinson, J.M. 291,302
Momsen, W. 56,66

SPECIES INDEX